EMPIRE OF KNOWLEDGE

EMPIRE OF KNOWLEDGE

The Academy of Sciences
of the USSR (1917-1970)

ALEXANDER VUCINICH

UNIVERSITY OF CALIFORNIA PRESS
Berkeley   Los Angeles   London

University of California Press
Berkeley and Los Angeles, California

University of California Press, Ltd.
London, England

Library of Congress Cataloging in Publication Data

Vucinich, Alexander, 1914-
  Empire of knowledge.

   Bibliography: p. 409
   Includes index.
   1. Akademiia nauk SSSR—History. I. Title.
AS262.A68V79 1984   354.470085′5   83-3484
ISBN 0-520-04871-7

Printed in the United States of America

1 2 3 4 5 6 7 8 9

TO VARTAN GREGORIAN

CONTENTS

PREFACE

The mammoth proportions of its institutional base and activities make the Academy of Sciences of the USSR a scholarly organization without historical precedent. In 1969 it embraced 240 research bodies, including gigantic research institutes, laboratories, observatories, learned societies, scientific councils, and experimental stations. Centered in Moscow, its institutional components were located in over fifty communities scattered across the vast country. The academic outposts ranged from isolated observation stations to complex science centers and "academic cities." In 1969 the academic presses printed 150 scholarly journals, edited by individual research institutes, and added 2,000 new titles to the national book output. In 1970 the Academy was the source of primary employment for close to 35,000 persons engaged in scientific and humanistic scholarship, and it was still growing at a rapid pace. The Academy was, and remains, the institutional summit of Soviet scholarship and the main mechanism of Soviet involvement in high science. One of its basic functions has been to select and train the future elite of Soviet scholarship; it has been the stage on which the electrifying drama of modern Soviet science has been cast, directed, and performed. To understand the dynamics of this institution is to understand the interplay of the political, social, and cultural forces that have shaped the realities of Soviet science.

This study concentrates on the history of the Soviet Academy of Sciences from 1917 to 1970. In October 1917 the Academy embarked on a historical course that made it an integral part of the Soviet system. By 1970 it had completed the first phase of its post-Stalinist history, a period dominated by the search for new science policy—a unique institutional accommodation to the twentieth-century revolution in science and technology and to the spirit of anti-Stalinism in Soviet politics and ideology. During this phase the authorities worked assiduously to achieve an effective recentralization of academic administration and a more systematic geographical deconcentration of academic research facilities. The post-Stalinist thaw—its effects on the relations of science to Soviet ideology and on the nature and magnitude of academic autonomy—had gone as far as the authorities had intended it to go. T. D. Lysenko's dominion, the critical force in the Stalinist attack on the moral code and intellectual integrity of science, was toppled in 1965; but it was not until the end of the 1960s that the last institutional outposts of Lysenkoism were removed and Lysenko's attack on science had ceased to be a topic of public debate.

This study stays close to developments directly related to the Academy, but a few notable exceptions to this rule will become apparent. A limited number of "digressions" into the larger field of Soviet science and science policy have helped to cast a broader light on the Academy; they have been included to provide a more general understanding of the Academy as an authoritative resource for acceptable solutions to organizational problems, as well as to demonstrate how the Academy acted as an institutional tool for bringing science and political ideology into harmonious relationship. According to a Soviet writer: "The Academy of Sciences of the USSR provides the general scientific guidance for research in the most important areas of the natural and social sciences conducted by the academies of sciences of Soviet republics, institutions of higher education and other research centers of the country."

I am deeply grateful to the Guggenheim Foundation, the University of Pennsylvania, and the Russian and East European Center at the University of Illinois in Urbana for generous support of this study. I am thankful to my friend Patricia Johnson for dedicated and expert help in preparing this study for publication. To Dorothy Vucinich, my wife, I am indebted for inspiring help and astute criticism.

A. V.

INTRODUCTION

The Academy of Sciences of the USSR is an integral part of the Soviet state—it is an agency of the Soviet government. It is governed by a complex system of administrative links based on the principles of modern management and opened to structured controls by appropriate agencies of the central government. Its basic social function is to ensure a constant flow of scientific knowledge to the multiple domains of national technology and to safeguard the unity of science and ideology. It performs the key role in the management of the national effort in science on the basis of authority vested in it by the Council of Ministers of the USSR. One of its main delegated administrative functions is to coordinate the national effort in basic research and to guide the growth and activities of the academies of sciences of individual Soviet republics.

The Academy is also part of the scientific community—it is a functioning microcosm of the community of scholars. As a general sociological category, the scientific community has two major characteristics: it transcends the limits of nation, race, and religion; and its members are not bound together by the rigid ties of structured authority and formal organization but by spontaneously evolved shared goals, consensus of scholarly opinion, tacitly agreed-upon professional ethics, values built into science as a subculture, and community sentiment. A scientist is a scientist only inasmuch as he belongs to the scientific community, and he belongs to the scientific community only inasmuch as he adheres to the principles that forge the intellectual and moral unity of his profession. History contains a record of individuals who worked within the framework of the scientific community but who were not part of it.

The intellectual unity of the scientific community is based on shared "metaphysical" assumptions that translate the diversity of scientific facts into a unified world outlook and create a strategy for the systematic search of knowledge and for the interpretation of new discoveries. These assumptions, never too rigid, define the main avenues of the scientific effort to unravel the mysteries of nature and to provide generalized answers to scientific questions.

The "metaphysics" of Newtonian science, which, in its time, reigned supreme in the scientific community, supported an intellectually unified view of the universe that gave scientists a community of purpose and a clearly delimited area of operation. In this view the atoms were structureless and indivisible, time and space preceded sensory experience, absolute causality was the reigning principle of scientific explanation, mechanics was the only legitimate source of models for the development of theory and methodology in all branches of scientific knowledge, and the universe was infinite. Post-Newtonian science rests instead on a "metaphysics" that recognizes the complex structure and fissionability of atoms and atomic nuclei, makes time and space empirical notions, challenges the omnipotence of Laplacian causality, and makes wide use of the idea of an expanding universe. By narrowing the vision of scientific concerns, the shared metaphysics of Newtonianism helped the scientific community to coordinate professional activities and to create a more expeditious and efficient way of marshaling scientific talent for work on problems of social import.

Thomas Kuhn has shown clearly and convincingly that the "systems of belief" that bring concert to work in science and give it a community basis are transitional phenomena: once their intellectual potential has been exhausted by the practitioners of science, they retreat before avalanches of new assumptions that describe the structure of the universe or its individual domains. The Einsteinian revolution in modern science provides a most graphic example of such a frontal onslaught on an established conceptualization of nature. It has been subtler than "the changes from geocentrism to heliocentrism, from phlogiston to oxygen and from corpuscles to waves," but it has been equally destructive of the reigning views of nature.[1]

The moral pillar of the scientific community—the ethos of science—consists of rules that define the "conscience" of scientists and set up the guiding values of their profession. These rules—clearly enunciated since the time of Francis Bacon—are products of informal scholarly consensus. They are not legislated by authorities external to science, and they are not coordinated by formal organizations of the scientific community. In the words of Robert K. Merton:

> The ethos of science is that affectively toned complex of values and norms which is held to be binding on the man of science. The norms are expressed in the form of prescriptions, proscriptions, preferences and permissions. They are legitimatized in terms of institutional values. These imperatives, transmitted by precept and example and reinforced by sanctions, are in varying degrees internalized by the scientist, thus fashioning his scientific conscience or, if one prefers the latter-day phrase, his superego. Although the ethos of science has not been codified, it can be inferred from the moral consensus of scientists as expressed in use and wont, in countless writings on the scientific spirit and in moral indignation directed toward contraventions of the ethos.[2]

Among the ethical norms guiding and holding together the scientific community, four are particularly significant. First: the scientific community rejects every notion of private property in scientific discovery; it holds that every new idea or artifact is deeply indebted to preparatory work by earlier and contemporary scholars. It takes to heart Newton's statement that he could reach the heights of scientific achievement because he "stood on the shoulders of giants." To dwell on the problem of national priorities in science is to flirt with some of the more questionable professional practices. In the words of Max von Laue, "Priority problems constitute an unfortunate chapter in the history of every science."[3] Second: the scientific community is guided by the idea that, in principle, all phenomena of nature and society are open to scientific inquiry. It alone has the right to define the scope and to set up the limitations of the scientific search for knowledge. Political—and other—ideologies have no legitimate right to demand the exclusion of specific natural and social phenomena from scientific scrutiny. Third: the scientific community recognizes no authority impervious to the ravages of history. In 1843 A. I. Herzen reminded both professional scientists and science dilettantes that science had "its own autonomy and its own genesis," and that, as a "free activity," it did not rely on any authority and did not try to impose its authority on anyone.[4] The scientific community does not deny the existence of authority in science altogether, as Kuhn's study has shown in detail, but it does operate on the assumption that every authority must be built on scholarly consensus and must be recognized as a transitory and historical phenomenon. Fourth: the scientific community alone has the right to certify scientific knowledge. Efforts by external (political or religious) forces to certify knowledge undermine the moral foundations and kill the free spirit of science.

This study deals with the Soviet Academy of Sciences as both a government agency and a microcosm of the scientific community from 1917 to 1970. It examines the role of political authority in making the Academy an integral part of the central government apparatus and a bastion of Soviet ideology. It analyzes the historical record of efforts on the part of academic personnel to protect the intellectual and moral foundations of the scientific community—to safeguard and to widen the base of academic autonomy necessary for a successful pursuit of professional activity. It examines the role of the Academy, thought of as a community of scholars, in protecting the internal logic and internally generated momentum of the growth of scientific knowledge. It deals with the evolution and dramatic expansion of the institutional base of academic science, and Soviet science in general, and the role of the Academy in the formulation and execution of the national science policy and in the intricate web of relations between Soviet science and Marxist ideology.

The word *ideology*, frequently used in this study, requires an explanation. As used here it denotes a politically sustained and manipulated system of beliefs that describe the work of the universe, establish the hierarchy of cultural values, and direct activities of individuals and groups either toward the achievement of predetermined social goals or toward the maintenance of the existing social order. Beliefs are ideological inasmuch as they are socialized, sustained, and articulated for the benefit of social conformity and social unity. Ideology is a historical phenomenon. Although its axioms persist in their basic orientation, they are subject to changing interpretations. At times, these interpretations may work to enhance the orthodoxy and rigidity of basic axioms; at other times, they may make these axioms more flexible and more tolerant of new ramifications. Soviet ideology is rooted in Marxist sociology, and political authorities are its codifiers and official interpreters.

Philosophy is the major branch of knowledge devoted to explaining the continually growing ramifications of ideology. One of its basic functions is to serve as a bridge between ideology and science—to channel the close axioms of ideology into the open world of scientific theory. In performing their ideological functions in the Soviet Union, the philosophers operate officially from *within* the scientific community, for Soviet ideology, as presented by its architects, is one with and inseparable from science. The most dramatic pages in the history of the Soviet Academy of Sciences present the conflict between the scientists, as the guardians of scientific legacy, and the philosophers, as defenders of ideology. This conflict is neither accumulative nor linear; its scope and intensity vary with changes in political climate. The effect of this conflict on the inner dynamics of the Academy occupies a pivotal position in this study.

I

ANCESTRY (1725–1917)

THE LEGACY OF PETER I

The St. Petersburg Academy of Sciences opened its doors in 1725, soon after the death of Peter I, its founder and spiritual father. On November 12, 1725, the first academicians, all Western European scholars (some with established reputations), held the first recorded discussion under the aegis of the fledgling institution. The discussion concentrated on mathematics and was highlighted by a dispute between Jacob Hermann, an avowed Newtonian, and G. B. Büllfinger, a strong Cartesian.[1] The Academy came into being at a time when Russia did not have a single institution of higher education. Its founding marked the beginning of modern Russia—a decisive break with medieval perspectives. It was founded on the principles of Baconian philosophy: it emphasized science as the prime source of social well-being and the most powerful weapon in the war against superstition and ignorance.

The founding of the Academy was the crowning point of Peter's inspired efforts to make "enlightenment and science" the major tools of Russia's economic and military development and the main index of Russia's readiness to join the civilized world. After his return from the Grand Embassy—which took him to Holland and England in 1697–1698 to observe Western technology and scientific workshops in operation—Peter worked assiduously to give Russia its first institutions devoted to the enrichment and diffusion of scientific knowledge. In 1701 he founded the Mathematical and Navigational School in Moscow and in 1715 the Naval Academy in St. Petersburg, although neither one matched at that time the level of a typical secondary school in the West. He then established a museum *(Kunstkamera)*, with a strong emphasis on natural history, and a public library, the nucleus of a rapidly growing depository of Russian and Western books. Peter also founded the first Russian press for publishing material in civil script. In 1719 Euclid's *Elements* appeared for the first time in a Russian translation, as did Christian Huygens's *Kosmotheoros*, which gave a popular presentation of the mainstream of scientific ideas connecting the ages of Copernicus and Newton. Indeed, it was the first book in the Russian language to endorse the Copernican astronomical theory.[2] Huygens presented Copernican ideas as "commonly accepted" knowledge, "very agreeable to the frugal simplicity nature shows in all her works."[3] The growing stream of translated works covered such diverse subjects as morality, navigation, geography, Western history, natural law, and Aesop's fables. The lively publication activity compelled the sparse and often inept ranks of translators either to invent Russian equivalents for new scientific words or to add to the language Russianized versions of such new terms as "logarithm," "sine," "tangent," and "secant."[4]

In the evolution of Peter's ambitious efforts to make Russia an active contributor to scientific knowledge and to create an institutional base for scientific research, the influence of Leibniz must be accorded a place of prime importance. Leibniz sought an intimate acquaintance with Peter primarily because Russia, by virtue of its vast unexplored Asian areas, loomed as one of the major challenges and invitations to the rapidly growing pursuit of exploratory work in natural history, a key source of empirical data for a long string of burgeoning sciences. Leibniz was particularly interested in Russia as a bridge between the Occident and the Orient; Russia, in his view, could help enrich Western science by facilitating a scientific study of the hidden wealth of large sections of Asia and could accelerate the flow of modern scientific ideas to Asian nations. Leibniz recognized two major bodies of contemporary science: *philosophia naturalis,* based on an alliance of mathematics and laboratory experiment; and *historia naturalis,* based on empirical facts obtained by a trained observation of the infinite variety of natural objects. He was convinced that Russia's path toward science must begin with *historia naturalis,* close to and often inseparable from folk science.[5]

Leibniz appealed to Peter for a variety of reasons. The Berlin Society of Sciences, founded by Leibniz in 1700, was much more attractive to Peter than the two older models: the Paris Academy of Sciences and the Royal Society of London. Peter could not but favor the Berlin Society's dedication to establishing a close union of science and the national economy and to coordinating the advancement of science with the state's goals in formal education. Leibniz presented a philosophy of science that was fully acceptable to the Russian monarch, for it viewed science as a new pivot of social integration rather than as a weapon of social revolution. Carefully and dramatically, Leibniz articulated a philosophical view based on the complementarity of religion and science—of truth produced by divine revelation and truth produced by the powers of human reason.

Peter and Leibniz corresponded extensively and met several times—in Torgau in 1711, in Carlsbad and Dresden in 1712, and in Pyrmont in 1716. All the dialogues and correspondence—some of the latter between Leibniz and Peter's advisors—contained the same refrain: Russia must establish a "general institution for the advancement of science and the arts"; attract competent Western scholars to St. Petersburg; obtain books and instruments from the West; send young Russians to various Western countries for advanced studies; organize a system of formal education; bring together distinguished men of science, arts, and modern industrial crafts; expand publishing and translation activity; and create museums and botanical and zoological gardens. Leibniz talked also about the urgent need for a systematic study of the natural resources of the vast empire, a survey of languages, and an appropriate station for regular magnetic observations.[6] He manifested uncertainty in his plans for bringing enlightenment to Russia on only one occasion, when he wondered whether Russia might benefit more by opening several universities in various parts of the country than by creating a central academy of sciences.

Leibniz died in 1716. In his thoughtful *éloge* to the great man, Fontenelle, permanent secretary of the Paris Academy of Sciences, did not fail to mention the nobility of Leibniz's assistance to Peter in his effort to bring the light of modern science to Russia. Peter received from Leibniz not ready-made and comprehensive projects but a wealth of diverse, fragmentary, and free-flowing ideas on the general problem of bringing science and science education to Russia. After Leibniz's death, Peter wasted no time in establishing a contact with Christian Wolff, a renowned professor of physics and mathematics from Halle University (and later from Marburg) and one of Leibniz's closest disciples. It was at this time that the idea of a Russian academy of sciences became fully crystallized. The proposed institution no doubt owed its direction and underlying philosophy to the selections Peter made from the recommendations of his Russian and foreign advisors. Wolff's main contribution was to help Peter iron out organizational difficulties and embark on the ambitious plan of attracting the first contingents of foreign scholars to the newly created academic positions in St. Petersburg.

Throughout this period, the scattered voices of skeptics, in Russia and abroad, found the idea of establishing such a highly elevated learned body as an academy of sciences in Russia highly unrealistic. Wolff belonged to this group. He contended that Russia should concentrate on building a university rather than an academy of sciences, for, in his words, it would be much easier to attract excellent teachers than established scholars to St. Petersburg. Wolff pointed out that unrealistically ambitious plans had made the Berlin Society of Sciences—later renamed the Berlin Academy of Sciences—a scholarly institution only in name.[7] Numerous skeptics agreed that Russia was not ready for such an ambitious venture into higher learning. To silence the critics, Peter undertook many activities to show that Russia's involvement in scholarship was not a dream but a reality. Russian work in natural history and cartography became a serious matter after Peter's visit to Paris in 1717, a visit clearly undertaken to give the tsar a closer look at the work and the organization of the French scientific community. Peter visited the Collège des Quatre Nations, the Sorbonne, and the Paris Academy of Sciences, and he was greeted by the leaders of the scientific community—from Abbé Bignon, president of the Academy, to G. Delisle, the Academy's leading geographer—not as a man in search of institutional models for organized scientific research but as an established "scholar." A map of the Caspian Sea, prepared in 1716, made a particularly strong impression on the members of the Academy, even though it represented only the beginning of major departures from previous maps. The Paris Academy honored Peter by making him the first monarch to be elected to its membership (he was proclaimed a member *hors de tout rang*). For reasons unknown, Peter waited until 1721 to accept the honor.[8]

Peter, on his return to Russia, hired D. G. Messerschmidt, a Danzig physician, to lead an exploratory team to Siberia to collect unique cultural artifacts and rare specimens of flora and fauna.[9] He commissioned Carl von Verden to prepare a new map of the Caspian Sea and appointed a team of cartographers to undertake a preliminary survey of the Azov and Black seas.[10] In 1719 he appointed the geodesists Ivan Evreinov and Fedor Luzhin to study the geography of Kamchatka and determine whether there was a land link between Siberia and North America. In the early 1720s St. Petersburg acquired a new stately building to house the rapidly growing library and museum. In all these activities Peter clung closely to the idea that science was both a primary source of national strength and an infallible builder of national prestige. An academy, Peter said, "would earn us a place of trust and honor in Europe by showing that we are engaged in science and that we are not barbarians disrespectful of science."[11] The Spiritual Regulation, made public in 1721, stated in no uncertain terms that the church must absorb the spirit of new science in order to become a force of social and cultural progress. Modern learning, the regulation noted, made it possible for Peter to create a strong army and to bring modern architecture, medicine, and government administration to Russia.[12] The church would have to overcome the ancient fear of scholarship as a source of heresy and recognize the benefits of modern learning, for otherwise it would be unable to perform the role of a useful social institution. It must never forget that "our early teachers studied not only the Holy Scriptures, but even natural philosophy." A. N. Pypin was correct in noting at the end of the nineteenth century that the Spiritual Regulation signaled a turning point in Russian history, for it substituted "rationalism" and "utilitarianism" for medieval thought steeped in superstition and belief in miracles.[13]

On January 22, 1724, the Senate approved a project for the founding of an academy, prepared, under Peter's guidance, by L. L. Blumentrost, a court physician and a highly educated person, and J. D. Schumacher, the head of the public library and a college dropout. The choice of these two men was logical: they were Peter's chief aides at home in maintaining contact with Western scholarship and in gathering information on the organization of Western scientific institutions.[14] The "project" defined the Academy as a scholarly body devoted to the advancement of knowledge and to providing expert answers to practical questions raised by the government. The disciplines treated by the Academy fell into three classes: mathematics and sciences grounded in mathematics; theoretical disciplines dependent on the experimental method, including physics, anatomy, chemistry, and botany; and the humanities (with a heavy emphasis on history, politics, law, and ethics).

The new academicians came in small numbers but included several distinguished scholars who were responsible for the generally favorable acceptance of the St. Petersburg institution by the Western world of scholarship. Fontenelle, in his *éloge* to Peter I, had remarked that Russia, which only two decades earlier was unaware of the existence of Euclid's geometry, had risen to the position of one of the leading contributors to the most advanced branches of contemporary mathematical analysis. At first, however, the sailing was not smooth. Peter's successors on the throne—Catherine I and Peter II—were unwilling to bolster the financial resources of the fledgling organization. The new regimes had little affinity with the spirit of Peter's reforms, and the advocates of Old Russia, now in power, succeeded in moving the royal residence to Moscow. Blumentrost, the first president of the Academy, went to Moscow as court physician, leaving behind his friend J. D. Schumacher, a government bureaucrat with no deep commitment to science, to administer academic affairs.[15] The latter made no effort to resist the rising opposition to the Academy as an expensive luxury.

In 1730 Anna Ivanovna ascended the throne, and the royal house returned to St. Petersburg, signaling a defeat for the antireform party. The status of the Academy thus improved financially, but Schumacher's rigid reign continued unabated. To add to the woes of the Academy, some academicians proved to be useless charlatans, and a core of reputable scholars—including Daniel Bernoulli—wasted no time in finding their way out of Russia.[16] There was also the problem of censorship. In 1728 J. Delisle's discourse on the rotation of the earth was denied publication in a Russian translation. In 1731 the government prevented the Academy from publishing a Russian translation of Fontenelle's *Conversations on the Plurality of Worlds*, which took Copernicus's view of the universe for granted. Although these difficulties contradicted the moral code of science, they did not threaten the survival of the Academy.

The Academy brought science and scientific activity to Russia, but it did not immediately make Russia—and the Russians—fountains of scientific wisdom. During its second year, the Academy began to publish the periodical *Commentarii Academiae Scientiarum Imperialis Petropolitanae*, which quickly became one of the premier publications devoted to scientific scholarship. All the learned papers published in the early years, however, were written by foreign scholars. Nonetheless, from the very beginning many steps were undertaken to enhance the Russian side of the academic equation. The *Brief Description of the Commentaries of the Academy of Sciences,* published since 1725, provided abridged Russian translations of scholarly articles from the Academy's main journal; the opening number provided the first systematic discussion of the concepts of causality and motion to be presented in the Russian language. Soon the Academy began to prepare "Notes" appended to the *Vedomosti*, the St. Petersburg newspaper, discussing modern scientific themes in a popular vein. In 1731 the Academy undertook a careful survey and publication of chronicles necessary for a critical and comprehensive study of Russian history. In 1735 it founded the Russian Council, which consisted of a growing corps of translators who pressed for a systematic study of Russian grammar and for an extensive effort to develop a Russian literary language unhampered by the ossified forms of Church-Slavic expressions. The council gave intellectually oriented personnel associated with the Academy a legitimate and attractive forum for the exchange of ideas.

Adapting to Russian realities, the St. Petersburg Academy acquired a feature that it did not share with similar bodies in Western Europe: it established an academic university and an academic gymnasium. It was through these institutions that the Academy helped make Russians active participants in scientific creativity, and science an organic part of Russian culture. In a country with no national network of educational institutions—lower, secondary, or higher—this was an enormously arduous task. Compounding the difficulties was the fact that most academicians were exceedingly slow—and, in the case of such individuals as G. S. Bayer, unwilling—to learn Russian at a time when Russian students were unprepared to communicate in foreign languages. Small wonder that, according to a recorded complaint in the early 1730s, the academic university had more professors than students, and in 1743 an academic administrator complained to the Senate that professors gave no public lectures and had no Russian students.[17] In due time, however, the Academy played the major role in selecting promising young scholars for training in the leading universities abroad. The founding of Moscow University in 1755 marked the first solid step toward the creation of a network of national centers of higher education.

During the first fifteen years of its existence, the St. Petersburg Academy produced enough solid work to become an internationally respected body of scholars. Its achievements were strong in the two major domains of contemporary science: *philosophia naturalis*, concerned with making mathematics and experimental techniques the supreme methodological tools of science, and *historia naturalis*, concerned with describing and systematizing the infinite ramifications of nature and encouraging exploratory work in previously ignored parts of the world.

In the first area there were several persons of prominence, including the mathematicians Jacob Hermann and Daniel Bernoulli, whose *Hydro-dynamics* became a classic of Newtonian science. But the brightest star in this constellation—and in all of eighteenth-century Russian science—was Leonhard Euler. He came to St. Petersburg in 1727, at the age of twenty, with two short publications by the Paris Academy of Sciences to his credit; by 1741, when Frederick II lured him to join the Berlin Academy of Sciences, he was universally regarded as one of Europe's most brilliant mathematicians. His *Mechanica*, published in 1736, was the first successful, though incomplete, effort to use infinitesimal calculus in interpreting and codifying the laws of mechanics. In pursuing this work he not only perceived the need for order in the house of mathematical analysis—in the new world of mathematics unveiled through the discovery of calculus by Newton and Leibniz—but also undertook to do something about it, a task that preoccupied him for the rest of his life. One result was the monumental, two-volume *Introductio in analysin infinitorum* (1748), the first systematic development of the mathematical notion of function. A modern biographer of Euler has noted that while *Mechanica* was the first "analytical mechanics," the *Introductio* was the first "theory of functions."[18] In his later work Euler made a major contribution to the advancement of the calculus of variation and to the emergence of modern textbooks in basic mathematics.[19]

Natural history, the second great domain of eighteenth-century science, established a strong foothold in the new Academy. In 1733 the Academy was asked to contribute its contingent of experts to the Great Northern Expedition (known also as the Second Kamchatka Expedition), led by Vitus Bering.[20] During the ten years of exploration, academic personnel—particularly J. G. Gmelin—collected valuable information providing the first systematic surveys of Siberian flora, fauna, topography, and history. The Kamchatka Expedition included a small group of young Russians, some of whom had transferred to the Academy from the Moscow Slavo-Graeco-Latin Academy for the sole purpose of bolstering the ranks of native participants in the Siberian enterprise. According to B. N. Menshutkin, Lomonosov's most prominent biographer:

> The Slavo-Graeco-Latin Academy, founded in 1684, consisted of eight classes: four lower ones (grammar, syntax), two middle classes (poetry, rhetoric), and two higher (philosophy, religion). The instruction in the lower classes was concentrated chiefly on the study of Latin, learned so thoroughly that at the end of the fourth year the students could read

and write it, and on the study of Slavonic languages. In addition, they
had to know geography, history, the catechism, and arithmetic. In the
middle classes, the students were already obliged to speak Latin. [21]

Exploratory work in the distant regions of the vast empire provided
valuable apprenticeship for many generations of young Russians aspir-
ing to become professional scientists. This work produced mountains
of empirical data that laid the foundations for the pronounced concern
of the Russian scientific community with the problems of ecology and
the classification of natural phenomena—and that helped make the scientific mode of thinking and inquiry a dynamic component of Russian
culture. Nonetheless, a typical Russian natural historian of this time had
no illusions about the relatively low place of his work in the hierarchy
of sciences; under the spell of Eulerian contributions—and in the spirit
of Newtonian philosophy—he knew that only mathematics could take
him to the heights of scholarly achievement.

During the first period of its existence, the Academy acquired enough
valuable experience both to support a cautiously optimistic view of its
future development and to become fully aware of serious barriers on
the road to its full integration into Russian society. Critics were quick
to point out that the "republicanism" (or associative autonomy) of the
Academy was alien to the patriarchal and autocratic fabric of the Russian
polity. "Friends" of the Academy, however, argued that Russia should
follow the example of France, which relied on the official policy of
enlightenment as a bridge between the absolutism of the state and the
republicanism of the Paris Academy. V. N. Tatishchev, the first Russian
to undertake a systematic survey of the history of his country, seems to
have found this policy workable. He was eager to point out that the
reign of Peter I was a period of fully consolidated autocracy and rapidly
developing science; this was the best proof that "freedom" (vol'nost')
was not an essential condition for the diffusion of scientific knowledge. [22]

The general political atmosphere of post-Petrine Russia made the
Academy an agency of the state rather than an autonomous body. Until
1747 the Academy was without a charter; it was administered by the
chancellery, made up of bureaucrats responsible only to the government.
The administrators—all foreigners—played the academicians against
one another and used arbitrary measures in dismissing or promoting
them. They inaugurated research projects without previous consultation
with academic experts, and they showed no reluctance in compelling

individual academicians to undertake research tasks that were outside their professional competence.

To add to the woes of foreign academicians, the government chose to keep them outside the official table of ranks, the system of officially specified privileges granted to the graded elite. The academicians were thus de facto without social status. The existing situation had three negative results: it made the Academy an ambivalent institution without a strong anchor in the government hierarchy or in society at large; it accentuated the agonizing feeling of statelessness shared by the academicians and militated against the integration of Western scholars into the mainstream of Russian life; and it worked against the academic university, for, in a society unaccustomed to considering the level of education a criterion in the assignment of ranks, a certificate of higher education brought no clearly defined social advantages. Small wonder that the academic university found it extremely difficult to attract more than a handful of students and, in particular, made no appeal to aristocratic youth.[23] According to Tatishchev, the academic university did not attract young gentry because it offered no instruction in Orthodox theology, Russian law, introductory sciences, and such "noble sciences" as horsemanship and modern dancing.[24]

THE AGE OF LOMONOSOV

In 1741 the Academy lost Euler, the main source of its international prominence. Together with P. L. M. de Maupertuis from the Paris Academy of Sciences, Euler went to Germany to help Frederick II in his ambitious designs to shake the Berlin Academy of Sciences out of its chronic slumber and hopeless drifting. Euler did not cancel his membership in the St. Petersburg Academy and continued to publish his papers in its journals and to provide countless services to Russia, always acting as an academic emissary in the West.[25] In 1766 he returned to St. Petersburg, where his death seventeen years later brought his remarkable career to an end.

During Euler's absence the St. Petersburg Academy was blessed with the rise of another star, a star belonging to a different order but emitting equally powerful rays. Euler brought Russia to the forefront of Newtonian science. Mikhail Vasil'evich Lomonosov showed clearly and convincingly that the Russians were ready to become active contributors to high science. There is no great scientific law associated with his name; nor is there a method or a new domain of inquiry to which he could claim priority rights. Lomonosov's achievement was to consider a wide range of theoretical questions that troubled the leading scientific minds of his time and arrive at competent and often refreshingly original—and unexpected—answers. He was not a strict Newtonian, but he was the first Russian to meet the challenges of the key theoretical positions that formed the basis of Newtonian science.[26] In doing this, he raised new questions that anticipated the main course of the development of theoretical science. For example, he anticipated the law of the conservation of mass and the limitless potentialities of a new synthesis of physics and chemistry (he coined the term "physical chemistry"). He combined the wave theory of light with the corpuscular theory of color. He organized the first Russian chemical laboratory, expressed doubts about the scientific validity of the phlogiston theory and the fluid theory of electricity, gave a rudimentary presentation of the molecular theory of heat, worked on the kinetic theory of gases, and experimented with atmospheric electricity. In many of his views in physics and chemistry, he was clearly ahead of his time. A corpuscular concept ran conspicuously through his theories of heat, air, gas, atmospheric pressure, chemical solutions, electricity, and magnetism. He was, in the words of S. I. Vavilov, the builder and consistent follower of a grand theory of mechanical atomism.[27] He ventured into Russian history, grammar, rhetoric, poetics, mineralogy, geology, and geography.

Lomonosov was firmly convinced that mathematics held the key to the advancement of science across the full spectrum of its specialized branches, but he personally was in command only of the most elementary mathematical skills and made virtually no use of them. The strength of his approach to science lay not in his command of methods of quantitative measurement but in his unmatched erudition in the theoretical achievements and dilemmas of contemporary science and his logical mind. He was capable of detecting both the inner flow of theoretical developments in the key sciences and the deeper meaning and anatomy of persisting controversies and inconsistencies in high abstractions. "We may or may not accept the ideas of Lomonosov," wrote a nineteenth-century Russian scientist, "but we cannot fail to see in his explanations a robust and ingenious logic."[28] His contributions lay in charting the lines of scientific development rather than in elaborating the empirical substratum of abstract theory.

Lomonosov represents more than a dramatic affirmation of the Russian scientific mind: he was the standard bearer of a growing struggle against forces inimical to the development of the national sources of Russian science. He fought the foreign academicians who spread the word that the high theory of science was alien to the Russian mind; he announced optimistically that the time was fast approaching for Russia to produce its own "Platos and Newtons." He fought G. S. Bayer, A. L. Schlözer, and G. F. Müller, who laid the foundations of the Norman theory of the origin of the Russian state, based on the assumed premise that the Russians were incapable of creating a politically organized society without the "help" of outside conquerors. He helped lay the groundwork for the founding of Moscow University and argued that Moscow should become a university center because it was close to the pulse of national life and not overly dependent on foreign scholars as a source of "Russian science." The Academy of Sciences, in his opinion, was too heavily controlled by foreign scholars to be transformed into a national institution within the foreseeable future. In this respect he was correct; it was not until the 1860s that the Academy, under the heavy pressure of an awakened national intelligentsia, began to take on the form and the spirit of a national institution.

Despite initial setbacks, history was kind to Lomonosov. It made him a culture hero—a national legend—of the first order. The evolution of the idea of Lomonosov as the father of Russian science was a long and tedious process. By the twentieth century it was the center of an elaborate and much-justified national mythology that viewed Lomonosov as a genius of limitless scope, a most authoritative spokesman for the pristine purity of the national commitment to rationalism, an embodiment of the noblest virtues of human existence and the most sublime expression of Russian patriotism. But in his own time Lomonosov was slow in rising to prominence. He was elected to full membership in the Academy in 1745, five years before he began to publish his first scientific papers. When his first papers did appear in the *Novi Commentarii* in 1750, they were noted in Western European journals without much enthusiasm and, in isolated cases, with biting criticism. Most of his subsequent scientific papers were buried in the archives;[29] it was not until the beginning of the twentieth century that they were retrieved, translated into Russian, commented upon, and published by B. N. Menshutkin. Menshutkin unveiled not only the full scope of scientific theses treated by Lomonosov but also the originality of his theoretical formulations and insights.[30] But the main outlines of Lomonosov's gigantic stature were drawn early; succeeding generations could only put the finishing touches on the grand picture of Lomonosov as a pristine expression of Russia's irreversible entrance into the age of modern science.

Lomonosov impressed his contemporaries by his relentless war on the despotic administrators of the Academy—J. D. Schumacher and J. K. Taubert—who worked against making the Academy a truly Russian institution.[31] He wrote odes to science, instituted public lectures on topics in natural science, advocated closer ties between science and the national well-being, and articulated an optimistic view of the growing role of science and the scientific world outlook in Russian culture. He was the first to show convincingly that the Russian language could express the most subtle thoughts of abstract science. His thinking was a synthesis of Baconian philosophy, which viewed science as the most effective weapon of technological progress, and the Newtonian view of the world as a mechanism governed by the laws of motion and the invincible logic of absolute causality.

Summing up the power of the Lomonosov legend, V. I. Vernadskii, the stalwart of the Academy in the first half of the twentieth century, stated: "He always stood for the application of science to man's practical activities. The task of science, according to him, was to improve the human condition. Along with philosophical abstractions, he was preoccupied with applied science. . . . In this he saw an ethical dictate."[32] D. I. Mendeleev stated that Lomonosov was the first Russian scientist, in the European sense of the word.[33] T. I. Rainov, a perceptive historian of science, brought the Lomonosov legend closer to reality when he stated that

> more than any other person in the eighteenth century, he contributed to building the intellectual environment in which his learned successors lived and worked; he gave that environment a strong sensitivity for natural science; he made the Russian language adaptable to the needs of natural science; he fought for the social recognition and dignity of the scientist; and he was the first to show to the world that Russia was capable of creating "her own Newtons."[34]

In its mythmaking role, history played down some of his more disagreeable traits: an exaggerated and passionate animosity toward the "German professors," a lordly attitude toward his Russian colleagues in the Academy, a self-centered assertiveness, a violent temperament, and an excessive restlessness. Numerous enemies were, of course, inclined to exaggerate Lomonosov's more unflattering traits. In 1743 he was arrested for participation in drunken brawls and was kept in jail for one month.[35]

In his unbounded belief in the power of science-based marine technology, Lomonosov wrote about the possibility of nagivation in the Arctic Ocean, a scheme that would have opened an additional trade route to the Orient. This view was challenged by more cautious scientists of the eighteenth century.[36] Benjamin Franklin, ever alert to new developments in science and technology, took note of Lomonosov's endeavor, but he limited his comment to stating that all known efforts to establish a navigable route in the Arctic were completely unsuccessful.[37] P. S. Pallas went even further by observing that "there can be no greater folly than the attempts to navigate the North Pole or the north coast of Siberia."[38]

Lomonosov was the first Russian to become a full member of the Academy. But he was not the only Russian academician at this time. Among his contemporaries, S. P. Krasheninnikov attracted much attention in the scientific community as well as among the slowly growing ranks of general readers. The leading Russian participant in Vitus Bering's Great Northern Expedition, he produced *A Description of Kamchatka*, a four-volume study presenting the first systematic natural-history survey of the geography, ethnography, and history of Kamchatka. In volume 1 (chapter 10), covering the physical geography of Kamchatka, he made use of material gathered by G. W. Steller, an adjunct of the Academy, that described Bering and Steller's voyage to "the shores of America." Published in 1755, the study was remembered as the first monograph on the natural history of a Russian area to be undertaken and completed by a native scholar and to be written in the Russian language. Krasheninnikov's meticulous study of the culture and society of Kamchatka natives earned him the title of the founder of Russian ethnography. He was the first scholar to prepare a comparative dictionary of Siberian languages, a work that was rapidly translated into German, English, and Dutch.[39]

The dictatorial regime instituted by the chancellery—the administering office of the Academy—coupled with continued feuding among various groups of professors and unfavorable finances, made it difficult to attract established foreign scholars to St. Petersburg. There were, however, notable exceptions. One was Franz Ulrich Theodosius Aepinus, who came to St. Petersburg in 1756 at the age of thirty-two and who produced *An Essay on the Theory of Electricity and Magnetism,* an ingenious effort to make electricity and magnetism strong and carefully ordered components of Newtonian *philosophia naturalis.* The historical value of this work has been clearly depicted by R. W. Home:

> Aepinus's *Essay on the Theory of Electricity and Magnetism,* first published in St. Petersburg in 1759, was one of the outstanding achievements of eighteenth-century physics. Previously, the sciences of electricity and magnetism had always been investigated in a wholly qualitative and non-mathematical way. In Aepinus's hands they acquired for the first time something like their modern rigorous and highly mathematical form.[40]

Franklin, who "found electricity a curiosity and left it a science," greeted Aepinus's work as a successful effort to apply his own principles of electricity to the magnetism theory;[41] perhaps Franklin recognized in Aepinus's work a bold effort, completed by J. Clerk Maxwell more than a hundred years later, to create a mathematical theory of the unity of electricity and magnetism. Aepinus quickly adjusted to life in St. Petersburg and was ready to take on many educational and popular assignments that endeared him to the royal court. R. W. Home gives the following glimpse into the growth of Aepinus's popularity:

> Aepinus's rise to imperial favor may have begun with his public lecture before the Empress Elizabeth and her court on 7 September 1758. But whether it was then or on some subsequent occasion that he first attracted attention, what is certain is that only a few months later, in early 1759, he was commissioned to prepare an exposition of the "system of the world" for the edification and amusement of the then Imperial Grand Duchess, soon to become the Empress Catherine the Great. His essay was extremely well received—so well, indeed, that he was appointed there and then by Catherine as her personal tutor in natural philosophy. And from this time onward, we are told, he was always treated by Catherine with ranks of the most exceptional favor.[42]

All the royal attention did not free Aepinus from occasional entanglements in academic squabbles. As to his scientific theories, they were too advanced to attract as much immediate attention in the scientific community—in Russia or abroad—as they deserved. During the waning decades of the eighteenth century, Aepinus's theories were commented upon, often with a mixture of admiration and slur, by such giants of contemporary science as Joseph Priestley, Charles Coulomb, R. J. Haüy, and G. Buffon.[43]

During much of the 1740s the Academy was impoverished and in disarray. Salaries of academicians were often delivered with considerable delay. From 1741 to 1746 not a single volume of the *Commentarii* was published.[44] It was not until the end of the 1740s that academic salaries began to improve and the "Russian Party" in the Academy found increasing numbers of highly placed government officials ready to listen to their grievances. Lomonosov acquired a place in the table of ranks, first as a "collegial counselor" and then as a "state counselor," positions generally closed to members of the Academy. He received this honor, however, thanks to the patriotic sonority of his poems rather than to the power of his science. But the chancellery showed no signs of relinquishing its despotic powers or of tolerating any search for academic autonomy. In 1747 the Academy became a legal institution in the full sense of the term, for it acquired its first charter, stipulating its internal organization and duties. But the charter legalized the status of the chancellery as the unchallengeable center of all authority in the Academy, ratified the existence of serious institutional limitations on free scientific work in Russia, and prompted Euler to state, in effect, that the West was becoming accustomed to hearing much more about the evil deeds of the chancellery than about the positive achievements of scholars gathered in St. Petersburg.

The charter, furthermore, divided the Academy into "the Academy proper" and the university. Until that time, the members of the Academy had been called "professors"; from then on they became "academicians." The title of "professor" henceforth went to university instructors, who might or might not have been "academicians." The Academy proper was divided into three classes, all dealing with the natural—primarily physical—sciences. The humanities and the social sciences became a domain of the university, exclusively as teaching subjects. This, of course, contradicted the intent of Peter I to empower the Academy to cover the full spectrum of scholarship. Neither the government nor the academic chancellery made an effort to explain the reasons for such a radical departure from the project of 1724, approved by Peter I, which was designed to give the St. Petersburg Academy a scholarly jurisdiction broader than that of the Royal Society in London and, particularly, the Paris Academy of Sciences. Perhaps the new policy was a nationalist response to the emergence of the Norman theory, advanced by foreigners in the Academy and much resented by Russian scholars.

At the end of the 1750s a basic change took place in the academic administration: Empress Elizabeth made Lomonosov one of the three managers of the Academy. The royal intent was to strengthen the role of Russian scholars in the administration of academic affairs without seriously impairing the supremacy of foreign academicians. Lomonosov, a bitter foe of the chancellery, was now a full member of it. As a chancellor, he in fact manifested the same autocratic disposition as did the administrators he once subjected to scorching criticism.[45] Lomonosov's ascent to the administrative summit had a considerable symbolic significance: it was the crowning point of the cultural achievements of the age, and it marked a solid beginning of the long process of strengthening the Russian side of the academic equation.

In 1761 Lomonosov recommended that the chancellery be reconstituted so as to include an equal number of foreign and Russian scholars, and in 1764 he prepared a project for a new charter for the Academy, which recommended that the chancellery be abolished and replaced by a representative academic council.[46] He fought on two fronts: against foreign rule in the Academy and against the government forces that prevented the Academy from advancing the mechanisms of self-administration. His hostile actions brought him more enemies than friends in high government circles close to the Academy. The anti-Lomonosov forces in the Academy received help from Euler (then in Germany), who, in a letter to G. F. Müller, expressed concern about the alleged destructiveness of Lomonosov's behavior. Then on May 2, 1763, Catherine II signed a decree forcing Lomonosov to retire. But before the Senate could act on the decree, Catherine changed her mind and abandoned her plan to remove Lomonosov from the Academy.[47] Catherine's wavering was the best example of the general confusion in the Academy, a battlefield for many wars.

During the reign of Elizabeth, which coincided roughly with Lomonosov's rise to prominence, the Academy pursued new interests and directions. In 1739 the Geography Department came into being as a permanent and inordinately active component of the Academy; before his departure for Berlin, Euler was its major consultant.[48] In 1745 the department published the *Atlas of the Russian Empire*, in Russian and in Latin, the first cartographic work produced in and about Russia, built upon mathematical and astronomical foundations.[49] The achievement was all the more remarkable because until that time the country did not have reliable latitude and longitude measurements for a single community—not even for St. Petersburg. In this age the Academy also undertook a more ambitious effort at the dissemination of knowledge:

from 1755 to 1764 it published twenty volumes of *Monthly Works,* a popular journal of encyclopedic scope. The Academy, however, was unsuccessful in preserving the university, which languished until 1786 and then faded out of existence. (The academic gymnasium lasted until 1803.)[50]

Lomonosov died on April 14, 1765, the third year of the reign of Catherine II. Faced with difficulty in finding an academician to deliver an *éloge* to the great man, the administration welcomed the offer of N. G. Leclerc, who only four days earlier had been made an honorary member of the Academy. Leclerc, a physician and a popular writer about Russian history in the Russian language, used a good part of the *"éloge"* to thank his lucky stars for his own election to the prestigious body. When he descended to his main task, he saved the kindest words for Lomonosov's talents and accomplishments as the poet of the age. He compared Lomonosov to a bird that flew "above the clouds." The Academy made no effort to preserve Leclerc's oratorical masterpiece; more than a century later it was discovered by P. P. Pekarskii in the archives of the Ministry of Foreign Affairs.[51]

THE ACADEMY AND THE ENLIGHTENMENT

Catherine II, who ascended the throne in 1762, opened Russia to the wholesome influence of the French Enlightenment. It was because of her direct initiative that Voltaire was commissioned to write a biography of Peter I, that Diderot visited St. Petersburg—where he attended the meetings of the Academy of Sciences—that special commissions were set up to translate Georges Buffon's *Histoire naturelle* and selected articles from the *Encyclopédie* into Russian, and that the Russians had the opportunity to read Rousseau in their own language.[52] The names of Jean D'Alembert, Montesquieu, and Pierre Bayle circulated far beyond the narrow domain of the Academy.

A combination of luck and adroit moves helped the Academy make up for the loss of Lomonosov through the acquisition—or reacquisition—of three masters of eighteenth-century science. In July 1766 Euler returned to St. Petersburg, where he completed his illustrious scientific career. He returned a different man. Maintaining his scholarly pursuits, he expanded his activities to give vent to his philosophical and religious ideas, to write a textbook in algebra, to work on number and probability theories, and to add new tools to the complex world of mathematical analysis. His *Letters to a German Princess*, highly literate and trenchant, was, in the words of Valentin Boss, "the most exhaustive and authoritative treatment of natural philosophy to be written by any major scientific figure in the eighteenth century."[53] In his long *éloge* to Euler, who died in 1783 at the age of seventy-six, Condorcet, permanent secretary of the Paris Academy of Sciences, remarked that "all the noted mathematicians of the present day were his pupils, for all were educated with the primary help of his works. All had learned the art of mathematical analysis from his books and all were directed and supported by the genius of Euler and his discoveries."[54]

Neither idealism nor nostalgia led Euler to return to Russia. According to an authoritative source, the persistent rumors that Frederick II was trying to persuade D'Alembert to become the president of the Berlin Academy made Euler, who was in open conflict with the famous mathematician and Encyclopedist, more than eager to explore the possibilities of new employment.[55] He was all the more willing to return to Russia because the St. Petersburg Academy met most of his demands. His salary was raised to stratospheric heights. His son Johann-Albrecht, a mathematician of limited promise, was "elected" a full member of the Academy. Another son was given a high position in the Artillery Department, where he rose to the rank of general, and the youngest son was appointed to a high position in the medical organization and served as a personal physician of Catherine II.

Euler's demand that he be made a vice president of the Academy was not met. He thought that this position would have assured him of a strong hand in shaping academic policies and would have given him entrance to the world protected by the table of ranks. The government, for its part, was not ready to give Euler prerogatives of power that would have weakened bureaucratic control of the Academy; nor was the government ready to make scientific achievement an official criterion of social status. The authorities were accustomed to viewing the Academy as a nonaristocratic institution; the available biographical data show clearly that only an insignificant number of academicians and University of Moscow professors came from noble ranks. Most native scholars came from the clergy or from the motley aggregate of "service people," such as soldiers in the royal-guards regiments, artists, artisans, and free peasants.[56] To a typical member of the Russian nobility of this time, learning was neither a necessity nor an ornament.

Peter Simon Pallas joined the St. Petersburg Academy in 1767, one year after Euler's return from Berlin.[57] While Euler was the widely acclaimed master of the most advanced branches of *philosophia naturalis*, Pallas was the prince of *historia naturalis*, equaled only by Alexander von Humboldt, the great explorer of the Americas. Born in Berlin and educated in Germany, England, and Holland, Pallas was an accomplished scholar before he accepted the invitation to move to St. Petersburg. His *Elenchus zoophytorum*, a study of sponges and coral polyps, published in the Hague in 1766, attracted wide attention not only through its richness of detail but also because of its theoretical insights of general biological import. In 1768 he left St. Petersburg for an eastern expedition that took his team to Samara, Orenburg, Ufa, Cheliabinsk, Tobol'sk, and the Baikal area, and, on the way back, to the lower Volga region. The expedition, which lasted six years, is generally considered the largest and most productive scientific endeavor of its kind to have been undertaken in Russia, before or after Pallas. It served as a model for several similar expeditions, some of them organized and led by Russian naturalists and all contributing to making the age of Catherine II "the age of exploration." In addition to having enriched the knowledge of Russia's natural resources and multiple ethnic components, these ventures widened the base of the existing sciences, helped to create new disciplines (such as ethnography), and produced theoretical insights of general scientific significance.

By removing dilettantism from natural history, Pallas brought this branch of science closer to natural philosophy—to a theoretical science reaching far beyond simple summations and systematizations of empirical data. For an entire generation, expeditionary work in natural history provided young Russian scholars with the best apprenticeship in theoretical work. Pallas's concern with theory was best expressed in two public lectures delivered in St. Petersburg in 1777 and 1780. In "Observations on the Formation of Mountains," he presented a general scheme of the major strata of the earth's crust, placing special emphasis on the role of intraterrestrial thermal forces in geological evolution. In 1813 Georges Cuvier noted that this paper "gave birth to a completely new geology."[58] In "A Memoir on Variation in Animals," Pallas presented a brilliant summary of the views of Linnaeus and Buffon on the transformation of living forms.[59] Darwin was particularly impressed with Pallas's views on animal hybridization—on "artificial selection." Although Pallas quickly retreated to an antitransformist position in general biology, his scientific legacy placed him among the forerunners of Darwinian theory.[60]

Caspar Wolff, the third pillar of Russian science in the Catherinian age, classified his work as an effort to combine a "philosophical" view and a "historical" view in embryology, a biological science to which he gave modern theoretical grounding. Trained at the Berlin Medical and Surgical Academy and at Halle University, he attracted the attention of the world of scholarship first in 1759 with the publication of *Theoria generationis* and then in 1764 when the first copies of *Theorie von der Generation*, a German translation and popular reworking of the ideas presented in the Latin book, reached the book market. These volumes made Wolff one of the most conspicuous and most consistent eighteenth-century forerunners of Darwin and his theories. According to an early biographer, Wolff's work was a polemic against the dominant physiological methods and ideas of the age.[61] His embryological theory of epigenesis was a direct attack on the reigning theory of preformation, according to which the growth of an embryo was merely an enlargement of predetermined and precreated parts of an organism. Wolff's theory of epigenesis placed heavy emphasis on the role of environment in modifying embryonic growth. Wolff attracted attention not because the scientific community was ready to absorb his heretical ideas but, on the contrary, because it was determined to wage a bitter war against them. No less an authority than Albrecht Haller, the master of physiology, took on the task of combating Wolff's heresy. In the process, however, he made a scrupulous effort to give an accurate presentation of Wolff's thought, thus helping to plant the seeds of biological evolutionism.

Wolff's heterodox ideas prevented him from obtaining academic employment in Germany; this made the invitation by the St. Petersburg Academy all the more attractive. Once in Russia he abandoned embryology and moved to anatomy; but even here his extensive study of monsters was clearly designed to stress the role of environment in the development of organisms. Despite his failure to concentrate on one problem, he fathered the strong Russian tradition in embryology, which included such eminent figures as Christian Pander, Karl von Baer, and A. O. Kovalevskii, each making a distinct contribution to the victory of the evolutionary view in biology.

During the reign of Catherine II, thirty-three scholars were admitted to the major categories of active academic membership; fourteen of these scholars were born in the Russian Empire and most were native Russians. M. I. Sukhomlinov, a nineteenth-century historian, created the myth that Russian academicians of this era comprised a succession of generations united by a common origin in—and dedication to—Lomonosov's scientific legacy. Sukhomlinov stated:

> S. Ia. Rumovskii, S. K. Kotel'nikov and A. P. Protasov received their scientific education under the guidance of Lomonosov. I. I. Lepekhin and P. B. Inokhodtsev were students of Rumovskii and Kotel'nikov. N. Ia. Ozeretskovskii, N. P. Sokolov and V. M. Severgin were educated under the beneficial influence of Lepekhin, etc. These generations of Russian scholars from Lomonosov to Severgin were related to each other by the basic principles of their scientific activity and the ties of a literary tradition that reflected the realities of the time and the historical development of Russian enlightenment.[62]

This position cannot be defended. Lomonosov was indeed an inspiration to the growing ranks of Russian naturalists, but these scholars did not necessarily follow the theoretical and methodological dictates of the great master. Collectively they represented the full arch of scholarly endeavor. Some—S. Ia. Rumovskii, S. K. Kotel'nikov, and M. E. Golovin—came from Euler's school; others—led by the naturalists I. I. Lepekhin, N. Ia. Ozeretskovskii, and V. F. Zuev—were closely identified with Pallas's approach to natural history; and still others were either "independents" or owed a heavy intellectual debt to a combination of scientific orientations.[63] None of these groups produced true giants whose names graced the annals of world science. All, however, formed a vital link in the evolution of scientific thought in Russia. They brought the Academy—and high science—closer to the burgeoning intelligentsia, which viewed science as a combination of positive knowledge and antiautocratic ideology. They played a major role in stepping up the popularization of science and in uniting the Academy with Moscow University, newly formed learned societies, and the expanding institutional base of public education. And they helped to give a distinctly Russian meaning to the utilitarian value of science.

In one important respect, Russian naturalists from Kotel'nikov to Severgin were followers of Lomonosov. All were encyclopedists who ranged over wide areas of unrelated scholarly activity. It was not unusual for an Eulerian mathematician to combine work in high algebra with ethnographic fieldwork, for an astronomer to concern himself with the details of natural history, or for a specialist in natural history to tackle the puzzling questions of philology. All, without exception, found it important to speak and write in defense of science as a cultural force of prime significance and to emphasize the compatibility of science and religion.

The age of Catherine II produced two new bodies dedicated to the advancement of scholarship: the Free Economic Society, founded in 1765, and the Russian Academy, founded in 1783. The first organization undertook systematic studies of problems related to the national economy, particularly to agriculture. Its aim was to propose ways of avoiding uncritical acceptance of Western economic institutions and practices, and of organizing a search for technological innovations and institutional mechanisms that would meet the unique conditions of the Russian land, climate, mineral resources, and social traditions.[64] The second institution, organized on the model of the *Académie française*, concentrated on preparing a Russian dictionary and, indeed, on creating a vernacular free of the ossified vestiges of the Church-Slavic language. The spirit of the time brought the two academies into a complementary relationship. Both responded to the national urge for "self-understanding." The Academy of Sciences studied the natural resources of the vast empire, the basis of material culture. The Russian Academy became the "treasure house" of national language, the basis of nonmaterial culture.[65] The membership of the Russian Academy included all the scholars of the Academy of Sciences who took part in expeditionary surveys. As new research centers emerged, they strengthened the role of the Academy of Sciences as a guiding force of the first rank.

The internal organization of the Academy underwent a succession of drastic changes.[66] The liberal air of the early years of Catherine's reign encouraged members of the Academy to search for larger domains of autonomous action. In 1764 Lomonosov submitted a project for a new charter that would reemphasize the preeminent role of the Academy as a research center and an educational institution. Two years later, Euler submitted a new project that focused on the Academy as a community of internationally recognized scholars, a proposal that would have slowed down the advancement of Russian scientists to the highest rungs of the academic ladder. While Lomonosov tried to make the Academy responsive to the critical needs of the nation, Euler was concerned primarily with the immanent interests of science, uncomplicated by specific practical considerations. The two projects had one common feature: both demanded the abolition of the chancellery and the creation of administrative bodies with a strong representation of academic personnel. Each proposal was considered separately, and each was rejected by the government authorities.

Leading scientists pressed for a firmer institutional footing for academic autonomy. The government, in turn, was both willing to listen to the arguments advanced by academic circles and determined not to make the Academy an oasis of "republicanism" in the system of autocratic institutions. The result was a succession of changes that disencumbered the Academy of some of its more disagreeable administrative components without expanding the institutional safeguards of academic autonomy. These changes began on October 5, 1766, when a government decree acknowledged the "decline" of the Academy and ruled that it be brought under the "direct control" of the empress as the first step toward restoring its "former state of florescence."[67] At the same time, the Academy was placed in the charge of V. G. Orlov, brother of a favorite of the empress.[68] The office of president was not abolished; however, K. G. Razumovskii, president since 1746, was assigned other functions in the government and was de facto separated from any responsibility in the Academy. The chancellery was abolished and replaced by a special commission containing a heavy representation of academicians.

The new body, estimable in design, did not succeed in preventing Orlov from becoming the unchallengeable authority in all matters of primary significance. The special commission was given the task of drafting a new academic charter, and it took on this new task with utmost seriousness; but its fully formed draft was lost in the maze of unspecified government channels. Orlov was succeeded by S. G. Domashnev, who acted as if the academic commission did not exist. A self-styled philosopher—and "a laughingstock of the salons"—he squandered the Academy's funds on pet projects and amended academic protocols to suit his own whims.[69]

The futility of efforts to make the Academy a self-governing body reached a peak when the academic commission was abolished by E. R. Dashkova, a literary figure and intimate friend of Catherine II, who in 1783 became the president of the newly founded Russian Academy and the new director of the St. Petersburg Academy of Sciences. The Pugachev uprising and the perceptible growth of rebellious intelligentsia who identified science with antiautocratic ideology made it increasingly more difficult to press for academic self-government. Instead, Dashkova concentrated on improving the material conditions of the Academy—

on raising the salaries of scholars, modernizing laboratories, expanding library and museum holdings, and rejuvenating the academic gymnasium. Under her influence the Academy acquired a new building, an architectural jewel. A strong personality unaccustomed to democratic processes, Dashkova went so far as to assume full authority in selecting and confirming new members of the Academy. Single-handedly, she dismissed V. F. Zuev from his position of adjunct in the Academy because she was displeased with his educational activities, which did not fall under her jurisdiction.[70] Pallas, who after the death of Euler in 1783 was the most illustrious member of the Academy, protested against her high-handed actions and came to the defense of Zuev, who was one of his most eminent Russian disciples. Soon Pallas moved on, first to the Crimea—but not until he had completed a much-noted exploration of that region—and then to Berlin.

The situation deteriorated rapidly as echoes of the French Revolution began to reach Russia. Catherine II rendered an ironic disservice to the Academy by not only instituting rigid censorship on imported foreign books but also appointing academicians Kotel'nikov and Inokhodtsev to serve as censors.[71] Under her successor, Paul I, there was no need for competent censors since he banned the import of all books and music published in the West. The exchange of academic publications with foreign scholarly institutions came to a standstill. The Academy's Geography Department was reduced in staff and budget to a position that forced the termination of most of its activities. The government did not trust the Academy with such a militarily and politically sensitive activity as cartography.[72] The addition of new scholars to the Academy was drastically limited. For his role in the French Revolution, Condorcet was stripped of honorary membership in the Academy. Even Princess Dashkova came under suspicion and was dismissed as director of the Academy of Sciences and president of the Russian Academy. In 1796 she was exiled to her estate. Heinrich Ludwig Nikolay replaced Dashkova as "president" rather than "director," bringing the thirty-year reign of directors to an end. Although Nikolay had difficulty communicating in Russian and had no background in—or relationship to—scholarship, he saved the Academy from full collapse during the closing years of the eighteenth century.

During the eighteenth century the Academy played a major role in shaping the institutional base of Russian science. It occupied a central position in the onrush of scientific thought, which changed the texture of Russian national life and cultural values. Science became broader in compass and deeper in meaning. In the age of Peter I science was viewed solely as a source of technical advancement and national prestige. The age of Catherine II did not overlook the technical value of science. Nor did it fail to acknowledge the prominent place of scientific knowledge—and the pursuit of science—in the development of a harmonious and intellectually balanced personality and in the evolution of an optimistic view of the future development of mankind. Turgot's view of science as the most reliable index of social progress was an important factor in shaping the thinking of Russian scientists and the new intelligentsia.

As Karl von Baer pointed out, while the age of Peter I placed primary emphasis on the universal theory of *philosophia naturalis*, particularly as applied to shipbuilding, navigation, artillery, and basic metallurgy, the age of Catherine II shifted the emphasis from the high theory of *philosophia naturalis* to the descriptive methods of *historia naturalis,* applied specifically in the study of Russian resources. Peter I was concerned with the flow of Western science to Russia; Catherine II wanted to ensure a constant flow of empirical knowledge on Russian topography and economic and human geography to the mainstream of world science. The history of the Academy, however, despite periodic wavering, was dominated by the clear effort to create a symmetrical relationship between the two major scientific activities.

The members of the Academy took different positions on the relationship of science to religion. The large majority of academicians avoided the question, for it inevitably led to disputes of an ideological nature. Euler represented the exceedingly small group willing and ready to emphasize the full harmony of science and religious thought; to Euler, the pursuit of science was a moral obligation, for it revealed the perfection of divine creations. The third group, represented by Lomonosov and a small number of other Russian academicians, recognized science and religion—the truth discovered by human reason and the truth given by divine revelation—as two completely different and incomparable but equally important approaches to the universe. Both modes of inquiry deserved full recognition, but they should not be allowed to interfere with each other. In general, Russian academicians avoided a deeper and more detailed analysis of the relationship of science to religion. The

academic translators of Buffon's *Histoire naturelle* into Russian were careful to omit the passages likely to anger the more vigilant members of
the Synod.[73]

Outside the Academy, the sparse ranks of the intelligentsia—typified
by N. I. Novikov and A. N. Radischev—viewed Newtonian science as
humanity personified and as a new ideology, democratic in essence. In
the time of Peter I, V. N. Tatishchev saw no reason why science could
not prosper in an autocratic state. In the age of Catherine II, A. N.
Radishchev had no qualms about professing a view of science and autocracy as realities totally alien to each other. He showed no hesitation in
echoing Condillac's views on the portable nature of Newtonian method
and theory—on their easy and advantageous transfer to the study of
human values and social realities.[74] A. N. Pypin was referring to the
cultural values of the Russian people, interpreted in the spirit of democratic liberalism and the ideals of Westernism, rather than to the political
principles of Russian autocracy, when he asserted that "in the era of
Peter the Great and during the entire eighteenth century no rational
being had entertained the slightest idea of a possible conflict between
science, borrowed from the West (for it could not have been borrowed
from anywhere else), and our national spirit."[75]

REFORMS AND RETREATS UNDER ALEXANDER I

Many Russian historians identified the beginning of the age of Alexander I, who ascended the Russian throne after the assassination of Paul
I in 1801, as a continuation of the policies of Catherine II, inspired by
the Enlightenment. The new tsar was heralded with unparalleled enthusiasm; the poet Derzhavin and the historian Karamzin greeted him as
a man destined to make Russia a haven for universal justice, full citizenship, and untrammeled enlightenment. From his teacher Frederic-
César la Harpe, selected by Catherine II, Alexander "learned about the
views of the Encyclopedists on human rights and duties, on the origin
of civil society, and on the relations between the rulers and the ruled."
From la Harpe he also learned that the greatest glory of the ruler was
in the fulfillment of his obligations as a man and a citizen. He wrote:
"To ask whether a future ruler should be a philosopher is to ask whether
he should be aware of his obligations and be ready to carry them out—
whether he should be a worthy citizen."[76]

Catherine II was unprepared to face the French Revolution and its onslaught on monarchical values. In panic, she betrayed the ideals of the Enlightenment and reverted to the tested techniques of despotic rule. Alexander I came to power after the French Revolution had run its course and had triggered far-reaching social and cultural changes throughout Europe. The flames of revolution precipitated unique changes in each country. But behind distinct national adjustments all countries shared the beginnings of basic transformations in social relations and cultural values. In no other area of social dynamics were these changes more visible than in education and science. The French Revolution—and Napoleon I—made science the universally recognized foundation of national strength. In the leading countries of Europe, the natural sciences and mathematics were made the mainspring of the educational process. As formal education became increasingly available at all levels to all social classes, it led to wide-ranging social changes. The professionalization of scientific work and the growth of academic autonomy acquired a wider base and a deeper meaning.

Despite its geographical remoteness and its unique social realities and political institutions, Russia did not escape the effects of the French Revolution. As soon as Alexander I became the monarch, a series of radical reforms produced a basic transformation of the institutional base of public education and national engagement in science. In 1805, in addition to a thoroughly reorganized Moscow University, the country had four universities: in Kazan, Khar'kov, Vilnius, and Dorpat. In 1819 the St. Petersburg Pedagogical Institute, founded in 1803, was transformed into St. Petersburg University, which played a major role in bringing the Academy closer to the institutions of higher education. The temper of the time helped to make the Medical-Surgical Academy, founded at the end of the eighteenth century, a lively center for research in the biological sciences directly related to medicine. It must be noted, however, that the experimental work Bichat had initiated in France found fewer followers in Russia than the metaphysical constructions of the German *Naturphilosophie*. The spirit of science—and the emphasis on scientific curriculum—was so pervasive that in 1819 the Moscow Theological Academy added calculus to its curriculum.

The universities, now under the jurisdiction of the newly founded Ministry of Public Education, acquired a new charter in 1804.[77] The Academy of Sciences played a major role in shaping the internal organization and curriculum of the unversities and, particularly, in making the universities both teaching and research centers. The charter had three distinguishing features: it guaranteed academic autonomy, as this had developed in the leading German universities (which made high administrative positions elective); it made the exact sciences—particularly mathematics—the core of the university curriculum; and, at least in theory, it opened universities to all social classes. The *raison d'être* of the universities was bolstered in 1809 when the government issued a strict ruling that a large block of top-level administrative positions in the state administration would be made accessible only to persons with certified university education. Russia was not too far behind France, Germany, and England in responding to the forces of change unleashed by the French Revolution; however, it was far behind these countries in making these forces more than a passing and shallow-rooted phenomenon. Academic autonomy was only a flicker of hope, rapidly extinguished by the vested interests of local administrations. The aristocracy hastened to open special lycées admitting only students of noble origin. The youth of gentry origin enrolled in foreign universities in rapidly increasing numbers.

It was in this atmosphere of conflict between the liberal law and conservative tradition that the Academy of Sciences acquired a new charter in 1803.[78] The new document noted that the previous charter was outdated and the budget of the Academy inadequate, and these shortcomings had caused a perceptible decline in scientific work. It also stated that the primary task of the Academy was to expand all the domains of knowledge, to champion science, to spread enlightenment, and to use scientific knowledge for human well-being. The Academy was ordered to conduct a systematic study of the country's natural resources and to make recommendations for their rational utilization. As the highest scientific body in the country, the Academy was made the supreme authority in certifying new theories and inventions. It selected scientific topics for papers competing for national prizes; it was authorized to counsel universities on curricular matters and organized research and to maintain close contact with learned societies.

The new charter was the product not only of the liberal inclinations of the new monarch but also of pressure mounted by the Russian members of the Academy. In 1801 academicians N. Ia. Ozeretskovskii, S. E. Gur'ev, and A. F. Sevast'ianov, in an unprecedented move, wrote a collective letter to Alexander I urging him to help check the decay of the Academy and to restore it as the blossoming institution it had been a generation earlier. The letter was an overt expression of awakened nationalism: its basic demand was that the Academy become a truly Russian institution—dominated by native Russian scholars and devoted to the idea of making the Russian language the basic vehicle of scientific communication.[79] The Academy, the letter stated, was in full disarray. It had three chemists but not a single chemical laboratory; it had an astronomical observatory but no astronomer to make use of it; "the botanical garden was in such disarray that it did not deserve to go under that name"; and the library had discontinued the purchase of new publications.[80] The writers of the letter demanded that the position of president be filled by individuals elected by the members of the Academy and that the real authority be vested in the academicians as a collective body, rather than in the president as a representative of the monarch. Furthermore, they demanded a drastic reduction of government administrators working in the Academy.

The temper of the time favored an Academy that would be democratic in inner organization and national in human composition and linguistic background. National preference, however, did not deter the Academy from trying to persuade K. F. Gauss, the rising star of nineteenth-century science, to join its staff. This happened immediately after Gauss published his celebrated *Disquisitiones arithmeticae* (1801), which revealed the unmatched depths of his mathematical mind. After an initial interest in the offer—and efforts to secure better working and living conditions for himself in St. Petersburg—Gauss decided to remain in Germany.[81] Later on he became one of the chief Western supporters of N. I. Lobachevskii and his non-Euclidean geometry.

At this time, N. Ia. Ozeretskovskii and N. Fuss, two influential members of the Academy, were commissioned by the government to draft a new censorship law more in tune with the liberal sentiments of the day. In remarks appended to their draft, the two scientists viewed censorship as an outright and crippling attack on both the freedom of thought and "the spirit of the time." Censorship, they wrote, threatened "to uproot candor, to weaken rational faculties, to stifle the love for truth, and to check the spread of enlightenment." The free exchange of ideas was, according to them, the basic vehicle of human perfectibility and the best school of citizenship.[82]

The charter recognized explicitly that the advancement, diffusion, and practical application of science was the major source of national well-being and that the Academy was the supreme scientific body in the country.[83] By having established close ties with the universities, the

Academy became an important vehicle in the transmission and diffusion of "useful knowledge." Through two popular journals—*Intellectual Exploration (Umozritel'nye issledovaniia)* and *Technological Journal*—it reached an audience beyond university professors and students. The chief contributors to *Intellectual Exploration*, published from 1808 to 1819, were the Russian members of the Academy: S. E. Gur'ev wrote original articles on mathematical themes; V. V. Petrov reported on his physical experiments in the general field of electricity; V. M. Severgin wrote about new advances in paleontology and mineralogy; P. A. Zagorskii covered anatomy; and N. Ia. Ozeretskovskii presented natural-history surveys of selected Russian regions. Although aimed at a general audience, all papers met the technical standards of contemporary science. They gave the readers a representative picture of the main lines of scientific endeavor, from the most obscure branches of mathematics to the simplest naturalist accounts of individual geographical regions. Above everything else, they served as solid testimony to the rapidly widening scope of scientific interests that Russian scholars handled with professional ease and remarkable originality. Gur'ev, one of the most accomplished contributors to the journal, may rightfully be called the first Russian scientist to make original contributions to high mathematics. His work in differential and analytical geometry made him a key link in the evolution of Russian mathematical tradition.[84]

The charter introduced the position of permanent secretary, whose duties included keeping a record of academic meetings, corresponding with other learned bodies, supervising publication activities, and controlling the archives. The Academy was granted the right to conduct its own censorship. It ceased to be directly involved in formal education; this, however, did not exclude the recruitment of promising young students for advanced studies under the guidance of individual academicians. The new charter created the impression that the Academy was now a self-perpetuating institution relieved of heavy bureaucratic entanglements and firmly identified with research in the leading sciences. By assigning the academicians definite positions in the table of ranks, the charter helped integrate the scientific elite into the structure of Russia's high society and made it easier for foreign scholars to orient themselves in the realities of Russian existence. Napoleon I showed his appreciation of scientific work by decorating scholars of high distinction with the Legion of Honor. Alexander I, for similar reasons, decorated many members of the Academy with the Medal of St. Anna.[85]

Strongly influenced by the unusually rapid advancement and popularity of economics in England and France, the new charter added political economy and statistics to the established research areas in the
Academy. Since political economy combined economic theory with
moral philosophy, it had an appeal beyond the narrow circles of academic professionals. It quickly found a place in the university curriculum
and attracted the attention of the intelligentsia in search of models for
a new society. Although the Academy deserved much credit for the
growing interest in economic theory—as well as in the areas where
economic thought spilled over into sociology and social philosophy—
its descriptive work was much more impressive than its involvement in
theory. The new field of statistics found its most significant expression
in the monumental work of A. Shtorkh and F. Adelung, who assembled
and analyzed numerical data on books published in Russia from 1801
to 1806. The statistics clearly demonstrated the country's heavy dependence on translated books and on books published in Russia in foreign
languages. For example, of 106 books placed under the category of "natural science," only 14 were Russian originals; of 53 books in mathematics,
only 19 were Russian originals.[86]

During the reign of Alexander I, the Academy made an effort to
strengthen the Russian wing of the academic roster. Of the seventeen
scholars who were elevated to the rank of full academician during this
period, eleven were born in Russia. But to understand the real ethnic
composition of this group of honored scholars, it should be noted that
seven of the eleven "Russians" came from Western European—mostly
German—families settled in Russia or from the German community in
the Baltic area. A typical member of the new group of academicians was
an average scientist with no distinguished record. Among all academicians, including those who had been elected earlier but who had reached
their peak of scholarly accomplishment during the first quarter of the
nineteenth century, the Russian historians of science gave the most
lavish accolades to V. V. Petrov, who transformed the Academy's physical "laboratory" from total chaos to a modern and well-equipped scientific workshop. Here he conducted experiments in galvanism and
various physicochemical aspects of light, heat, electricity, and magnetism. Petrov was caught squarely in the middle between the generation
of Lavoisier, which viewed heat and light as products of distinct chemical
elements, and the generation of Oersted and Faraday, which, by establishing the unity of electricity and magnetism, was well on the way to

unifying physics—and to rising above chemical reductionism in physics.

The triumph of the "Russian" orientation in the Academy made it possible to rectify a grave omission of the earlier era: the presentation of some kind of encomium to Lomonosov. The task went to V. M. Severgin, who in 1805 delivered a long and passionate oration recounting Lomonosov's contributions to science in general and to "Russian science" in particular. Having done his homework with impeccable thoroughness and involvement, Severgin succeeded in showing not only the wide scope and originality of Lomonosov's work but also the temper of the first decade of the nineteenth century, when it was popular, in the words of M. I. Sukhomlinov, to equate the "love for science" with the "love for Russia."[87] He gave new life to Lomonosov's dictum that although "truth" and "belief" were "daughters of the one supreme parent"—and as such were not in conflict with each other—it would have been an unpardonable error to apply the tools of science in the examination of religious beliefs or the tools of religion in the search for scientific truth.[88] Lomonosovian tradition, like that of Peter I, called for a full separation of science and religion—an institutional separation no less than an intellectual separation.

Severgin, one of the most widely known and popular academicians of this era, represented a new step in the evolution of Russian scientists as professionals. During the eighteenth century all Russian scientists, Lomonosov no less than Krasheninnikov, Rumovskii, Kotel'nikov, Ozeretskovskii, Lepekhin, and others, were "naturalists" with encyclopedic interests. Severgin was the first Russian specialist in science. His *Foundations of Mineralogy* was the first original Russian work of its kind.[89] His orientation, however, was purely descriptive; at the basis of his classification of minerals were the chemical attributes of component elements. R. J. Haüy's ground-breaking effort in structural crystallography was too complex and removed from descriptive procedures to attract Severgin's attention.

At the beginning of the nineteenth century, the Russian intelligentsia in general evinced a lively interest in Western philosophy, which at that time concentrated heavily on the nature and limits of scientific knowledge. The Academy, however, with its interests extending in many directions, found the spirit of philosophical fermentation less attractive. The universities, heavily populated by Western professors, brought to Russia every major orientation in current philosophy. Most remarkable was the widespread concern with the philosophy of Immanuel Kant, which found both enthusiastic supporters (particularly among the German professoriat) and determined critics (led by Professor T. F. Osipovskii from Khar'kov University). To supporters, Kant's critical philosophy represented the opening of new horizons in the evolution of thought and a new justification of the complementary relations between science and metaphysics. To critics, particularly to Osipovskii, it represented a revival of ancient philosophy that was clearly discredited by the great scientists of the seventeenth century.[90]

Popular, too, at this time were the empiricism of Locke and Condillac and the utilitarianism of Jeremy Bentham. Schelling's *Naturphilosophie*, a critique of the mechanistic footing of Newtonian science, reached a peak in popularity in Russia in the 1830s, at a time when Schelling had drifted to the even more abstruse terrain of philosophical irrationalism. While the followers of Schelling considered Bacon's philosophy of induction and experiment a blinding force in science, critics of the *Naturphilosophie* treated Bacon's ideas as the true fountain of modern natural science.[91]

The Academy's isolation from philosophical thought established a precedent generally adhered to during the remainder of the tsarist era. Individual academicians were known for making statements of a philosophical nature, but none made an effort to elaborate a system of philosophical thought. Philosophical aloofness was part of a carefully cultivated effort to isolate the Academy from ideological involvement. Outside the Academy, philosophy, according to M. I. Sukhomlinov, attracted a wide audience concerned with differences between "ideals" and "reality," and with human rights. Philosophy provided a wide umbrella for discussions on the cultural attributes of Russian nationality in their historical context rather than in their official definition.[92] Despite its apparent detachment from philosophical controversies, the Academy resorted to many techniques in expressing its unsympathetic attitude toward attacks on the Baconian legacy in the methodology of modern science. One such technique was the differential treatment of published works competing for academic prizes: the academicians seldom failed to favor the studies written in the spirit of the philosophical dictates of the great English advocate of the inductive method in science. The Academy never struck a blow against the Baconian legacy.

Meanwhile, Alexander's attitude toward science and science policy changed markedly. His earlier recognition of both the power and the enlightenment value of science gradually gave way to grave doubts

concerning the broader enclaves of culture in which science was
anchored. This change was compounded by the influence of Joseph de
Maistre on A. K. Razumovskii, minister of public education, which gave
encouragement only to the foes of liberal reforms. The Sardinian plen-
ipotentiary in St. Petersburg, de Maistre was a bitter critic of Baconian
"materialism" and the leading articulator of a philosophy opposed to
the ideas unleashed by the French Revolution. The Napoleonic Wars,
the Holy Alliance, the triumph of conservative policies in German universities, and the founding of the Russian Bible Society turned the tide
in favor of new political forces that ushered in an age of blind reaction,
pathological mysticism, and intellectual repression. German professors
were dismissed from Russian universities for the simple reason that they
were Protestants.

Articulated by M. L. Magnitskii, curator of the Kazan school district,
and D. P. Runich, curator of the St. Petersburg school district, the new
policy sought to establish a new science—a science anchored in a mys-
tical version of Christianity bitterly opposed to Newtonian mechanistic
theory and to all scientific theories seen as challenging the supreme
authority of divine powers and scriptural interpretations. Proliferating
government instructions demanded that university professors teach a
physics that had no room for the law of causality, a mathematics that
was imbued with the spirit of religion, and a physiology without experiments.[93] The new strictures brought an end to the teaching of geology
and demanded that the teaching of moral principles be closely tied with
religion, that theology be an obligatory course for all students, and that
natural law not be allowed to go beyond the official interpretation of
Russian law. To symbolize the new unity of university education and
religion, the Ministry of Public Education, the spark plug of liberal
reforms, was replaced by the Ministry of Religious Affairs and Public
Education.[94] The religion of the new authorities had little in common
with the Russian Orthodox church: it was rather a unique blend of
abstract mysticism and elaborate ritualism. Opposition by the leaders of
the church shortened the life of the new heresy—and the national nightmare it had created—and brought its downfall two years after the death
of Alexander I in 1825.

Uninvolved in formal education, the Academy of Sciences did not feel
the full impact of the deadening blows of the new policy. N. Fuss,
permanent secretary of the Academy, was the only member of the Scholarship Committee of the Ministry of Religious Affairs and Public Edu-

cation who openly criticized Magnitskii's attack on the "materialism" of science and contemporary education. Fuss attacked Magnitskii's groundless condemnation of "universities" and "entire estates" as fountains of atheism and then quickly retreated from overt criticism and devoted his time to isolating the Academy from contemptuous and staggering attacks on science.[95] During the ten years of the Magnitskii era (1817–1827), only six full members were added to the Academy. That all these scholars were either foreigners or Russian Germans was proof that the Academy had managed to escape closer scrutiny by the new authorities.

S. S. Uvarov, the new president of the Academy, was conspicuously silent during this period. A minor scholar in ancient history and literature, Uvarov wrote all his papers in French and preferred to speak in French rather than in Russian. While recognizing the paramount social and cultural value of science, he advocated an acceptance of "Western science" on "Russian terms." Magnitskii and his group strove to create a Russian science and, in articulating their views, made a crude attack on Newtonian theory as the mainstream of contemporary science. Uvarov, in contrast, wanted to make Western science adaptable to the needs of Russian society. Unlike Magnitskii and his allies, he placed primary emphasis on the utilitarian aspect of science. His policies—particularly under Nicholas I—showed that he favored separate roles for the Academy and the universities: while the function of the Academy was to bring "Western science" to Russia, the function of the universities was to give this science a Russian meaning. He wanted the Academy to stay outside the domain of ideology, and he wanted the universities to harness science in support of autocratic beliefs.

THE ERA OF "OFFICIAL NATIONALITY"

In 1833 Uvarov became minister of public education, a post he held until 1850, without surrendering his position as president of the Academy. In full command of Russian educational and scholarly institutions, he could now act with more authority in making the Academy an integral part of Russian life as he perceived it. As the ideologue of Nicholas I's conservative regime—the regime of "official nationality"—he made censorship an elaborate and far-reaching activity carried out by a complex system of professional agencies. He did not hesitate to abolish ideologically deviant journals or to impose drastic limitations on their circulation.[96]

Uvarov's first task as the new minister was to dismiss university professors suspected of political disloyalty. He kept philosophy out of the
Academy and pursued a policy of favoring Orthodox priests as teachers
of philosophy on the university level. At first he was suspicious even
of the Slavophiles, who, relying on Schelling's philosophical ideas,
preached the purity of Russian soul, expressed in the social virtues of
the patriarchal *obshchina* and in the psychology of Orthodox Christianity,
and he ordered school authorities to wage a war against them. Uvarov
contended that theoretical science was an ideological liability, for it
opened sacred values to critical scrutiny; but he did not undervalue
applied science as a national necessity of prime significance. His aim
was to effect a functional reconciliation of the two extremes. His rhetoric
favored the purity of autocratic ideology; his deeds showed that he was
willing to make the Academy a scientific forum of maximum utility.
P. H. Fuss called his academic reign "the epoch of the regeneration of
the Academy," an epoch unmarred by "violent and quite insensible
crises" and dominated by efforts to bring order into the complex world
of academic activity.[97]

In 1836 the Academy received a new charter, a minor revision of the
1803 charter. The office of vice president was instituted for the first time.
The most obvious—and the most practical—intent of this office was to
give some relief to Uvarov, who was deeply involved in the affairs of
the Ministry of Public Education. Like that of president, the position of
vice president was nonelective; indeed, the vice president was appointed
directly by the president. From 1835 to 1852 the position of vice president
was filled by M. A. Dondukov-Korsakov, a man with no affinity with
scholarship.[98] The new charter set the number of full academicians
at twenty-one, three more than allowed by the previous statutes. It provided for the creation of an administrative committee to manage
the internal affairs of the Academy not directly related to research. The
committee consisted of six members, including two academicians.[99]
When the committee members voted against the wishes of S. S. Uvarov
as president of the Academy, the matter was referred to S. S. Uvarov
as minister of public education, who had the authority to produce irrevocable decisions.

Although the members of the Academy were not too unhappy about
the charter, they never ceased pressuring for wider autonomy. With the
exception of occasional participation in the collective effort of expeditionary research, which involved a relatively small group, academicians

were involved primarily in personally initiated research projects. This did not exclude the participation of individual scholars in various government committees. Despite the desirability of this kind of practical appointment—and the extra income that it brought—most academicians were advocates of "pure science," a science guided by its own internal logic rather than by practical dictates emanating from government agencies. To boost their income, a good number of academicians took on teaching duties at St. Petersburg University, the Medical-Surgical Academy, and various professional schools. Some academicians, typified by the mathematician M. V. Ostrogradskii, became so involved in teaching duties that they abandoned research altogether.

The problem of the social—or practical—role of science received much more attention from the intelligentsia, who operated outside the institutional framework of science, than from the country's scientific elite. A typical academician was satisfied with unadorned and unelaborated statements on science as an autonomous and self-propelled cultural force. A. I. Herzen started his long career as a leader of the intelligentsia with a critical analysis of the interrelations of the scientific community and the society at large. In "Dilettantism in Nature," published in 1843, he noted that although science did not have open enemies, it had many false friends, sapping its vitality and preventing it from performing its social duties. Among the false friends, two categories stood out in bold relief: the dilettantes and the "Buddhists of science." In the process of anchoring their ideological views in science, members of the intelligentsia—as dilettantes in science—were often guilty of advocating a world view based on superficial and often untested scientific ideas. Included in the ranks of dilettantes were also university professors, typified by M. G. Pavlov, who in his course on physics disguised his ignorance of current achievements in science by elaborate retreats into Schellingian metaphysics, completely isolating philosophy from experimental science.[100] The Schellingians sought to "correct" science by relying on a system of philosophical ideas; Herzen, under the temporary spell of Auguste Comte, advocated a new philosophy fully congruent with the experimental data of science.[101] By misinterpreting the state of science, Herzen thought, the dilettantes rendered a disservice to both science and society.

The "Buddhists of science," according to Herzen, were members of the scientific community who strove to become a special estate living in the narrow confines of an ivory tower, in isolation from the realities of social existence. A product of excessive specialization, they isolated themselves also from the total substance and spirit of science. "The Buddhists of science," in Herzen's words, "who have risen to the sphere of the universal in one way or another, never depart from it. There is nothing that could tempt them into the world of realities."[102] The "Buddhists," he concluded, were "fiercer champions of science than science itself"; they would rather die than remove science from the heights of intellectual absolutism. Herzen made no specific reference to the Russian situation, but this did not prevent him from giving a new focus to the intellectual life of his age: he shifted emphasis from the unending discussions of the relations of science to religion, morality, and metaphysics to the relations of science to society as a living and dynamic reality. He attacked the cultivated effort of the Academy elite to ignore the social moorings of science and to preserve the tranquillity of the inner sanctum of their profession. From now on, the charges that the Academy was dedicated more to enriching "Western" scientific theory than to meeting the acute needs of Russian society were coming from many sides. Herzen was the real architect of the movement of Russian intelligentsia that viewed the Academy as a "German institution" detached from the realities of Russian life. While conceding that the dilettantism of the intelligentsia was curable through sober acquisition of knowledge and cultivation of philosophical realism, he was uncompromising in his belief that "Buddhism in science" was a "grave and incurable malady."[103]

During this era the ethnic composition of the Academy underwent significant changes that helped both the non-Russian and Russian groups. The influx of German academicians, born in or outside of Russia, was no doubt part of Uvarov's plan to staff the Academy with persons too far removed from Russian realities to be effectively involved in the ideological stirrings of the time. These scholars raised the reputation of the Academy and made it one of the leading scientific institutions in Europe. H. F. E. Lenz received immediate and widely heralded recognition for the formulation of what became known as Lenz's law, which indicates the direction of the flow of an induced electric current. The scholarship of the embryologist Karl von Baer was held in such high esteem everywhere that Darwin was eager to have him on his side of the evolution controversy. F. G. W. Struve not only founded the Pulkovo Observatory in 1839 but made it one of the leading astronomical centers of the world. H. H. Hess was subsequently recognized by Wilhelm Ostwald as the founder of thermochemistry. M. H. Jacobi's work in galvanoplastics and successful experiments with constructing electricity-powered riverboats received wide acclaim; Michael Faraday read a report on them at a meeting of the British Association for the Advancement of Science. A. Th. Kupffer was a vital link between the scientific legacy of R. J. Haüy and modern structural crystallography; he played an important part in building a strong Russian tradition in this discipline, which reached a high point in the work of E. S. Fedorov, who, at the end of the century, relied on mathematical procedures to establish 230 basic crystalline forms. Non-Russians—mainly Germans—produced a new generation of explorers of Siberia who were modern specialists in individual disciplines (for example, G. P. Helmersen in geology and paleontology, and A. T. Middendorff in zoology). The work of these scholars marked the transfer of regional exploratory work from undifferentiated natural history to specialized disciplines. Indeed, their work marked the end of natural history as a major activity of the Russian scientific community.

"Foreigners" in the service of Russian scholarship formed a tightly woven community with carefully circumscribed ties with the society at large. They maintained closer relations with the Western scholarly world than with any group in Russia. The Academy, which depended mainly on the French and German languages for internal communication, recognized that non-Russian scientists needed to maintain this relative isolation. During the reign of Nicholas I the situation began to change at a slow but steady pace. The scope of nonacademic activities by members of the Academy became ever wider. "Foreign" academicians began to search for broader contacts with the general public, concentrating on giving public speeches on scientific themes and publishing articles in popular journals—such as the *Library for Reading*—that gave special attention to modern developments in science. They helped the *Mining Journal,* established by a government department in 1825, to become a highly successful publication. They helped the Mineralogical Society, established in 1817 in St. Petersburg, to become a going concern and to gain enough public support to publish a scientific journal. Karl von Baer, supported by the famous explorers F. B. Lütke and Ferdinand Wrangel, initiated a series of quickly unfolding actions that led to the founding of the Russian Geographical Society in 1845. During the first few years of its existence, the society was a haven for foreign scholars; by the end of the decade, however, the Russians had succeeded in establishing a dominant position without alienating von Baer and other non-Russian scholars. The society wasted no time in pursuing two main activities: organization of expeditionary teams to various parts of the huge empire and publication of new materials on the geography of Russia and neighboring Asian lands.

The Russian wing of the academic corps expanded drastically in 1841 when it absorbed the members of the recently abolished Russian Academy, founded by Catherine II in 1783 to study Russian literature and language. With one stroke Russian scholars became the major group in the Academy. This, however, did not change the situation in the natural sciences, where the "foreigners" continued to be a dominant factor. The new members bolstered the humanistic studies of Russia's past without weakening the Academy's primary commitment to the natural sciences. The defunct Russian Academy became the Department of Russian Language and Literature—usually referred to as the Second Department— with its own presiding officer. The Department of the Physical and Mathematical Sciences was known as the First Department and the Department of History and Philology as the Third Department. The new organization did not allow for a precise separation between the Second and Third departments.

The numerical superiority of the Russian wing was also enhanced by the increasing number of Russian scholars elected by the First and Third departments, although few of these received recognition beyond their native country.[104] Among those who attracted international attention, two stood out: M. V. Ostrogradskii and P. L. Chebyshev, both mathematicians. A student of A. L. Cauchy, from whom he received high praise, Ostrogradskii worked in mathematical analysis, the theory of numbers, and theoretical mechanics. He was the first to present a mathematical explanation of the theory of the distribution of heat in fluids, and J. Clerk Maxwell took note of his contribution to the variation principles in mechanics. In 1845 Chebyshev started his lifelong preoccupation with random magnitudes and limit theorems of probability theory, which placed him in the forefront of the modern work on approximative mathematical methods. The Paris Academy of Sciences made Chebyshev an *associé étranger*, the first Russian academician to be so honored. In addition, Chebyshev took part in the Academy's first serious effort to collect and publish Euler's works.

The third great Russian mathematician of this era—N. I. Lobachevskii, the creator of the first non-Euclidean geometry—was not elected a member of the Academy. Swayed by criticism from Ostrogradskii, the Academy refused even to publish Lobachevskii's mathematical papers. V. Ia. Buniakovskii, elected to full membership in the Academy in 1841, carried on the fight against all efforts to recognize Lobachevskii's contributions. The Academy was among the last Russian scientific institutions to attach any value to the first non-Euclidean geometry. Lobachevskii, isolated at Kazan University, wandered a long distance from the safe paths of the paradigmatic unity of science, the guardian of harmony and deep conservatism in the Academy. Nevertheless, like Ostrogradskii and Chebyshev, he was inspired by the brilliance of Leonhard Euler's accomplishments that played a major role in making mathematics highly attractive to the Russian scientific community.

The era of Nicholas I ended on a sour note. Overreacting to the revolutionary waves in Western Europe in 1848, the government moved expeditiously to suppress the branches of intellectual effort considered a breeding ground of ideas inimical to autocratic ideology. In the universities, the teaching of philosophy and constitutional law came to an end. Young Russians were forbidden to attend Western European universities. In 1850 the government invalidated paragraph 33 of the 1836 Academy charter, which stipulated that "scholarly books and journals, subscribed to by the Academy or individual full members of the Academy, are not subject to censorship." The censorship machinery at this time was so elaborate that, at least in some cases, individual publications went through as many as seven agencies located in various ministries.

THE ACADEMY IN THE ERA OF REFORMS

The end of the Crimean War in 1855 was the beginning of the glorious "epoch of the 1860s," dominated by basic reforms that emancipated the serfs, ushered in a period of extensive experimentation with local self-government, reshaped the judicial system, intensified and deepened cultural contact with the West, and liberalized education policies. Russian naturalists of this epoch were "the luckiest generation" of Russian scientists, because the "spring" of their professional activity coincided with a "general spring" that embraced the entire country, awakening it from an intellectual slumber that marked the previous quarter of a century.[105] Science thereby established deep national roots and strength. K. S. Veselovskii, permanent secretary of the Academy, observed that "the great changes" that "immortalized" the reign of Alexander II and gave new life to the nation found the clearest expression in an extraordinary dedication to science, "the branch of knowledge without which humanity could not progress." He said: "Under the influence of the irrepressible and constantly expanding force which we call spirit of the time, the study of natural science has penetrated all phases of life."[106] Such creative men as I. M. Sechenov in physiology, A. M. Butlerov in organic chemistry, D. I. Mendeleev in general chemistry, and A. O. Kovalevskii and I. I. Mechnikov in evolutionary embryology very rapidly entered the ranks of the great masters of modern scientific thought. The corps of new Russian chemists grew sufficiently to supply a steady stream of articles for the *Zeitschrift für Chemie*, a leading German scientific journal.[107]

At the same time, a group of philosophers who called themselves realists but were generally known as nihilists posited science as the essence and the moving force of national progress.[108] The physiologist I. P. Pavlov noted many years later that under the influence of D. I. Pisarev, the leader of the nihilist movement, the natural science departments of the universities became the main attraction to the swelling ranks of students. But Pisarev did more than inspire young Russians to study natural science: he made science the nerve center of a world view opposed to the dominant values of the autocratic system. He gave the Russian intelligentsia a new meaning for science—a vision of science through the prism of ideology. Pisarev accepted F. A. Pouchet's theory of spontaneous generation not because it was sustained by tested experience and laboratory experiments but because it meshed with the scientific materialism of his heroes Ludwig Büchner and Karl Vogt.[109] In siding with Pouchet, Pisarev turned against Pasteur, Pouchet's most eminent opponent; in turning against Pasteur, Pisarev turned against the "pontiffs in science"—against the social detachment of academic science. In the process he provided ample proof that "science" and the "scientific world view" did not always work in unison.

K. S. Veselovskii, in his annual reports, and other academicians, in their popular writings, made only scanty references to the growing public awareness of science as a social and cultural force of the first magnitude. They left this job to university professors and, particularly, to the new breed of philosophers, typified by Chernyshevskii and Pisarev, whose scientific world view did not always jibe with the facts of current science. A. P. Shchapov, a professor at Kazan University until he was dismissed for his radical views, undertook to write a new Russian history, centered on the transition of Russian thought from the Byzantine legacy of mysticism to the rigorous method and rationalism of modern science. Lomonosov, he wrote, represented a turning point in the intellectual history of Russia, for he was the first Russian to create scientific theories as guides to experimental research and to contribute to the most advanced areas of contemporary science.[110] But he also noted that Lomonosov's scientific legacy was slow in taking root in Russian culture. The turbulent atmosphere of the 1860s had created conditions favoring a transfer of the general study of history from philosophy and jurisprudence, an aristocratic and clerical approach, to the natural sciences, a vehicle and expression of democratic ideals. While Shchapov adduced voluminous information in support of his thesis that science—particularly the empirical variety—was showing signs of integration into Russian culture, Ia. K. Grot, a member of the Academy, spoke for a rapidly dwindling group of skeptics when he claimed that to nurture science in Russian society was the same as to grow tropical trees in Russian soil.

The spirit of the reforms encouraged a search for national "self-understanding" (*samopoznanie*) for the national identity of Russian culture. This development did not favor the Academy, which more than ever before stood out in public eyes as a "German institution," fundamentally alien to the pulse of Russian life. The attacks came from both the Left and the Right. The radical group viewed it as a bastion of intellectual conservatism, devoted to science shorn of social obligations. This group made the scientific world outlook the prime mover of antiautocratic ideology, and it considered the ideological neutrality of the Academy a deliberate effort to protect the outmoded political and social institutions of the tsarist system. The conservatives, however, resented the detachment of the Academy from nationalism based on the unity of "autocracy" and "orthodoxy." The commemoration of the one-hundredth anniversary of Lomonosov's death in 1865 gave the conservative press an attractive opportunity to voice its critical views on the "uselessness" of the Academy. But the conservatives were by no means united. Fearful of excessive and destructive criticism, M. N. Katkov made a plea for helping the Academy to take a "correct" course, instead of "destroying" it. To "correct" the Academy meant to transform it from a "German" to a "Russian" institution.

Ia. K. Grot, an academician in full standing, took on the job of defending the Academy. In 1861 he stated in the conservative journal *Russian*

Thought that the orientation of the Academy reflected the external conditions under which it operated rather than the fiat of academicians and that only a change in these conditions could place the Academy on the right track.[111] Grot took the opportunity to note that the Academy had made substantial progress on the road to becoming a Russian institution. He observed, for example, that for the first time the permanent secretary of the Academy was a native Russian; the percentage of Russian academicians, he noted, was higher than ever before; for the first time, academic reports of the First and Third departments were presented in Russian rather than in French, "as was the case until recently."[112] In 1862 the Academy published the first number of the *Memoirs (Zapiski)*, consisting of papers selected by the Academic Conference for the purpose of giving a representative picture of the types of research and other activities carried on under the auspices of the Academy. Printed in Russian, the *Memoirs* was meant to reach the general reading public; free copies were sent to all high schools.[113]

To show that it was an institution with deep roots in national culture, the Academy made an ambitious effort to present a detailed picture of its own past. On December 29, 1864, K. S. Veselovskii delivered a long speech on the history of the Academy as a Russian institution. Recounting the glories of academic achievement, he lashed out against critics who contended that the Academy was more a bureaucratic appendage of the state than a dedicated and functional component of Russian society.[114] His argument was that the government alone had sufficient resources to lift the Academy to the heights of scholarly achievement. Thanks to the alliance of the Academy and the state, he said, Russia had become not only an important link in the equilibrium of European politics but also an independent contributor to a civilization built on science.

The commemoration of the one-hundredth anniversary of Lomonosov's death in 1865 triggered a vast number of studies on the life and work of the great pioneer in Russian science.[115] The power of Lomonosov's science attracted considerable attention, but not as much as the cultural significance of his literary output, his contributions to creating the institutional base of Russian science, and his role as a source of inspiration to several generations of Russian scholars. For the same reason, the Academy transformed the celebration of the fiftieth anniversary of Karl von Baer's scientific work in 1864 into an auspicious public occasion. At the same time, P. P. Pekarskii was commissioned to write a comprehensive history of the Academy; this endeavor produced two huge tomes covering the period from Peter I to Catherine II. These and many other works had two common themes: the transformation of the Academy into a full-blooded Russian institution and the rising stature of the Academy as a contributor to the global pool of scientific knowledge.

In 1865 the Academy established the Lomonosov Prize of 1,000 rubles, to be given annually for the best publication in the branches of knowledge that the great man had claimed as his research domain. Because of Lomonosov's great versatility, the competition for the prize was open to a wide range of scholars working either in the natural sciences or in the humanities.[116] The Lomonosov Prize, the most coveted national recognition of scholarly achievement, had three characteristics. First, it expressed the nationalistic focus of the time by considering only the works of Russian scholars published in the Russian language, a move clearly designed to strengthen the position of the Academy as an organization of true national orientation. Second, it excluded members of the Academy from competition, a move designed to establish a living bond between the Academy of Sciences and the Russian scientific community. Third, it was intended to enhance the role of the Academy as the supreme judge of national contributions to both science and humanistic scholarship.

In 1868 the Department of Physics and Mathematics, which covered all the natural sciences represented in the Academy, consisted of an equal number of Russian and "foreign" members. In the Department of History and Philology, the academicians with Russian names constituted less than one-third of all members.[117] The "Russification" of the Academy during the 1860s and 1870s was a slow but irreversible process. It was not until the early 1880s that native Russians established a majority in the Department of Mathematics and Physics and the Department of History and Philology. During the 1860s several non-Russian members of the Academy—including Karl von Baer, F. G. W. Struve, A. Th. Kupffer, and A. T. Middendorff—had died or retired, and most were replaced by native Russian scholars. This helped to accelerate the process of Russification but also pushed the Academy a few notches below the level of scholastic quality it had reached in the era of Nicholas I. The numerical superiority of Russian scholars in the 1880s did not silence the Academy's critics, who continued to disparage it as a "German" institution.

The lingering "foreign" scholars responded to the flames of nationalism by intensifying their efforts to reach the greater Russian society outside the narrow confines of the Academy. Karl von Baer, before his retirement in 1862, founded the very practical Entomological Society in St. Petersburg, began to publish articles in Russian aimed at reaching a wider audience, wrote papers on the projected reorganization of the universities, graced the pages of the *Bulletin* of the Russian Geographical Society with articles on geographic and ethnographic themes related to Russia, participated in the discussion on the practicality of the current project to build a canal connecting the Caspian Sea and the Black Sea, and contributed an article on anthropology to the *Encyclopedic Dictionary*, edited by P. L. Lavrov.[118] The extraacademic activities of most other "foreign" members of the Academy were equally broad and diversified.

Despite the broadening of the social framework of their activities, the academicians—both "foreign" and Russian—defended an ideology that viewed "pure science" as the proper domain of the Academy's activities. K. S. Veselovskii gave this view full expression when he stated that "the duty of the Academy is to safeguard the interests of pure science." He added: "This duty is sacred particularly at the present time, when ephemeral impulses have given rise to political passions . . . that often silence the voice of justice and truth."[119] Obviously, here Veselovskii was rebutting the nihilists, who looked to science as a vehicle for a war against autocracy. He was also contesting the strong movement among the intelligentsia to bring science closer to the immediate needs of Russia. Veselovskii made it clear that the Academy treated science as "one and indivisible for the entire world." The time had come, he said, to recognize that the most abstract theoretical discoveries were no less practical than the study of national history and natural resources. There were no sciences, regardless of how abstract they might be, that were useless to Russia. Karl von Baer went a step further when, in his address inaugurating the Entomological Society, he asserted that science in its ascent to the higher levels of abstraction could ally itself more profitably with philosophical idealism than with materialism.[120]

The emphasis on pure science did not prevent the Academy from pursuing a wide range of activities that brought it closer to Russian society. The academic museums were enriched and beautified for the sole purpose of attracting more visitors, who came not only to observe exotic specimens but also to hear lectures on themes from natural history. In 1864 the Zoological Museum was attended by 13,369 visitors, almost twice as many as seven years earlier.[121] The Lomonosov ceremony in 1865 broke all records of public attendance of meetings sponsored by the Academy: it attracted 1,200 persons and made academic administrators jubilant over the growing public awareness of the Academy as a national institution of the first order. On the educational front, the role of the Academy was even more impressive: it served as the prime mover in sending promising graduate students to Western universities for advanced studies. In 1870 I. I. Mechnikov noted that he did not know a single scholar of his generation who had not benefited from research experience in foreign universities.[122]

During the 1860s the full members of the Academy were encouraged to invite promising young university scholars to submit papers to academic journals. This was a way of establishing closer ties with the scientific community outside the Academy and of scouting for future academicians. The contributions of these scholars were acknowledged in the annual reports on the achievements of the Academy. A. O. Kovalevskii's embryological work on the transitional links between invertebrates and vertebrates attracted the attention of Darwin because they were published by the Academy.

Despite these and similar activities, university professors had serious doubts about the role of the Academy as a propelling force of scientific thought in Russia and as a genuine Russian institution. This became particularly clear in 1865, when the university councils were asked to comment on the project for a new charter for the Academy. The main purpose of the projected charter was to bring the Academy in tune with the fiscal policies of the central government, to expand the professional staff, and, in particular, to regulate the salaries and pensions of academicians. Moscow University issued a special memorandum charging that the Academy continued to be a "foreign" institution alien to the "democratic" principles characteristic of university life.[123] The council of St. Petersburg University resented the Academy's policy of keeping the public out of most of its meetings.[124] In a review of responses by all the universities, the popular journal *Fatherland Notes* recorded in 1866 that there was unanimity in criticism of the Academy. Almost no criticism was addressed to the quality of academic scholarship; rather, it was generally limited to the privileged status of the Academy in the national network of scholarly bodies and to the reluctance of the Academy to adjust itself to the needs of the day.[125] The new charter was not approved by the government, not because it failed to appeal to the universities but because the authorities decided to wait for calmer times before acting on such a sensitive matter.

The Academy continued to produce notable work in many fields of scientific endeavor. F. V. Ovsiannikov in physiology, A. V. Gadolin in structural crystallography, P. L. Chebyshev in probability theory, and N. N. Zinin and A. M. Butlerov in chemistry are only a few of the Academy's scholars who attracted the attention of the scientific community in and outside of Russia. But the striking feature of the age was the strong competition the universities, for the first time, gave the Academy as centers of scientific research. The difference between academic science and university science was both clear and deep rooted. Academicians were conservative, generally living on the glories of the past achievements that brought them into the Academy in the first place. Dominant among university professors was the rapidly growing group of young scholars eager to establish their reputations. Academicians were generally reluctant, or unprepared, to commit themselves to research in the new branches of knowledge opened by the scientific revolution of the 1850s, spearheaded by the discovery of spectral analysis, the triumph of experimental physiology and cellular pathology, the emergence of modern bacteriology, the establishment of thermodynamics, and the formulation of Darwinian evolutionary theory.

The Academy produced the chief opponent of Darwin's ideas—Karl von Baer—who used every opportunity to combat the streams of ideas unleashed by the new theory and who supplied material for an entire generation of critics of the new biological orientation. Von Baer's attack was carefully conceived: he claimed that his criticism was not directed at the idea of transformation in the living world but at the inadequacies of Darwin's evolutionary theory. The universities, in contrast, were ready to go along with Darwin and to carry his ideas to such diverse fields as paleontology, embryology, physical anthropology, and ethnography. With no vested academic interests, the new generation of university professors absorbed the onrushing scientific ideas with great expediency and enthusiasm. The new naturalist societies—such as the Society of the Amateurs in Natural Science, Anthropology and Ethnography, anchored in Moscow University—adopted Darwin's theory as a sure guide to future advances in science, both natural and social. Devotion to Darwinian thought was a form of rebellion against conservatives in the Academy and in older societies, typified by the Moscow Society of Naturalists.[126] Proponents of Darwinism were at the same time proponents of using Russian as the prime language of Russian science.

At the end of the 1860s, the Academy was ready to recognize the great Darwinian influence on modern thought, but it did not produce a scholar willing and ready to challenge von Baer's anti-Darwinian crusade. In the annual report for 1869, K. S. Veselovskii had Darwin's contributions in mind when he stated that the new developments had brought zoology and botany into closer relationship with paleontology and had made it possible to compare the presently existing forms of organic life with the fossils discovered in various layers of the earth. All this, he noted, had paved the way "for a fuller understanding of general genetic laws governing the gradual transformation of living forms, from the first appearance of life on earth to its present-day diversity and abundance." "In the same way," he continued, "the ancient history of man is beginning to merge with geology and paleontology, and, thus enriched by the new methods of inquiry, to carry the torch of knowledge far beyond the points in time where the history of many countries was thought to have started."[127]

In 1862 von Baer retired and moved to Dorpat, Estonia; in 1867 the first Karl von Baer Prize, established by the Academy, went to two young evolutionists: A. O. Kovalevskii and I. I. Mechnikov. In retirement, von Baer intensified his criticism of Darwin's theory.[128] He also published a lengthy paper aimed against A. O. Kovalevskii's study of the ascidians as a transitional living form between invertebrates and vertebrates. It was Darwin who, in *The Descent of Man*, drew the attention of the scientific world to the great promise of Kovalevskii's evolutionary approach in embryology.[129]

Prone to dramatic expression, K. A. Timiriazev, a noted student of photosynthesis, described the state of the Academy:

> During the period under consideration, there was a strong movement against the German domination in Russia, in general, and in the Academy, in particular. This unfavorable attitude toward the "German" Academy, as the St. Petersburg Academy was called sardonically, was best expressed in the passionate and impressive protest made by A. N. Beketov, and in the defense of the Academy by Permanent Secretary K. S. Veselovskii. But the subsequent transformation of the "German" Academy into the "Russian" Academy did not strengthen the role of this institution in the development of Russian science. The "German" Academy, with such great names as von Baer, Lenz, Struve and

Hess . . . did not represent Russian science. But, in any case, it represented the status of science in Russia. This was not true in the case of its successor, the "Russian" Academy. Now the most famous scientists—Tsenkovskii, Mendeleev, Sechenov, and Stoletov—were conspicuous by their absence from membership in the Academy. Under such conditions, the Academy could not be a true indicator of scientific achievement in Russia. Such an indicator was the universities. They were the centers for the training of new scientists, and they brought science in direct contact with society.[130]

Timiriazev was ready to admit that the institutions of higher technical education, rather than the universities, played the leading role in the development of modern scientific thought. Represented by the Academy of Military Medicine and various technological institutes, this category of schools acquired more advanced laboratories and more favorable conditions for experimental research. Most of these institutions published scientific periodicals. It seemed, wrote Timiriazev, that the blossoming of science, which gave the 1860s its most distinctive feature and its most fervent expression, had become the "imperturbable trend" in the evolution of Russian thought. But this did not happen. The forces of political reaction dampened the enthusiasm of the scientific community and reduced the intensity and the daring of scholarly endeavor. In a paper published much later, Timiriazev consoled himself by assuming that "the reactionary spell of the last quarter of the century" did not fully paralyze the development of science; "the healthy seeds planted in the Russian soil during the 1860s did not die, despite the unfavorable odds." The creative impulse had been slowed down, but advances continued to be steady.[131] The numbers of scientists and scientific institutions grew continuously.

THE AGE OF POLITICAL CRISIS

In the early 1880s the Academy became a target of new attacks by the leaders of the scientific community, mainly university professors. As in the 1860s, criticism centered on the remoteness of the Academy from Russian realities. The most bitter criticism was sparked by the Academy's unfavorable treatment of D. I. Mendeleev, at that time Russia's most eminent scientist. In 1880 Mendeleev was a candidate for membership in the Academy, only to be turned down by a majority vote of the Academic Conference.[132] Dismayed by this alarming turn of events, A. M. Butlerov, a pioneer of structural orientation in organic chemistry and an academician since 1874, delivered a scorching public attack on the Academy, calling it "imperial but not Russian."* He stayed with the old argument: the Academy was too much under the control of foreign academicians and their conservative Russian allies to perform a useful role for the nation. An "alien" institution, the Academy selected new members by depending on foreign referees and ignoring the judgment and consensus of the Russian scholarly community.

Butlerov's article came out in the wake of bitter denunciations of the Academy occasioned by the Mendeleev incident. Protests came from all directions—from individuals and from universities and learned societies. "Almost without exception," Russian chemists sent letters and telegrams expressing their high respect for Mendeleev's achievements in science and condemning the "academic majority," which worked steadfastly against the best interests of the nation. A petition signed by fourteen members of the faculty of physics and mathematics of Moscow University stated:

> The history of many academic elections has shown clearly that the dark forces in the Academy have been engaged in suppressing the voice of men of science and have regularly closed the doors of the Academy to Russian talent. . . . In the name of science, in the name of patriotism and in the name of justice, we consider it our duty to voice our disapproval of the [Mendeleev] action, which is incompatible with the dignity of the learned corporation and is insulting to Russian society.[133]

V. I. Modestov, professor of St. Petersburg University and a liberal critic of the policies of the Ministry of Public Education, stated that the Mendeleev affair had shown that in Russia the interests of science had not yet become a matter of public concern, a change that would be necessary to prevent groups of "intriguers" from transforming universities and the Academy into their private domains.[134] The famous Russian novelist Fëdor Dostoevsky, disturbed by the Mendeleev vote, suggested that there was only one way out of the academic dilemma: the creation of a Free Russian Academy, supported by voluntary contributions, and the abandonment of efforts to reform the existing Academy, which he considered a hopeless task.

Mendeleev, however, thought that the situation was not hopeless. He did contend that only a major recasting of the Academy's organizational structure and functional orientation could make it a useful institution.

---

*The official title of the Academy was the Imperial Academy of Sciences in St. Petersburg.

Guided by the dictum that "science in its entirety" was a "free activity," he stood for expanded institutional guarantees of academic autonomy. But he also wanted stronger institutional ties between the Academy and other national centers of learning: he proposed, for example, that the universities and selected learned societies be given the right to nominate candidates for academic membership.[135] Mendeleev saw no good reason why all the members of the Academy should be stationed in St. Petersburg. What the Academy needed most of all, he thought, were precise and reliable methods of selecting new members. Scholarly contributions of candidates for membership in the Academy must be open to public inspection and judgment. An outstanding engineering feat must be recognized no less than a triumph in scientific theory. Furthermore, Mendeleev felt that the Academy should end its practice of being a "kind of sinecure" and a source of automatic pensions for its members and that, instead, it should become an active "scientific center" in which the academicians were remunerated only for undertaken and completed research assignments. Clearly, Mendeleev wanted an Academy that would differ radically from the existing institution. He demanded far more than either the government or the Academy was willing to support.

In 1882 D. A. Tolstoi became president of the Academy; at the same time, he held the position of minister of internal affairs. During a time of growing political turbulence, when the government made a last-ditch effort to preserve the rapidly eroding autocratic system, Tolstoi's dual function symbolized the growing rift between the police and the educated strata of the population. The government moved fiercely in checking the flow of "dangerous" ideas, and the disenchanted segments of the population responded with equal vehemence. In 1884 the new university charter wiped out all the attributes of academic autonomy that had been guaranteed by the 1863 charter. The positions of rector and dean ceased to be elective. In order to control the content of professors' lectures, university examinations were now administered by special government boards.[136] While students joined various revolutionary movements, ranging from anarchism to Marxism, professors formed the backbone of a rising campaign in favor of parliamentary government. The physicians, united in the Pirogov Society, declared openly that the existing political system could not be part of a future Russia. The government ordered the closing of the Moscow Juridical Society, known for its intensive engagement in a critical inquiry into the fabric and inner dynamics of Russian society. The Free Economic Society, traditionally apolitical, did not hide its newly acquired liberal views and its skeptical attitude toward the reactionary course of the government.

In this milieu the Academy could not escape political entanglement. At the end of the century, a strong wing of liberal academicians joined the critics of autocratic government. The liberals were at first clearly recognized by their allegiance to the *zemstvo* movement and by their critical views on the key problems of modern society and government administration. They did not preach revolution, and most of them did not go beyond demands for a government that was both monarchical and parliamentary. In the early twentieth century they were found typically in the ranks of Constitutional Democrats (Kadets). Some academicians, typified by the mathematicians A. A. Markov and A. M. Liapunov, were "liberal" in response to their conscience rather than to the dictates of an articulated ideology.

In 1899 a group of academicians added their signatures to a protest note against police brutality at St. Petersburg University. Two years later, a similar note was signed by academicians A. N. Beketov and A. A. Shakhmatov, along with ninety-seven other persons engaged in education and literary work. Beketov refused to meet the request of Grand Duke Konstantin Konstantinovich, president of the Academy since 1889, to submit a written explanation of his reasons for signing the petition. In 1902 a protest of larger proportions was related to the academic fortunes of the writer Maksim Gorky. On February 25, 1902, the Academy's newly founded Section of Fine Arts elected Gorky an honorary member, but a week later Nicholas II invalidated the election by refusing to sign a document to that effect. The voices of protest against this reversal came from several individuals. The writers V. G. Korolenko and Anton Chekhov showed their displeasure by resigning as honorary academicians.[137] The mathematician A. A. Markov considered the incident a flagrant attack on the intellectual and moral integrity of the Academy and threatened to resign from his prestigious post.[138] He claimed that the decisions of the Academy in matters of professional nature should be considered final and irrevocable.

The revolution of 1905 brought more politics—and overt expressions of discontent with the general situation in the country—than any previous events or sequence of events. After the disastrous outcome of the war with Japan, the major cities swarmed with angry groups demanding radical changes in the structure of political relations. The Academy took an active part in the rebellion of university professors against the institutional mechanisms that disenfranchised both teachers and students and isolated the educational process from cultural and economic needs of the nation. In January 1905 a group of academicians—including, among others, the historians A. S. Lappo-Danilevskii and A. A. Shakhmatov, and the mathematicians A. A. Markov and A. M. Liapunov—signed the "Note of 342 Scholars," which stated that the policies of the tsarist government were incompatible with the true spirit of education and science.[139] This document proclaimed:

> Science can flourish only in the countries where it is free, where it is protected from external encroachments and where it can work constantly on illuminating all sides of human existence. Where these conditions do not prevail, the institutions of higher education, as well as secondary and primary schools, can only barely exist. The perilous situation in public education compels us to express our profound conviction that academic freedom is incompatible with the state system in present-day Russia. To try to achieve this freedom by piecemeal changes in the existing system would not be sufficient; what is needed is a fundamental transformation of the system. At the present time, such a transformation can no longer be postponed.[140]

The Academy protested the use of its premises for stationing troops that policed the beleaguered students of St. Petersburg University. Its representatives signed petitions demanding the freedom of the press and the right of the Ukrainian people to speak and write in their own language. The academicians were represented in the newly founded Academic Union, an alliance of local groups of university professors, all operating without a charter ratified by the government. Pressured by a group of angry academicians, the government abandoned its effort to make the Academy the central agency in a much-expanded censorship apparatus. When the revolutionary waves receded, the Academy retreated from the battlefield and, in general, from overt activities contradicting government policies.

But the new academic tranquillity was not complete: it was disturbed by the voice of V. I. Vernadskii, who continued to battle against the myopia of the Ministry of Public Education. Vernadskii, a longtime Moscow University professor and a full member of the Academy since 1912, was the country's most distinguished mineralogist and the founder of a complex and influential orientation in geochemistry. He was the first

Russian scientist to advocate, and to undertake, a systematic study of national resources in radioactive elements. He also was the man who, at the end of 1904, initiated the activities resulting in the formation of the ill-fated Academic Union.[141]

One theme ran through all of Vernadskii's writings on the social dynamics of science: science is an expression and a product of creative individuality in search of truth. Only societies that encouraged individuality of thought, he stated repeatedly, and therefore challenged authority, provided favorable conditions for the development of scientific thought. To be engaged in science, he argued, is to be engaged in a democratic process. He kept alive the argument advanced during the 1905 revolution that Russia's autocratic system and science were incompatible. The future of Russia, as he saw it, was in a state system based on political parties expressing diverse views on the fundamental principles of governmental organization.[142] Scientific research must be an organic component of the activity of university professors, for without it teaching could not be a creative process.[143]

When Vernadskii stated that "every institution of higher education is not only a school but also a scientific organization involved in research of a large magnitude," he depicted what he thought should have been the main orientation of the future development of Russian universities. Behind his argumentation was the idea that the Academy of Sciences could not ascend to greater heights in its scholarly dedication as long as the universities continued to place insufficient emphasis on scientific work. (According to the 1884 university charter, university professors received salaries based on the number of students in their classes, fully disregarding the quality of their contributions to scholarship.) Vernadskii's reasoning was as accurate as it was simple: without outstanding university professors there could be no outstanding members of the Academy.

Vernadskii survived unscathed, because the manner of his writing was inoffensive and ideologically subdued, and because the authorities found at least some of his plans for the reorganization of the national system of scientific centers untainted by political considerations of a subversive nature. He did not blame the government alone for the relatively slow growth of science; he placed equal blame on inadequate support from the society at large. Traditionally, Russia did not have a social class or estate that could be relied on as a steady source of scientific manpower.[144]

The Academy greeted the February Revolution with jubilation and strong hopes for a future alliance between science and democracy. In May 1917 it realized one of its oldest dreams: it elected its president. For the first time the president was not appointed by the government, and for the first time he was a distinguished scholar. A. P. Karpinskii, the new president, was a leading Russian expert in the tectonics, stratigraphy, and paleogeography of European Russia. S. F. Ol'denburg, permanent secretary of the Academy since 1904, was appointed minister of public education in the provisional government, and A. S. Lappo-Danilevskii became a ranking member of a group working on election plans for the Constituent Assembly, whose task it was to decide the future course of democratic development in Russian politics. The name of the institution was at the time changed from the Imperial Academy of Sciences in St. Petersburg to the Russian Academy of Sciences. Several leading academicians joined Maksim Gorky in a popular effort to create a free association of scientists devoted to the advancement and diffusion of scientific knowledge.

The history of the Academy after 1884 was shaped not only by the growing political crisis, which brought the fall of the Romanov dynasty and the full collapse of autocratic principles of political organization in 1917, but also by rapid changes and advances in the national economy. During the period of political crisis, the country was deeply engaged in developing a modern economy dominated by large-scale industrialization. The economist James R. Millar has described the Russian economy as it had evolved prior to the Bolshevik takeover:

> The economy the Bolsheviks assumed control of was already quite large by world standards, and its growth performance was improving as well prior to World War I. Large-scale modern industrial establishments had been established in a number of areas. The necessary basic transportation network was essentially complete, and a permanent labor force was in the making. Moreover the Tsarist regime had introduced measures following the unsuccessful 1905 revolution designed to undermine traditional agricultural institutions, which were deemed resistant to modernization of production, and to encourage establishment of independent farmers. If successful, these reforms would have raised the growth of agricultural output, thereby increasing also over-all and per-capita growth of GNP to rates more nearly equal to those of the major world powers.[145]

Rapid industrialization of the country brought the social role of science into sharper focus and contributed to a broadening of the institutional base of research, all of which affected the Academy of Sciences. The emergence of privately endowed research centers was the most distinguishing feature of the time. Private organizations were of two types: those that combined higher education and research and those devoted exclusively to research. Typical of the first group were the Lesgaft Institute, which ran an elaborate laboratory in experimental biology; Bekhterev's Neurophysiological Institute, which combined physiological research with an unorthodox curriculum that made the study of personality the central theme of all courses in the social sciences; and the A. L. Shaniavskii University in Moscow, founded by a rich owner of gold mines in Siberia.[146] These schools accelerated the flow of scientific ideas to Russia and generated research activities untrammeled by restrictions imposed on university curriculum. They brought new financial resources to the world of scholarship and made higher education accessible to social and ethnic groups discriminated against by the state universities.[147]

The privately supported centers devoted exclusively to research assumed many forms and pursued diverse lines of interest. Dominating the field were the newly created foundations that did not engage in research directly but concentrated on extending financial help to institutes, laboratories, or individuals engaged in types of research that linked science and technology. The Kh. S. Ledentsov Society for Assistance to Experimental Sciences and Their Practical Application began to operate in 1909, thanks to an initial endowment of two million rubles. It helped such eminent scientists as the neurophysiologist I. P. Pavlov, the physicist P. N. Lebedev, and the aerodynamics pioneer N. E. Zhukovskii to improve their laboratories and to undertake research projects of larger magnitude and complexity. In 1911, in his report to the annual meeting of the society, Lebedev stated that his laboratory was involved in a spectroscopic study of atomic structure.[148]

The fact that the Ledentsov Society had no qualms about supporting such scholars as Pavlov and Zhukovskii, who were deeply involved in basic theory, demonstrated that it was not guided by narrow practicalism. Its records show that it supported the work of two kinds of useful innovations: those aimed at overcoming serious technological difficulties in industrial production and those invited by the advancing frontiers of scientific theory. The society was guided by a philosophy that viewed the advancement of scientific theory as the basic requisite for the advancement of industrial technology. It was ready to assist promising research ventures of individual scholars of proven accomplishment and group projects centered on complex scientific and technological problems.[149] Anchored in Moscow University and the Moscow Technical School, the Ledentsov Society symbolized the growing awareness in Russian society of the pressing need for a more rapid and predictable flow of scientific ideas to industrial technology.

In 1911 a large group of Moscow university professors resigned in protest against the oppressive policies of L. A. Kasso, minister of public education. Among the professors were many leading scientists, including P. N. Lebedev, who had acquired an international reputation through his experiments on the pressure of electromagnetic waves.[150] J. J. Thomson of Cambridge University called Lebedev's investigations "the most striking triumph of experimental physics" and a major contribution to "the general theory of radiation."[151] Svante Arrhenius, director of the Nobel Institute in Stockholm, invited Lebedev to join his research staff, but Lebedev was not ready to leave the country.[152]

Deprived of access to university laboratories, the displaced scholars made an appeal for private support. The most important immediate result of this appeal was the founding of the Society for the Moscow Scientific Institute in 1912, soon after Lebedev's death. An unnamed benefactor contributed 300,000 rubles, making it possible for the society to establish the Moscow Physical Institute, which moved to its newly built laboratory on January 1, 1917.[153] Excited about the lavish flow of private donations, the founders of the institute saw the makings of a private academy of sciences modeled after the Royal Society of London.[154] The society was "dedicated to the memory of February 19, 1861," the day of the emancipation of the serfs: just as the emancipation marked the creation of new domains of social activity untutored by the state, so the establishment of the Society for the Scientific Institute, according to its founders, marked the beginning of scientific work in Russia unhampered by government restrictions.[155] In late 1916 the industrialist G. M. Mark gave the society a sum of 1.2 million rubles for the purpose of organizing a chemical and a biological institute.[156]

The emergence of this and similar societies was widely acclaimed in the scientific community. Nevertheless, the typical scientist was convinced that the future of science in Russia depended not only on the creation of a strong private sector but also on expanded financial assistance by the state. He was interested in breaking the state monopoly in directing and controlling scientific research rather than in doing away with the public sector in science. Vernadskii was both the leader of the movement seeking a broader scope for the private sector and the chief advocate of an increased government role in creating a network of scientific institutions tied to the economic needs of the country and to the challenges of rapidly advancing science. He recognized the supreme authority of the government in establishing a national commitment to science and in leading a planned and unified effort.[157] In 1916 Vernadskii noted enthusiastically that the government's recognition of the growing role of science in the national life was the most impressive feature of the new era.[158]

The expansion of the institutional base of science took place on yet another front. The growth of industry necessitated the creation of new engineering schools modeled after the German *Technische Hochschule*. The St. Petersburg Polytechnical Institute, one of the most advanced schools of the new breed, was engaged not only in training modern engineers but also in organizing new research laboratories and in forging new ties between scientific centers and industry. Unencumbered by vested academic interests of long standing, these institutions were ready to absorb not only modern techniques but the newest theories as well. The St. Petersburg Polytechnical Institute was the first Russian institution of higher education to include Planck's quantum theory and Einstein's relativity theory in its research and seminar activities. A. F. Ioffe's informal seminar in this institution, organized in 1916, heralded Russia's entrance into the world of modern physics.[159] Several decades later, two members of this seminar—N. N. Semenov and P. L. Kapitsa—joined the distinguished group of Nobel laureates. This school became the sounding board for the most recent developments in chemistry, mathematics, and demographic statistics. According to James C. McClelland:

> One of the innovative features of the new institute was its independent faculty of economics (which enrolled approximately half of the student body) in addition to the more typical faculties of metallurgy, electromechanics, and ship-building. The teaching staff included not only famous scientists such as N. A. Menshutkin, but also controversial intellectual figures such as economists Petr Struve and M. I. Tugan-Baranovskii, who were barred from teaching at the regular universities because of their political views. Most teachers were interested in practical applications as well as theory, emphasizing laboratory work and discussion groups rather than lectures. Memoirs testify to close teacher-student relations and to an *esprit de corps* unusual for educational institutions at that time.[160]

This and related developments were special expressions of—and responses to—a national recognition that the scope of social and cultural values associated with science was rapidly broadening. The nation was searching for a broader basis for applied science. Science became an ideological force of major importance. The language of science became a symbol of social progress. Science acquired a more representative base in the structure of social classes. Never before were the avenues connecting the institutions of higher education and scientific training with the active needs of society so wide and so diversified. The world of scientific scholarship acquired a broader institutional base and a purer and more comprehensive ethic.

The new developments made science, as a cluster of values and a source of progress, a topic of many and often contradictory interpretations. The social value of science was articulated with more precision and in more detail than ever before. But, at the same time, the potential dangers of "science imperialism" were debated with equal vigor and popular appeal. The geochemist V. I. Vernadskii, the physicist O. D. Khvol'son, and the mathematician A. Vasil'ev were among the leaders of the group combining a broad optimism in the growing social role of science with a deep conviction that, in order to develop its full strength and to achieve maximum social usefulness, science must know not only its strengths but also its limitations. The scientist, according to Khvol'son, must recognize that the difficult tasks of drawing a line between life and inorganic matter, of choosing between the finitude and infinity of space and time, and of explaining the freedom of will were clearly outside his competence.[161]

While higher education worked on expanding its appeal and its institutional base, the government broadened its direct involvement in scientific research. In 1882 it founded the Geological Committee, which quickly became the national center of geological research. Its *Bulletin* and *Transactions* were the chief geological journals published in Russia. The pressure of industrialization compelled the government to become a direct and major participant in organized geological surveys, the first step in search of new mineral deposits. In 1892 the government founded the Institute of Experimental Medicine, a complex of modern biological laboratories related to medicine. Organized for the purpose of identifying and eliminating the causes of recurrent epidemics of plague, the institute quickly branched into many research activities. It was in the laboratories of this institute that I. P. Pavlov conducted his research in the neurophysiology of digestion that brought him a Nobel Prize in 1904; and it was here that S. N. Vinogradskii discovered the role of micro-organisms in the nitrification of soil.

The *Archives of Biological Sciences,* published by the institute in French and Russian, became the rallying point for scholars engaged in experimental studies of the structure and dynamics of living forms. In 1894 the Main Administration of Land Tenure and Agriculture established the Bureau of Applied Botany, which waited until 1907 to become a consolidated and blossoming research body. Its journal, in turn, became the rallying point for the growing ranks of experts in artificial selection. While the universities were steeped in Darwinian tradition, the bureau concentrated on a popularization and advancement of the ideas set forth by Gregor Mendel and modern genetics. It also invited the first Russian efforts to reconcile Darwinism and Mendelism.[162] These and similar institutions expanded the experimental base of Russian science without altering the educational focus of university laboratories. At the same time, they became preoccupied with research areas that, with notable exceptions, did not receive strong support in the Academy of Sciences. These institutions broadened the horizons of Russian science without challenging the academic strongholds.

Nor did the Academy stand still. To the contrary, it expanded its activities in several directions: it became a coordinating center of the national meteorological service; it joined an international academic organization working on a worldwide bibliography of scholarly publications; it made preparations for work on a general history of knowledge; and it undertook a systematic survey of science, scientific institutions, and scientists in Russia. In 1911, on the occasion of the celebration of the two-hundredth anniversary of Lomonosov's birth, the Academy unveiled an ambitious plan to build a new research center—the Lomonosov Institute—both to honor the great man and to acquire more efficient laboratories for pursuing the intricate lines of research revealed by the ongoing revolution in science.[163] Modeled after the Cavendish Laboratory at Cambridge, the new institute was to be open to all promising or established Russian scientists and was expected to broaden the base of biological research far beyond what was possible in the antiquated laboratories of the Academy and the universities.

In 1916 a group of academicians reported that work on the institute was still in the planning stage. The dream of establishing the Lomonosov Institute, in fact, did not materialize. But the work of the planners was not fruitless: they had shown clearly and convincingly that the future of the Academy was in building laboratories on much broader foundations than in the past and in establishing more effective and diversified relations with the scientific community at large. The plan showed that the Academy was ready to undertake the necessary measures to meet two new challenges: the twentieth-century revolution in science and the growing needs for direct ties between science and technology. The Academy thus became the leading force in the renewed national effort to create the economic foundations of national independence. At the same time, frequent voices were heard charging that academies were a thing of the past and that the future of science would depend on well-equipped and well-financed institutes concentrating talent in closely related fields.[164] In 1917 Vernadskii wrote that just as general academies answered the needs of Renaissance science, so specialized institutes, equipped with modern laboratories, were the best response to the needs of twentieth-century science.[165]

In 1915 the Academy made a bold and generally successful effort to add a new dimension to its institutional makeup: it established the Commission for the Study of Productive Forces (KEPS), directed to organize "scientific, technical, and social forces for a more effective participation in the war effort."[166] The commission organized the search for deposits of rare and noble metals, initiated general surveys of natural resources of industrial relevance, and worked on finding ways of producing industrial and scientific instruments whose importation from the West (mainly from Germany) had been discontinued because of the war. Its publication activity grew rapidly, and it initiated the organization of several new research institutes. As interpreted by V. I. Vernadskii, the commission pointed out important directions for the future development of science in Russia. Vernadskii observed that the new commission, born in a time of crisis, was only the first step in the evolution of a central organization serving as a scientific base for the national economy.[167] The commission provided ample proof of the advantages of creating large scientific centers whose members did not split their working time between teaching and research. It showed the advantages of cooperative and concentrated effort in science over isolated and diffuse activities. Furthermore, the commission pointed to the urgent need for a national science policy dedicated to achieving a maximum engagement of science in the economic and cultural life of the nation and, at the same time, to protecting the basic principles of academic autonomy. Above everything else, the commission showed that the emancipation of Russia from heavy dependence on the West could be brought about by establishing closer links between science and the national economy.

Select members of the Academy, led by V. I. Vernadskii, and of the universities, led by N. K. Kol'tsov, contended that a mere expansion of research facilities, expert personnel, and financial resources would not alone elevate Russian science to new heights and bring it closer to the needs of society. Russia needed not only more activity at the forefront of science but also a true scientific community. The work toward creating an effective scientific community proceeded in several directions. The first serious campaign was undertaken to point out the advantages of reduced government control in the rapidly growing domain of scientific activity. Vernadskii spoke for a majority of Russian scientists when he stated in 1917 that the government should continue in its effort to provide adequate finances for research, expand the institutional network of science, and determine national priorities in applied science, but that it should leave the management of research activities to the scientific community. The government must recognize the scientific community as the sole guardian and interpreter of scientific legacy. It must recognize science as a "free activity" that cannot tolerate external interference.[168] Vernadskii said: "The organization of scientific work must be entrusted to the free scientific creativity of Russian scholars, which cannot and must not be regulated by the state." The government, he argued, should do everything in its power to help Russian scholars establish more intimate ties with the scientific scholarship abroad.

The world of scholarship concentrated on creating two types of associations: specialized national scientific societies and a general national association of scientists. During World War I several specialized national societies were established: the Russian Botanical Society, the Russian Paleontological Society, the I. M. Sechenov Society of Russian Physiologists, and the All-Russian Astronomical Union.[169] The first president of the Botanical Society was I. P. Borodin, and the first president of the Sechenov Society was I. P. Pavlov, both members of the Academy.

In these and similar organizations, individual members of the Academy of Sciences were both charter members and the most influential leaders. All societies published professional journals in their specialized fields and organized national congresses. Both activities helped to consolidate the ranks of specialists—and to counteract the centrifugal effects of isolation—through intensified personal contact with active scientists and through coordination of scholarly effort. Professional journals and congresses also brought together specialists representing the key components of the institutional maze of Russian science, and they contributed to building national traditions in various sciences and to maintaining a record of the national commitment to science. These forums gave the scientific community effective means for expressing views on the most important questions of national science policy related to their specific fields. And the new journals relieved the scholars of the need to look abroad for publication of scientific papers.

On May 31, 1916, the Ministry of Public Education ratified the charter of the newly founded Association of Russian Naturalists and Physicians. This was the crowning event in the persistent effort of Russian scholars since 1869 to establish an organization on the model of the British Association for the Advancement of Science.[170] The government accepted the new association as a necessary aid in mobilizing scientific manpower for work on the critical problems of the national economy, gravely crippled by the war situation. The scientists greeted it as a defender of professional interests, particularly in relations with the government. War conditions, however, prevented the association from moving into action and from testing its professional strength and national appeal.

On the eve of the October Revolution, the Academy of Sciences occupied the most prominent place in the network of Russian scientific institutions. In this position it was more a symbolic expression of achievement in Russian science than a real workshop of new ideas. It represented and reflected the past successes of Russian science much more than the current national commitment to absorbing and adding to the streams of ideas released by the ongoing revolution in science. The Academy was clearly in no position to cope with the swelling research demands and perspectives of modern science or to serve as a supreme judge and coordinator of the national effort in science. Yet, as Loren Graham has pointed out, it was somewhat justified in clinging "to its claim to be the highest scientific institution in Russia": it was blessed with a heavier concentration of talent and a more glorious past than any other single institutional component of Russian science, and its members were the most astute leaders in the growing struggle for freedom of scientific inquiry. More than any other institution, it kept a careful record of the growth of science as an organic part of national culture and of Russia's contributions to the international pool of scientific knowledge.[171]

During the last two decades of its existence as a tsarist institution, the Academy moved cautiously in search of a new position in the expanding labyrinth—and unsteady hierarchy—of scientific institutions. Among new developments, one stood out in bold relief: the Academy had shown a definite preference for the humanities and the social sciences. Of thirty-seven new members elected during this period, only fifteen worked in the natural sciences. For every new physicist or mathematician, there were seven new specialists in Russian language and literature. The two academic elections during the Kerensky regime in 1917 produced an even more dissonant picture: of the five newly elected members, only one was in the natural sciences—and he was a physicist unaware of the critical scientific repercussions brought about by quantum theory and Einstein's revolution.

II

THE SEARCH FOR NEW IDENTITY (1918–1928)

SCIENTIFIC INSTITUTIONS: MAJOR SYSTEMS

In the years immediately following the October Revolution, the Bolshevik authorities moved cautiously and rather insecurely in laying the foundations for a national system of scientific institutions. In this unwieldy enterprise, they worked on three fronts: adapting the traditional scientific institutions to modern conditions, creating new institutions to respond to previously unattended needs, and building a bridge between science and ideology. The science policy of this period showed the first signs of government efforts to devise mechanisms of centralized control over scientific institutions that would fully reflect the new political philosophy. In carrying out this task, the government was seriously hampered by political uncertainties, ideological unsettledness, administrative inexperience, and the drastic depletion of economic resources. Scientific scholarship was bound to a network of five major institutional complexes: the centers of higher education, sectoral research centers, learned societies, Marxist organizations, and academies.

THE UNIVERSITY SYSTEM

The Soviet Union inherited from tsarist Russia an elaborate system of higher education, the institutional backbone of the national commitment to science. Moscow and St. Petersburg, in particular, boasted a complex system of universities, polytechnical institutes, medical schools, and specialized institutes, all with research laboratories and organized publication outlets. The Bolshevik government encountered grave difficulties in its effort to bring the professorial community under control. Indeed, "the formidable *esprit de corps* of Moscow University" had intimidated the Commissariat of Education so much that it "made no approach to the university until the middle of 1918." The Moscow academic community, writes Sheila Fitzpatrick, like that of Petrograd (Leningrad), was dominated by "prominent members of the Kadet Party."[1]

Professional resistance was widespread. When the Czech military units forced the Red Army detachments to abandon Kazan in August 1918, the academic council of the local university issued a public proclamation greeting the "downfall" of Bolshevism and pledged support to the White generals. When the Bolsheviks reoccupied Kazan a month later, many professors and a "surprisingly large" number of students retreated with the units of the White Army.[2] On October 26, 1917, a meeting of the Petrograd members of the Academic Union passed a resolution pleading with the professoriat not to side with the Soviet authorities.[3] The resistance reached a peak in January 1922, when professors of Moscow University, one of the country's most advanced and largest institutions of higher education, went on strike. The professors rebelled against unfavorable working conditions and a rapidly deteriorating standard of living. They joined the staffs of other universities in opposing the new education policy, which stressed the doctrinaire (and accelerated) training of Marxist professors, recognized Communist groups as the only legal representatives of the student population, and revamped academic councils to include representatives of the students and local government in addition to the teaching staff. They fought against the policy of "a revolutionary renewal of the teaching staff," which called for a concerted effort to dismiss professors unfavorably disposed toward Marxist ideology and the new political system.[4] They had little sympathy for the newly founded "workers' faculties" on university campuses, which were part of a broad campaign to "proletarianize" the student body and to "revolutionize" the methods of instruction and "the content of science."[5]

The termination of the Civil War did not bring the university crisis to an end. On the contrary, the hasty efforts to proletarianize the student body, the rising unpredictability of fiscal policies, and the increased frequency of faculty purges made the situation more critical and volatile. Having begun with efforts to strengthen the "proletarian" side of the university equation, the authorities now concentrated on advancing the institutional mechanisms of stringent government control. The general confusion was compounded by frequent, and often radical, shifts in the organization of faculties, the administrative and curricular centers of the universities.

The general crisis in higher education did not deter the government from founding new universities and teaching institutes. New universities sprang up in Minsk, Voronezh, Gor'kii, Dnepropetrovsk, Erevan, Perm, Rostov, Tashkent, Tbilisi, and Sverdlovsk. The number of universities grew from ten in 1916 to thirty-four in 1925.[6] An unusually large number of institutions of higher education—particularly polytechnical, pedagogical, and medical colleges—were established in areas under the control of the White Army, in most cases by professors fleeing from Bolshevik-held cities. Of the thirteen new institutions, only four managed to survive.[7] The unsettled situation in the country prior to the October Revolution started a steady migration of professors to the Crimea, where those who did not emigrate engaged in teaching and research. Having received strong local support, a group of distinguished scholars founded the University of Taurida in Simferopol. Because of its geographical position, the new institution was the first Soviet university to receive financial aid from the West, which helped it in acquiring modern laboratory equipment and Western scientific journals. The American Red Cross undertook to finance the new university's medical school.[8] After the initial outburst, the university became a typical provincial school, deeply steeped in the study of local archaeology, ethnography, history, and natural resources.

As in the past, the leading institutions in Moscow and Petrograd were the indisputable centers of the most intensive and innovative research. A magnificent coalescence of two generations of eminent mathematicians in Moscow brought the Soviet Union to the forefront of modern work in mathematics. By the middle of the 1920s Moscow University could boast of the work of P. S. Aleksandrov in topology, of N. N. Luzin in the theory of real functions, and of A. N. Kolmogorov in the foundations of mathematics and probability theory, all attracting international attention. The outstanding and most refreshing feature of the Moscow school of mathematics was its direct confrontation with philosophical issues. But the new philosophical involvement had nothing in common with the ideological problems raised by Leninist epistemology; it was limited exclusively to the logic of mathematical theory-building and operation, an area alien to and unattended by dialectical materialism.

As for developments outside mathematics, professors at Moscow and

Leningrad institutions of higher education presented the first system-atic—and critical—Russian surveys of quantum theory and the theory of relativity. In 1922 L. S. Berg of the University of Petrograd presented his theory of nomogenesis, which promptly attracted much attention in the West. This theory rejected natural selection as the motive force of organic evolution and combined modern genetics with a philosophy of purposive evolution of the living world. In the same year A. A. Fridman, also of Petrograd University, attracted the attention of Albert Einstein by his bold conception of the expanding universe. Fridman's work helped to accelerate the growth of Soviet physics both as a theoretical and an experimental discipline. The number of papers by Soviet pro-fessors published in such prestigious foreign journals as the German *Zeitschrift für Physik* was truly impressive. These papers so impressed Max Planck that in 1925 he readily admitted that German physicists could learn from their Russian peers no less than Russian physicists could learn from their German colleagues.[9]

Prior to the October Revolution, professors pursued personal research interests; institutes requiring coordinated group effort existed only in isolated cases. The Soviet government acted quickly in recognizing the need for special research units in the institutions of higher education. But it was apparent from the beginning that the new regime favored institutes independent of the universities, which it regarded primarily as centers for the training of future scientists. The government heeded the advice of V. I. Vernadskii, a distinguished geobiologist and member of the Academy of Sciences, that in order to give modern science a broader and firmer footing in national life, the country must create a system of specialized institutes, unencumbered by educational duties.[10] The dual system of universities and research institutes made it possible for the government to ease the conflict between the policy of proletar-ianizing education and the policy of preserving the roots of bourgeois science, without which, according to Lenin, a Marxist design for socialist society could not be translated into reality.

Most of the twenty-odd institutes founded by the universities during the 1920s worked under the most adverse conditions, lacking sufficient staffs, modern laboratory equipment, and adequate space.[11] There were, however, notable exceptions. The Physical Institute, the Zoological Insti-tute, and the Mathematical Institute, all at Moscow University, were among the most advanced research organizations in the country. This was true as well of the Physical Institute and the Chemical Institute at

Leningrad University. A relatively large number of leading professors associated with these institutes became members of the Soviet Academy of Sciences and, at the same time, continued to offer courses in their respective universities.

SECTORAL ORGANIZATION

Sectoral research centers—so called because of their identification with specific sectors of the national economy and culture and their direct subordination to individual people's commissariats—formed one of the most dynamic and rapidly expanding systems of research centers.[12] A strong contingent of these organizations was of pre-Soviet vintage. The contributions of such pre-Soviet government research centers as the Institute of Experimental Medicine, where I. P. Pavlov had one of his main laboratories, the Geological Committee, and the Institute of Experimental Biology were widely acknowledged. As a result of favorable governmental attention, this sector grew rapidly and in many directions. Its basic contribution was in carrying modern research to fields that previously had received little or no attention. Some centers became vital components of the institutional network of Soviet science; others followed an uncertain course and were either disbanded or subjected to radical transformations.

The Institute of Experimental Biology, founded in 1917, became a center for the most modern laboratory work in genetics, enabling Soviet scientists to make pioneering contributions to population genetics (S. S. Chetverikov) and to undertake an early study of the genetic implications of macromolecular chemistry (N. K. Kol'tsov). The latter efforts opened new research areas that three decades later led to the emergence of molecular biology.[13] Publication of the *Bulletin of the Institute of Experimental Biology* began in 1921. The new journal became the outlet for a novel orientation in biology, one advocating not only a synthesis of biological disciplines but also a systematic and extensive use of the methods of physics and chemistry in the study of life. Helping to counteract the rapid fission of biology into smaller subdisciplines, the *Bulletin* contributed to a modern integration of biological knowledge and methods.[14] It also helped push biology closer to the standards of exactitude prevalent in the physical sciences and aided Soviet scholarship in becoming actively involved in one of the most promising developments in modern science. After overcoming grave shortages in space and laboratory equipment, the Roentgenological and Radiological Institute—split into the Physico-Technical Section and the Medical-Biological Section—became the country's leading center for radiation studies. Headed by A. F. Ioffe, the Physico-Technical Section (which later became the Physico-Technical Institute) initiated an entire generation of young Soviet physicists in experimental work in quantum theory and related fields.

The State Optical Institute was founded in 1918 for the purpose of providing a scientific and technical base for the development of an optical industry, previously nonexistent in Russia. Prior to World War I, Russia depended on Germany for all its supplies in optical glass and optical instruments. Thanks to the talents of D. S. Rozhdestvenskii, the founder of the institute, molecular and atomic spectroscopic studies became one of the main interests of Soviet physicists during the 1920s.[15]

LEARNED SOCIETIES AND ASSOCIATIONS

Learned societies occupied a unique position among institutions dedicated to the advancement of science. They facilitated the exchange of scientific ideas by organizing congresses and conferences of specialists, founding special libraries, expanding the exchange of professional literature with similar institutions abroad, and providing new publication facilities. In particular, the emergence and growth of learned societies had made possible the growth of specialized journals. *Mathematical Symposium*, for example, published by the Moscow Mathematical Society, was one of the oldest and most respected specialized scientific journals originating in Russia.

In the pre-Soviet period, learned societies attracted private financial support for scientific research. These societies were either fully independent of or were anchored in individual universities. The Russian Geographical Society typified the first group, and the Moscow Mathematical Society (operating from Moscow University) was the best-known institution of the second group. Most tsarist societies survived the October Revolution, but all encountered serious difficulties in adjusting to the rapidly changing and inordinately fluid state of the national effort in science. For example, the membership of the Russian Geographical Society decreased from 1,318 in 1917 to 549 in 1927.[16] A few societies, overcommitted to liberal ideologies, found it difficult to operate and ceased to exist as legal entities. Some were abolished by the government because of their open opposition to Marxist views on the nature and social role of science.[17] The Free Economic Society, for example, founded in the age of Catherine II, was too favorably disposed toward the ideology of a free-enterprise economy to be acceptable to the new regime. The Pirogov Society, known for its extensive work in the study of public health and for its efforts to extend the benefits of modern medicine to the countryside, received the same unfavorable treatment and quickly ceased to exist.[18] The open campaign of its journal, *Social Physician*, against "the autocratic regime of Russian Bolshevism" was the most obvious reason for the failure of the society to survive the scourge of the Bolshevik vendetta.[19]

University professors were favorably disposed toward learned societies, which provided important avenues for the advancement of professional interests of individual groups of scholars and of the scholarly community as a whole. The old regime had been skeptical about voluntary associations created for the purpose of protecting the welfare of individual groups and serving as a forum for criticizing unpopular government policies. During the Kerensky regime, scholars took advantage of the generally relaxed atmosphere to intensify and modernize the process of creating scientific associations as national organizations uniting specialists in individual disciplines. These associations were the main sponsors and organizers of national congresses of experts in particular branches of scientific knowledge, who met periodically in selected cities. In tsarist days workers in individual sciences did not hold national congresses; they were limited to organizing special sessions at the general congresses of "naturalists and physicians," sponsored and hosted by individual universities. The new associations intensified the role of oral discussion of theoretical and methodological issues in individual sciences, broadened the contact between the scientific community and the general public, and created personal ties between scholars from different cities. The I. M. Sechenov Society of Russian Physiologists (with I. P. Pavlov as its first president), the All-Russian Astronomical Union (later the Association of the Astronomers of the Russian Soviet Federated Socialist Republic [RSFSR]), and the Russian Physical Association were the first such new organizations.

The Soviet regime recognized these bodies and did not discourage the formation of new ones. From the very beginning, it became clear that these organizations would not be allowed to serve as forums for the articulation of professional interests of specialists in various fields or as springboards for actions incompatible with the national science policy.

Scientific congresses became forums for public reports on national achievements in individual sciences and for open discussion of knotty theoretical questions that troubled the scientific community.

It was only proper and logical that physicists, whose discipline led the ongoing revolution in science, held the first congress of Soviet naturalists. Convened in Petrograd on February 4–7, 1919, the First All-Russian Congress of Physicists mainly attracted scientists from the two largest cities. Unstable conditions in the country, caused by the Civil War, allowed only a handful of scholars from Kiev and other cities to attend the Congress.[20] The First All-Union Congress of Geologists convened in Petrograd in 1922; 222 of 300 attending scientists were from Petrograd. By contrast, the Third All-Russian Congress of Zoologists, Anatomists and Histologists, which met in Leningrad on December 14–20, 1928, attracted 755 specialists from sixty-one cities.[21] Government authorities treated national congresses as public demonstrations of close ties between science and Soviet ideology, as handy vehicles for informing foreign scholars (who were often invited to attend these meetings) about Soviet achievements in science, as means for alerting the general public to the rapidly expanding social role of the scientific community, and as effective publicity for the advantages of the collective endeavor in science. The unusually large contingent of Western scholars who attended the Sixth All-Union Congress of Physicists, held in Moscow and Saratov in the summer of 1926, included among others the physicists Paul Dirac and Max Born, the mathematician Richard von Mises, and the philosopher Philipp Frank.[22] All this led S. F. Ol'denburg, permanent secretary of the Soviet Academy of Sciences, to note in 1926 that "scientific congresses are currently passing through a stage of significant transformation."[23]

The representatives of Marxist scholarly organizations took full advantage of the new opportunity to present their views on special questions of scientific theory at the meetings of non-Marxist scientists. Particularly after 1925 it was not unusual for the congresses of scientists working in the more sensitive areas to hold special sessions on dialectical materialism. As time went on, Marxist groups played an increasingly more important role in the committees entrusted with the task of organizing congresses and demanded a stronger concern with the four pillars of Marxist science: theoretical (ideological) unity, group effort, utilitarian orientation, and centralized planning.[24]

There were two general kinds of scientific congresses: those sponsored by individual learned societies or associations and those with no permanent institutional base. The latter were organized by host institutions—by a university, by a research institute, or by the State Planning Commission (GOSPLAN)—which changed from one meeting to another. Some of these congresses published their proceedings. In the mid-1920s all-Union congresses covered such fields as geophysics, geology, botany, hydrology, pathology, and natural resources. All-Russian and other regional congresses also took place, sometimes combining several related disciplines. In exceptional cases individual congresses established permanent formal organizations; such, for example, was the origin of the All-Russian Association of Zoologists, Anatomists and Histologists.[25] The rapid growth of the ranks of scientists, the fission of fundamental sciences into special sciences, and the rise of interdisciplinary branches of knowledge deeply affected the dynamics of congresses, associations, and societies. When old organizations became huge and unwieldy and ceased to be functional, they gave way to more specialized formations.

In a move to engage large segments of population in scientific work, the Bolshevik authorities encouraged and assisted a special type of nonprofessional learned society—the society for regional studies (*kraevedenie*).[26] The immediate aim of these organizations, which antedated the October Revolution, was to make work in science a cherished national preoccupation and to provide a firm foundation for the scientific world view built into Marxist ideology. Large sectors of the population were thus involved in gathering empirical data of scientific significance and in building a network of local laboratories, natural history museums, specialized scientific stations, and naturalist parks. According to the linguist N. Ia. Marr:

> For scholars working in research centers, regional studies are sources of new data indispensable for formulating abstract theories of universal scope; for local communities, regional studies are a fountain of social thought and social self-awareness. . . . For individual ethnic groups, they are the safest path to a national renaissance and the best guarantee for the growth of a national language as the mainstay of cultural identity.[27]

In 1919 the Academy of Sciences established the Central Bureau of Regional Studies to help young naturalist societies in their search for more advanced tools of scientific inquiry and to take the spirit of science to all corners of the country. The Leningrad Museum of Anthropology and Ethography counted on *kraevedenie* enthusiasts to supply some of the material for the journal *Etnografiia* and to contribute to the current national survey of regional cultures. The First Congress of Scientific Societies for the Study of Local Life, held in Moscow in 1921, relied on *kraevedenie* amateurs when it unveiled an ambitious plan for a study of population dynamics in Russia.

S. F. Ol'denburg noted in 1921 that the societies for regional studies existed even in "the most isolated corners" of the country and that some of them had become "large institutions."[28] In 1922, according to official reports, the army of *kraevedenie* enthusiasts numbered 40,000.[29] At first interested mainly in local history, ethnography, and archeology, *kraevedenie* societies gradually transferred their attention to "the study of natural riches, economy, and statistics." In 1926 a government report noted the existence of 1,299 organizations—including naturalist societies—engaged in the study of the natural resources of individual regions.[30] With the exception of a very small percentage of technical and administrative personnel, the huge membership of these groups contributed free labor to the propagation of science. Furthermore, they provided the clearest manifestation of a "democratic" organization of science, as distinct from the "aristocratic" principle on which the Academy of Sciences was built.

MARXIST ORGANIZATIONS

A network of special organizations dedicated to bringing Marxism and modern scientific thought into harmonious relation formed the newest complex of scholarly institutions. This segment was dominated by the Socialist Academy (later the Communist Academy) of the Social Sciences, founded in 1918, which undertook the task of propagating historical materialism as the theoretical and ideological base of the social sciences and the humanities. In the words of E. A. Preobrazhenskii: "Marxism in Russia is the official ideology of the victorious proletariat; the Socialist Academy is the highest scientific institute of Marxist thought. . . . It recognizes only the branches of social science which are anchored in Marxism." He added that "the theory of historical materialism is more important for the social sciences than the laws of Kepler and Newton are for physics."[31] The Academy was also given the task of maintaining active relations with Marxist theorists outside the Soviet Union and of training teachers in the fields of law, history, politics, national economy, sociology, and philosophy.[32] For several years the new organization was scarcely more than a library and a debating club, meeting infrequently and suffering from ambivalent goals and internal fragmentation. Its members represented the full gamut of Marxist thought, from loyal Leninists, such as M. N. Pokrovskii, to "Machists" and "empirio-critics," such as A. A. Bogdanov and V. A. Bazarov. The list of foreign members was also marked by a wide range of ideological coloration, including, among others, Romain Rolland, Franz Mehring, Klara Zetkin, and Karl Liebknecht. The membership included many "socialists" unwilling to cooperate with "communists."[33]

The Socialist Academy assumed the task of serving as the national center for the advancement of Marxist theory, because the Academy of Sciences had no organized research in this field. During the first twelve years of its existence under the Soviet regime, the Academy of Sciences had only one full member—the linguist N. Ia. Marr—who professed to be a Marxist. At the other extreme were isolated academicians, led by I. P. Pavlov and V. I. Vernadskii, who did not conceal their anti-Marxist leanings. Occupying the middle position, the vast majority of academicians pursued a line of "ideological neutrality," a derogatory label coined by Marxist critics.

Until 1923 the Socialist Academy achieved too little to justify its existence. To most leading members the work in Marxist theory was an extracurricular, and clearly secondary, activity. The young Academy was hampered by unrealistic plans, a small and generally incompetent research staff, limited publication outlets, and the pronounced absence of a consensus on the finer points of Marxist theory. The impotence of this new body was best manifested in its failure to produce a general treatise on dialectical materialism that could provide general guidance in the training of a new generation of Marxist theorists. And the same was true of the doctrine of historical materialism, the grand theory of Marxist sociology. Almost until the end of the decade, Marxist training centers depended mainly on Bukharin's book on historical materialism as a system of Marxist sociology, a work that was heavily attacked by the leading Marxists in the Soviet Union and abroad (for example, by Georg Lukács and Antonio Gramsci). Critics resented Bukharin's effort to create a "mechanical" model of the structure and dynamics of human society and to substitute the static notion of "equilibrium" for the dynamic notion of "dialectic" as the key concept in Marxist sociology. But Marxist schools relied so heavily on Bukharin's volume for one simple reason: it had no competition. The Communist Academy could not find either an individual or a team to prepare a comprehensive textbook on historical materialism.

After 1923 the Communist Academy—as the Socialist Academy was now renamed—rapidly expanded its staff and publications. By the middle of the decade its many research units included the Section of Natural and Exact Sciences and the Institute for the Study of Higher Nervous Activity. Although isolated scientific reports prepared by the Academy's staff or invited contributors did not raise, at least directly, the issues of ideology and Marxist philosophy, an overwhelming number of oral reports and published papers became part of a general effort to arrive at a unified Marxist interpretation of the theoretical foundations of modern science.

The members of the Communist Academy now undertook to criticize the leading scientists who either displayed philosophical aloofness or opposed Marxist thought. The distinguished members of the Academy of Sciences served as particularly attractive targets for denunciatory attacks. V. I. Vernadskii was accused of flirting with Bergsonian vitalism and of challenging the notion of the material unity of the universe, firmly built into Marxist ontology.[34] In general, however, the Marxist critics were favorably impressed with Vernadskii's exploratory work in the study of the biosphere. The famed neurophysiologist I. P. Pavlov was the target of numerous innuendos depicting him as the mastermind of recurring efforts to make the study of conditioned reflexes the only basis of scientific sociology and social psychology. The Communist Academy provided a forum for a group of Marxist theorists, led by Leon Trotsky, who contended that the excessive worship of Pavlov's theories worked against the burgeoning efforts to effect a fruitful synthesis of Marxism and Freudianism.[35] Pavlov's publicly expressed aversion to the use of revolutionary methods as a tool of resolving social conflict was directed against the political strategy of the Bolshevik party.[36] N. I. Bukharin, a member of the Communist Academy, wrote a lengthy article refuting Pavlov's claims that the October Revolution was a historical anomaly.[37]

Attacks on Vernadskii and Pavlov were seldom direct, particularly in the case of Pavlov, and never serious or concentrated enough to produce damaging results. With foreign scholars joining his laboratories for advanced studies, Pavlov's star now shone brighter than ever before. Vernadskii, too, attracted the attention of the international scientific community by his pioneering work in biogeology.[38] While the more critical Marxist theorists felt uncomfortable about the antidialectical views of these two eminent scientists, the government was eager to make them the backbone of the Soviet scientific community. Subsequent developments on the scientific front showed clearly that the government was much more tolerant of such illustrious "bourgeois" scientists as Pavlov and Vernadskii than of the stalwarts of Marxist "science" in the Communist Academy, who were extremely slow in giving up the idea of "Communist science."*

Nor did the Marxist theorists spare the Academy of Sciences as an institution. M. N. Pokrovskii, president of the Communist Academy, made no effort to disguise his view of the Academy of Sciences as a sanctuary of bourgeois thought and an institutional antithesis to Marxist plans for organized research. After 1925, when the government made the Academy of Sciences the officially recognized center of the national commitment to science, the members of the Communist Academy

---

*This does not mean that the government was not baffled by the behavior of the two scientists. Vernadskii gave the authorities much to be concerned with. At the time of the October Revolution, he and his family withdrew to a family settlement near the city of Poltava in the Ukraine, but on May 18, 1918, he was in Kiev—at that time not under Soviet control—where he helped organize the Ukrainian Academy of Sciences, of which he was the first president. In 1920, ill with typhus, he went to Yalta to live with relatives—and to be near the emigration routes. As soon as his health showed signs of improvement, he began to lecture on geochemistry at the newly founded Tauridian University. At this time he appealed to the British Association for the Advancement of Science to help him emigrate. Soon he received word that a British ship was on its way to take him out of the Soviet Union. In the meantime, he was elected rector of Tauridian University, a turn of events that persuaded him to abandon his emigration plans. In 1921 he returned to Petrograd and assumed his old position in the Academy of Sciences. But at the end of the year he received an offer to join the faculty of the Sorbonne as a visiting professor. The Academy of Sciences not only allowed him to go to Paris but also provided him with sufficient funds to visit the radium centers at Vienna, Heidelberg, London, and Cambridge. He was also given permission to take his family with him. In 1925, when the Academy celebrated the two-hundredth anniversary of its founding, he was still abroad; the published list of the members of the Academy given to the foreign dignitaries who attended the celebration did not contain his name (*Spravochnik dlia uchastnikov*, pp. 3–5). Vernadskii and his wife returned to Russia in 1926 (*Otchet-S:* 1926, vol. 1, p. 3); their son and daughter stayed abroad. During the long absence he continued to contribute papers to Soviet journals and to maintain close relations with the Academy. In addition to scientific work, he published two small volumes of essays and addresses (*Ocherki i rechi*) that had originally appeared before the October Revolution. The Soviet authorities had no quarrel with Vernadskii's basic ideas on the national organization of scientific work; however, they had little sympathy for his general philosophical and sociological views regarding science. At the same time, he published a popular booklet on "the beginnings and eternity of life," in which he stated that materialistic philosophy contradicted the revolutionary changes and gains of modern science (*Nachalo*, pp. 57–58). In mid-1927 he was back in Western Europe for a three-month tour of scientific centers. During the next five years, seldom did a year pass without a visit abroad (V. S. Neapolitanskaia, *Perepiska:* 1918–1939, pp. 81–97). He always came back—always ready to fight for a philosophical view that had little in common with Marxism.

changed their strategy from direct attacks to a recognition of the complementary functions of the two institutions. The new line recognized
the need for the continued work of the Academy of Sciences in advancing
"bourgeois science" as a source of tools and theories that could be har-
nessed for a full victory of socialism. It also recognized the need for the
Communist Academy to work on laying the foundations for "Soviet
science" as a unique system of research methods, organizational principles, and theories. But the notion of "Soviet science" was not clearly
developed; it was obscured in a maze of irrelevant comments, flowery
exhortations, and oblique hints. Its main defender was the historian M.
N. Pokrovskii, but he never looked beyond the social sciences, where
the question of ideology was both clear and acute. To most members of
the Communist Academy, the search for "Soviet science" was the search
for a Marxist interpretation of the revolutionary developments in modern
science.

The main function of the Communist Academy was to sponsor an
organized search for philosophical—or methodological—unity of all the
branches of knowledge. This unity, however, was a distant goal; during
the 1920s the Academy was torn by profound differences in the unceasing search for a Marxist interpretation of major developments in modern
science. According to Pokrovskii, the diversity of opinion was a result
of the Academy's mode of operation, which encouraged individual work
in research and nonconformism in matters of theoretical import. He
lamented the low development of collective effort in research, which
alone, he thought, could serve as a solid base for creating a body of
generally accepted postulates of Marxist philosophy of science. The
Academy, for example, supplied both philosophical critics and philosophical defenders of Einstein's theory of relativity.[39] It also provided
both proponents and opponents of quantum theory, psychoanalysis,
Pavlovian neurophysiology, and Mendelian genetics.[40] The Institute of
Philosophy, founded in 1928, inaugurated the era of Stalin's determined
efforts to standardize the basic premises of Marxist-Leninist thought and
to establish a unified Marxist philosophy of science. By an ironic twist
of fate, many leading articulators of Marxist philosophy operating from
the Communist Academy did not survive Stalin's sweeping and terror-
ridden moves to tie Soviet ideology to a unified system of Marxist
philosophy.

The Communist Academy was the guiding star in widely ranging
efforts to create new centers for the training of future Marxist scholars

and to advance Marxist theory as a most general science. These centers appeared at different times in distinctive roles. The Institute of the Red Professoriat was founded in 1921 for the purpose of supplying institutions of higher education with instructors in economics, sociology, and philosophy. Centered in Moscow, with branches founded first in Leningrad and then in other major cities, the institute flooded the academic world with interpreters of Marxist orthodoxy, working more as custodians of old doctrines than as explorers of new intellectual regions open to Marxist scrutiny. Although dismally undertrained in the natural sciences, they did not hesitate to criticize those philosophical views of the pioneers of modern physics, biology, and mathematics that did not mesh with the core postulates of Leninist epistemology. These "guardians of ideology" contributed greatly to the difficulties in bringing dialectical materialism in tune with advances in modern science and in making Marxist philosophy attractive to scientists.

The Russian Association of Scientific Research Institutes of Social Sciences (RANION), founded in September 1923 and closely associated with the Communist Academy, worked on two fronts: consolidation of Marxist theory and training new cadres of philosophy and social science instructors for the institutions of higher education.[41] As envisaged by its creators, RANION was expected to grow into a national "scientific" organization with branches in all large communities. This, however, did not materialize. Prior to 1925 it was dominated by non-Marxists and "formal Marxists" who knew little about theory of any kind. After 1925 RANION found it difficult to retain Marxists who preferred either to work in the theoretical centers of the Communist Academy or to seek employment in such "practical" agencies as the Supreme Council of National Economy (VSNKh).[42] While losing the battle on the "scientific" front, RANION intensified its involvement in mass campaigns of a political nature. In particular, it concentrated on harassing scientists and educators accused of active opposition to the reforms promoted by the Soviet government.

The aggressiveness of Marxist groups reached a peak in the activities of the Society of Militant Dialectical Materialists, founded in 1924.[46] Organized on a national scale, this body played a particularly active role in interpreting the advances of modern science from the position of dialectical materialism. Its members did not engage in scientific research; their main function was to oppose the theoretical views of modern science that were regarded as contrary to Marxist philosophy and Soviet ideology and to advocate a Soviet science policy emphasizing polytechnical education and central planning in research.

At the end of the 1920s, close to 100 different Marxist institutions, identified as research centers, learned societies, or schools, worked on building a bridge between science and Soviet ideology; but most of this effort concentrated on the social sciences and humanities. According to official statistics, only 8 percent of the total of 2,000 Communist party members engaged in scholarship in 1928–1929 were in the natural sciences (64.8 percent were in the social sciences and philosophy).[43] Philosophical uncertainty, ideological oscillation, and the grave scarcity of experts, particularly in the natural sciences, prevented these institutions from taking deeper roots and from playing a significant role in the development of scientific scholarship. The prodigiously wasteful effort to reconcile the iconoclasm of science with the conformism of ideology stood in the way of a broader and more intensive concern with scientific inquiry.

Militant Marxists, from A. M. Deborin and A. K. Timiriazev at the apex of the Communist Academy to the lowly teachers at newly established teacher-training schools, contributed more to generating skepticism toward the great achievements of modern science than to developing a healthy interest in research and devotion to science as a backbone of modern civilization. The Deborin group, in particular, preached a new creed that made science a servant of philosophy and proclaimed philosophy "the highest result of human thought at every stage of the development of science and society."[44] All Marxist theorists agreed that the Communist Academy should be the center for a systematic historical and sociological study of science; however, the proposed institute for the history of science and technology did not materialize. The members of the Academy, poised for ideological confrontation, favored the widely explored and simple dictates of Leninist epistemology over the little-explored and turbid perspectives of Marxist sociology as applied to the evolution of scientific thought. In the early 1920s A. Gol'tsman stated that much preparatory work in gathering documentary sources would be required before Marxist scholars could undertake a serious study of the social and political causes of scientific theories. He stated:

> To place a sociological base under Einstein's pure physical theory at the present time is as difficult as to give a sociological explanation of the conditions that led to the emergence of Newton's and Planck's theories.

> In order to explain the social causes of these theories and the development of science in full detail, we must first advance a microscopic Marxist approach. Unfortunately, at the present time we do not have precise measuring instruments in the field of sociology, and all efforts to give individual physical theories a sociological explanation would prove to be inadequate.[45]

The effectiveness of Marxists institutions was drastically curtailed by the glaring absence of a unified Marxist philosophy of science. Marxist theorists were split into two major feuding camps—the mechanists and the dialecticians—each with a distinct attitude toward the theoretical products of the ongoing revolution in physics and each accusing the other of gross misrepresentations of the philosophical legacy of the fathers of Marxist thought. In this bitter feud, which attracted public attention in 1926, dialecticians accused mechanists of giving preference to "mechanical causality" over dialectical "unities of contradictions," rejecting Lenin's "reflection" theory as the foundation of Marxist epistemology, considering "matter" a scientific but not a philosophical concept, treating measurable properties of natural phenomena as the only objectively existing reality of nature, and expressing a skeptical attitude toward some of the most revolutionary developments in modern physics, such as, for example, Einstein's theories of relativity.[47] In the thick of the debate—in 1928—the dialecticians were also ready to accuse their adversaries of philosophical inconsistency, pseudoscientific showmanship, and unfounded and pretentious claims.[48] A. M. Deborin, the chief dialectician, labeled the mechanists "a self-styled bloc of Freudians, former Machists, and a mixture of silent and vocal empiricists and mechanical materialists."[49] He went so far as to assert that the mechanist theory, brought to a logical conclusion, adhered to the "mathematical theory," which claimed that the smallest particles of matter were indistinguishable from numbers—that "number" was "the essence of matter."[50]

The mechanists, for their part, accused the dialecticians of questionable loyalty to Leninist elaborations of Marxist thought, a dogmatic attitude toward the idealistic underpinnings of Hegelian philosophy, and the uncritical and indiscriminate acceptance of the revolutionary claims of modern physics. While the dialectical orientation was dominated by old-line Marxist philosophers, the mechanist orientation was dominated by natural scientists who had joined the Marxist ranks after the October Revolution.[51] The bitter feud between the two camps was complicated by the clear tendency of each party to exaggerate and distort the major claims of the opposing group and by the existence of strong differences in philosophical views within each group.[52] The Deborin group won most skirmishes because it was more compact and more aggressive. In the words of David Joravsky: "Mechanism in various forms was probably the dominant philosophy among Soviet Marxists, but relatively few came to its defense when the Deborinites challenged it."[53]

Whether the Communist Academy should be concerned with the philosophical foundations of natural science in the first place was another key source of divisiveness in the ranks of Marxist theorists. A. M. Deborin led a strong group of leading Marxists who viewed dialectical materialism as both the supreme methodology of modern science and the safest link between science and Soviet ideology and who favored the Academy's primary concern with the epistemological foundations of revolutionary changes in twentieth-century science. D. B. Riazanov, in contrast, represented a relatively small group of theorists who claimed that neither natural science nor mathematics offered much help to the Academy in carrying out its major task: the study of the structure and dynamics of human society. According to Riazanov, the unwarranted preoccupation with the philosophy of natural science had contributed only to the infusion of positivist leanings into Marxist philosophy.[54] He resented the "contamination" of dialectical materialism by an excessive reliance on developments in the natural sciences. It should be noted that even though Deborin advocated substantial involvement in the dialectical-materialistic foundations of modern science, his personal writing dealt primarily with topics from the history of philosophy. The typical member of the Communist Academy was not much impressed with either Deborin's overemphasis on epistemology or Riazanov's overt efforts to move away from philosophical involvement. This typical member used philosophical metaphors as intellectual decor and ideological guidelines, but he made no effort to give them serious attention or to make them a solid component of his work.

An important function of Marxist "scientific" organizations was to advance the study of dialectical materialism in all institutions dedicated to scholarship. For this purpose, several Marxist institutions organized auxiliary societies or "collaborating societies"—such as "collaborating societies of mathematicians-materialists" and "collaborating societies of chemists-materialists"—to help in the war against "rightist" groups in the Academy of Sciences and other scientific bodies. These societies served as bases for training Marxist scientists and for bringing together scientists with clearly manifested Marxist philosophical inclinations. The members of these societies provided scientific arguments for endless philosophical and ideological campaigns led by Marxist theorists, and they accelerated the process of creating Marxist enclaves in the scientific community. An exceedingly small group of the members of these societies became either recognized scientists or prominent defenders of Marxist philosophy. During the first ten years of the new political regime, Marxist scholarly organizations made little impact on the growth of scientific thought in the Soviet Union. The Section of Natural and Exact Sciences of the Communist Academy, established in 1925, could survive only by accepting papers contributed by non-Marxist scientists.

THE ACADEMIC NETWORK

Academies of sciences, the fifth complex of scholarly institutions, performed a dual function: they honored the most distinguished scholars by giving them high titles and more favorable working conditions, and they served as advanced research centers dealing with the most critical areas of science. They trained the most promising young scholars, represented the scientific community at international scientific forums, and played a decisive role in guiding and coordinating the national effort in science. In the pre-Soviet period the Russian Academy of Sciences was the only academy dedicated to the advancement of science. After 1917 the situation changed dramatically. In 1918, before Soviet power was established in the Ukraine, a group of scholars, led by V. I. Vernadskii, founded the Ukrainian Academy of Sciences, at first heavily weighted in favor of the humanities and the social sciences.[55] This was the first step in a long process (the second step was made in 1929 with the founding of the Belorussian Academy of Sciences) that gave each republic a general academy of sciences. The new authorities opted also for the creation of specialized academies; however, the Communist Academy, the Lenin Academy of the Agricultural Sciences, founded in 1929, the Academy of Pedagogical Sciences, founded in 1943, and the Academy of Medical Sciences, founded in 1944, set the general limit in this category of academic expansion.

During the 1920s the evolution of the institutional network of Soviet science was dominated by two trends: the continual and unusually rapid, but not symmetrical, development of the five systems of scientific organizations; and the consolidation of the Soviet Academy of Sciences as the institutional summit of Soviet science. It is the latter development that deserves special attention, for it provides the clearest and the most comprehensive and dramatic picture of the unique structure of the institutional underpinnings of Soviet science and of the combination of remarkable successes and disconcerting dilemmas of Soviet scientific scholarship. By understanding the process through which the Academy became both an integral part of the Soviet polity and the nerve center of the institutional maze of Soviet scientific scholarship, we come to see Soviet science as a rapidly growing system of knowledge, a mode of inquiry, a cultural value, an instrument of national power, and a source of far-reaching ideological dilemmas.

THE TIME OF UNCERTAINTY

In 1917, the year of the October Revolution, the Russian Academy of Sciences was less than eight years short of entering the third century of its illustrious existence. The versatility of its interests and the measure of its achievements elevated it to an honored position in the highly competitive realm of world scholarship. The Academy gave Russian science a sense of direction and a cultural vibrancy of the first order, and its long history was dominated by scholars who added lasting pages to the annals of science. Over the long period of its dedication to science, its membership had included such world-renowned scientists as Leonhard Euler, the greatest mathematician of the eighteenth century, Karl von Baer, the founder of comparative embryology, and A. A. Markov, a noted pioneer in the modern theory of probability. Among its distinguished members were M. V. Lomonosov, rightfully called the father of Russian science, I. P. Pavlov, the master of neurophysiological experiments in digestive processes and conditioned reflexes, E. S. Fedorov, one of two scholars to produce the first complete mathematical picture of crystalline structures, and V. I. Vernadskii, a leading pioneer in modern studies of the biosphere.

The revolution of February 1917 marked a turning point in the history of the Academy. An impressive group of academicians welcomed the revolution as the beginning of an intensive and dedicated search for democratic processes favorable to free inquiry and academic autonomy. Several academicians, including S. F. Ol'denburg, occupied high positions in the new government.[56] The geologist A. P. Karpinskii became the first elected president of the Academy (previous presidents were appointed by the government), a clear triumph for the cause of academic autonomy. The Academy gave strong support to a group of eminent members of the scientific and artistic intelligentsia who paved the way for the creation of the Free Association for the Advancement and Diffusion of Science, which was dedicated to expanding the lines of free communication between the scientific community and the society at large and to promoting science as a distinct body of knowledge and an ideal of humanity.

Most members of the Academy, however, did not welcome the October Revolution.[57] On November 21, 1917, the general conference of the Academy approved a resolution drafted by a special committee—including S. F. Ol'denburg, chemist N. S. Kurnakov, and others—that called for unrelenting opposition to the new government and for continued support of the idea of a Constituent Assembly.[58] A few years later, I. P. Pavlov, the bulwark of the Academy's inner strength, called the October Revolution a "historical error" and a "national catastrophe."[59] The government, in turn, was in no position to give much attention to the Academy; the Civil War and the shattered economy demanded most of its attention. During the Civil War and a few years afterward, wrote V. P. Volgin, the government could give scientific and educational institutions only "a minimum of attention." Most of "the limited time and energy which could be devoted to cultural construction went to the founding of [Marxist] educational and scientific centers."[60] Volgin considered the founding of the Socialist Academy of the Social Sciences and the Sverdlov Communist University, both deeply involved in "Soviet and Party work," the major achievement of the government in the fields of education and scholarship during the first five years of the Soviet system.

When in 1919 *Priroda,* a journal devoted to the popularization of scientific knowledge, presented the plight of the Mineralogical Society in Petrograd, it gave a fair description of the plight of the entire scientific community. It stated: "The [critical] political situation in the country, chronic delays in government subsidies, prohibitive publication costs, and the shortage of printing material have brought scientific work in the Mineralogical Society to a standstill and have eliminated most of its publishing activity. . . . The trying conditions of personal life, hunger and heightened anxiety have left deep scars on the face of our Society."[61] The plight of science was given the most dramatic expression in 1921 when the famed mathematician A. A. Markov informed the Academy of Sciences that he could not attend the scholarly meetings because he did not own a wearable pair of shoes.[62]

It was at this time that Maksim Gorky organized the Central Commission on the Improvement of Living Conditions of Scholars (TsEKUBU), which operated mainly in Petrograd and Moscow. This organization issued scientists "academic rations," which had a slightly higher caloric value than the typical rations. It assisted scientists in finding housing and places in rest homes and sanatoria and provided small financial assistance to helpless families.[63] In Moscow it managed the House of Scholars, which provided space for popular lectures and ran a library boasting a good collection of current scholarly journals published abroad.[64] TsEKUBU also organized a "Sunday university" for workers. The government made no effort to tie the cultural activities of TsEKUBU to Marxist aspirations, and in fact the lectures given at the Moscow House of Scholars had no Marxist goals. Even representatives of philosophical idealism were given the opportunity to express their views on the freedom of will, scientific determinism, the nature of creativity in the arts and sciences, the human soul, spiritual activities, the uniqueness of religious consciousness, and the broader cultural context of science.[65]

At first the Soviet authorities did not know exactly what to do with the Academy, particularly how to define its place in the new political system.[66] Their attitude could best be described as ambivalent. On the one hand, they treated science as the most potent and most advanced method for unlocking the mysteries of nature and making social progress a reality.[67] As worshipers of science, they were certain to avoid the grave error committed by the French revolutionaries when they abolished the Paris Academy of Sciences and guillotined Antoine Lavoisier, the father of modern chemistry, because he symbolized, in their opinion, the most repressive features of the "ancien régime". The Bolsheviks did not abolish the Russian Academy of Sciences, and they did not look for a Russian Lavoisier. On the other hand, in their dealings with the Academy they did not conceal their distaste for many features of the time-honored institution. The Academy, they contended, was built on an aristocratic principle: not only did it occupy the summit of the pyramidal organization of scientific institutions but it was also a self-perpetuating body in the sense that it alone nominated and elected new members. As an "aristocratic" institution, the Academy stood at the opposite pole from the mounting outcries for the "proletarianization" of science—or the "proletarian socialization of science."[68]

Although the Academy represented the main stumbling block on the road to the proletarianization of science, the government refrained from subjecting it to drastic institutional restraints. Nor did the government overlook the Academy in its plans for a future system of scientific institutions. Lenin cautioned A. V. Lunacharskii, commissar of education, that all government plans for the future of the Academy must be based on "prudence, tact and extensive knowledge."[69] Lunacharskii contended that "the Academy must adapt itself to the general needs of the state and society," and that it must not be allowed to exist as a state within a state. "We must," he said, "bring it closer to us, be familiar with its activities, and give it some direction; but it would be premature to undertake a major reorganization at this time."

The Bolshevik authorities were also influenced by a strong liberal tradition in Russia that treated the Academy as a "German institution" because it was dominated for so many years by German scholars. The Academy was held guilty of not having elected such luminaries of national science as the chemist D. I. Mendeleev, the neurophysiologist I. M. Sechenov, and the physicist P. N. Lebedev to its membership.[70] During the early years of the Soviet regime, K. A. Timiriazev, a member of the once-influential group of liberal professors and one of the first leading scientists to express loyalty to the Soviets, continued to exploit the unfavorable image of the Academy as a "German institution." Thus the Academy was found guilty on two counts: it was painfully slow in organizing scientific research along the lines most suitable to the immediate needs of the country, and it did not fulfill its duty of bestowing high academic honors on the most deserving scientists. The Bolsheviks were quick to interpret these "aberrations" as specific manifestations of aristocratic detachment from the realities of national life.

The Bolshevik leaders found the unbalanced representation of various disciplines in the Academy particularly objectionable. They resented the overrepresentation of the humanities and the social sciences. They felt it was the natural sciences that held the key to economic progress and a successful recasting of the structure of society. Moreover, they thought that the academic representatives of the humanities and the social sciences were overcommitted to the ideology of the old regime. While the natural scientists, they thought, could be persuaded to work toward a practical future regardless of their ideological stance, the humanists and social scientists were too deeply steeped in the ideological past to be of much use to the new sociopolitical system. The Bolshevik leaders looked with great disfavor at the emigration of the economic historian P. B. Struve and the Roman historian M. I. Rostovtsev, who continued their bitter anti-Soviet campaign in the West. They resented the attacks by the medievalist P. G. Vinogradov, member of the Academy and Oxford University professor, who brought his practice of frequent visits to Russia to an end.

While the emigrated social scientists and historians quickly became anti-Soviet activists, the ranking natural scientists who decided to leave the country were in many cases unwilling to engage in anti-Soviet campaigns or to break all relations with the scientific institutions of the homeland. In 1923 the Academy bestowed honorary membership on S. N. Vinogradskii, the world-famous student of the role of microorganisms in the nitrification of soil and an associate of the Pasteur Institute in France. He had stayed in the territory controlled by the White Army until an opportunity for emigration was available. The chemist P. J. Walden, a full member of the Academy, emigrated to Germany after the October Revolution; however, in 1927 the Academy elected him an honorary member. Nor did the authorities look favorably at the unusually heavy concentration of political liberals in the Academy, who were interested in changing national politics along democratic lines rather than in the spirit of Lenin's prescriptions for a dictatorship of the proletariat.

For their part, the members of the Academy made no secret of their profound opposition to the science policy of the new government. They were particularly unyielding in resisting the new official emphasis on applied research. A typical academician evoked Mendeleev's counsel that only truth sought for its intrinsic meaning and "absolute purity" could become a true source of practical knowledge. The microbiologist V. L. Omel'ianskii, for example, endorsed K. A. Timiriazev's statement that in its development science depended on "the internal logic of facts rather than on the external pressure of practical needs."[71] He contended that all efforts to subordinate science to "external needs" inevitably imposed limitations on the freedom and breadth of creative research. Nor did the members of the Academy conceal their disaffection with the government's emphasis on officially approved research plans. They were particularly dissatisfied with the limited resources the government placed at their disposal. In his progress report for 1920, S. F. Ol'denburg complained about the critical shortage of printing material, poor lighting and heating of academic premises, substandard salaries for scientific personnel, and the drastic depletion of young scholars attached to the Academy.[72] He reported two years later that an American gift of 500 tons of printing paper had helped the Academy to expand and accelerate its publishing activity,[73] but he noted the injurious effects of the isolation of Soviet scholars from Western scientific centers. What disturbed him most, however, were official efforts to make science one with the reigning political ideology. Although he made no specific references, his message was clear and meaningful:

> Many periods in the history of mankind were dominated by misdirected actions to subjugate science to temporal currents of ideas regarded by their champions as incontestable. While these efforts came and disappeared quickly, science continued to march forward, following a straight and impartial path and helping mankind in its struggle to overcome delusions and uncertainties. Let us hope that the new Russian way of life will not repeat these old errors and will fully recognize the creative and unswerving power of free science and its unceasing search for truth.[74]

On another occasion, Ol'denburg recalled the statements of the great Russian historians S. M. Solov'ev and V. O. Kliuchevskii in favor of the full independence of scholarly pursuits from "dominant views" in society as the best guarantee for socially useful work in science.[75] The mathematician V. A. Steklov, vice president of the Academy, made an excursion into the history of science to give added strength to Ol'denburg's lamentations. In his biography of Galileo, published in 1923, he stated that no force could subordinate "the free mind of an exact scientist and thinker to predetermined and permanently fixed slogans of a party."[76] Political ideology and scientific creativity were mutually exclusive phenomena. In order to be productive, "the representatives of science must be allowed to work outside all parties and party commands."

Bolshevik critics were generally correct when they claimed that the members of the Academy represented the scientific intelligentsia in its negative attitude toward the new government.[77] This attitude, in the words of a Marxist scholar, ranged from "petty-bourgeois wavering" to "open enmity."[78] This scholar went on to speculate that the "neutrality" of most leading scientists represented a thinly veiled antagonism toward the new regime and was an excuse for retreat into the most isolated and abstruse research domains. In the years immediately following the October Revolution, he said, some academicians reduced their activity to a mere gathering and systematization of scientific facts. Most scholars considered the proletarian revolution a historical accident and felt no need to adjust their work and ideas to the demands and ideological dictates of the new authorities.

"Open enmity" of the members of the Academy toward the new system did not subside quickly. Indeed, it found a particularly strong expression in 1923 when the Academy brought out a new edition of A. S. Lappo-Danilevskii's noted *Methodology of History*, a critical assessment of modern approaches to the study of history. Lappo-Danilevskii, an academician of liberal convictions, had died in 1918. His *Methodology* included a long chapter on the Marxist theory of history that concentrated on the major flaws of Marx and Engels's efforts to make history a "nomothetic" discipline, concerned with general laws of social development and built on an axiomatic basis borrowed from the natural sciences. The basic weakness of the Marxist theory was that it considered historical events primarily as illustrative material for preconceived theoretical schemes and ideological dictates. It was also marred by inconsistencies and imprecise statements that invited an endless procession of revisionist theories. Lappo-Danilevskii showed much more sympathy for P. B. Struve and M. I. Tugan-Baranovskii, who tried to merge the theories of Marx and Kant, than for Lenin's efforts to protect the orthodoxy of classical Marxism.[79] The republication of Lappo-Danilevskii's work was one of the more dramatic ways in which the Academy resisted the growing external pressure to make Marxism the official philosophy of the natural and social sciences.

Lappo-Danilevskii figured in yet another effort of the Academy to assert its intellectual independence. The founding of the Socialist Academy in 1918 by a group of Marxist theorists prompted the Academy of Sciences to undertake various measures to protect non-Marxist social theory. The initial intent of the Socialist Academy was to serve as the supreme forum for setting up the methodological guidelines and substantive interests of all the social sciences—to articulate the general principles of Marxist social theory. Interpreting this move as a direct assault and encroachment on its involvement in social analysis and theory, the Academy of Sciences made a bold effort not only to protect itself from the threat of Marxist monopoly in the vast domain of social thought but also to make this domain a stronger and more diversified component of modern research. In June 1918 Lappo-Danilevskii prepared a plan for the establishment of the Institute of Social Sciences, designed to be an autonomous body closely associated with and modeled after the Academy of Sciences.[80] The professed purpose of the new organization was to employ the dispassionate tools of scientific analysis in a broad effort to understand the massive dislocations caused by the recent revolution and to anticipate future developments in Russian society. Undoubtedly, Lappo-Danilevskii's proposal was guided by the belief that the Bolshevik takeover was a passing phenomenon and that there was an urgent need to counteract the claims of Marxist doctrinaires in the social sciences. The plan for the new institute was submitted to the government, which channeled it to the Socialist Academy for examination and quick rejection.[81]

The historian S. F. Platonov, elected to the full membership of the Academy of Sciences in 1920, was even more direct in expounding a negative view of the new political order. The October Revolution, he said, represented a categorical refutation of historical materialism and its theory of social-class conflict as the main wheel of social history. The October Revolution, he noted with discernible sarcasm, was engineered by a small group of zealots who occupied no position in the system of social relations in production and were guided by mental images of a future society and personal courage rather than by the social-class realities of the day.[82] The Academy made no effort to disassociate itself from Platonov; on the contrary, in 1922 it sponsored a festschrift in his honor.

Until the end of the 1920s, the Academy remained unchanged in one important respect: it elected twenty-six new members, none of whom had the least affinity with Marxist thought.[83] Most academicians did what came easiest to them: they ignored the Marxist philosophy of science in its totality. On the rare occasions when individual academicians asked themselves broad questions, they answered them in philosophical metaphors alien to Marxist thought and often implicitly critical of the views built into dialectical and historical materialism.

V. I. Vernadskii is a prominent example. At a time when the call for a Marxist study of historical and sociological aspects of science was on the rise, Vernadskii published two volumes of papers written before and during World War I in which he expressed views on the history, sociology, and philosophy of science totally alien to Marxist thought.[84] A few years later, he wrote an essay on the evolution of science in which he argued that scientific revolutions were not related to basic restructurings of socioeconomic relations and resultant changes in cultural values. He maintained that it was not regular and predictable historical

processes, as the Marxists claimed, but irregular and unpredictable clus-
terings of talent in particular historical periods that held the key to
understanding revolutionary upsurges in science, as well as other intellectual and artistic activities. While Marxist philosophers claimed that
scientific knowledge was a specific "reflection" of "objective reality"
existing independently and outside the human mind, Vernadskii argued
that it was a specific construction of human reason. The progress of
science, Vernadskii reasoned, was a record of the human drive for individual expression.[85] High levels of scientific creativity could be reached
only in societies tolerant of unorthodoxy in the realm of ideas.

Vernadskii firmly rejected the Marxist view of science and metaphysics—and of science and religion—as mutually exclusive modes of interpreting the work of nature.[86] He agreed with Marxist writers that the
flourishing of the natural sciences was the dominant feature of modern
culture; but, unlike Marxist writers, he argued that science could realize
its full potential not by dominating and subverting other modes of
inquiry and sources of wisdom but only by entering into a symbiotic
relationship with them. The Marxists accepted in principle Auguste
Comte's philosophy of history, which saw the future of man in the total
intellectual emancipation from metaphysical and religious thought. Vernadskii thought that since science, philosophy, and religion satisfied
different human endowments, the progress of one did not lead to the
decline of the others.

Marxist theorists presented Soviet ideology as a world view not only
compatible with but fully based on science; they viewed Marxist ideology
as a science. Vernadskii, in contrast, argued that there is a pronounced
difference between science and the scientific world view—the latter
incorporating, in addition to scientific knowledge, all kinds of metaphysical assumptions, ethical dictates, and outdated scientific theories.[87]
Marxist philosophers accepted Engels's dictum that the scientific study
of nature was a study of the basic forms of "matter in motion"; Vernadskii
claimed that the modern empirical base of the study of symmetry had
nothing in common with the study of "matter in motion," although this
did not prevent it from becoming one of the leading preoccupations of
modern science.[88] He went even further: the modern revolution in science, he thought, was dominated by powerful efforts "to create a world
without matter."

Marxist writers criticized Vernadskii's tolerance of various "idealistic"
philosophies of science and treatment of the scientific world view as "a

unique expression of social psychology" transcending the logic of scientific inquiry and the reservoir of accepted scientific facts and theories.[89] In *Under the Banner of Marxism*, V. I. Nevskii was quick to discern in Vernadskii's "vitalistic" leanings a specific expression of the philosophical position that condemned materialism and treated it as an orientation "opposed to the discoveries of modern science." Vernadskii, in Nevskii's view, typified the scholars who exploited the crisis in science for the purpose of reviving idealism and "placing new foundations under the disintegrating bourgeois metaphysics."[90]

Vernadskii had a strong following among scientists who feared that the new worship of science might invite invidious attacks on traditional culture and make science a tool of ideology. The geographer L. S. Berg, whose attack on Darwin's notion of struggle for existence attracted international attention, showed particular interest in those cultural and epistemological attributes of science in direct opposition to the Marxist interpretation. In *Science: Its Content, Meaning and Classification*, published in 1922, Berg echoed the tone and the spirit of Vernadskii's views on the cultural matrix of modern scientific thought. He, too, emphasized the limits of scientific vision and the interdependence of scientific, metaphysical, and religious wisdom. He criticized not only cognitive absolutism in epistemology (as opposed to epistemological relativism) but also the intellectual narrowness of materialistic ontology. In Henri Poincaré's emphasis on "pure science" as the most sublime achievement of the human mind he saw the surest guidepost for a sound science policy. In an unusually provocative statement, Berg inferred that the study of the evolution of living nature required both a scientific study of causes and a metaphysical recognition of ethical purposiveness. Like Vernadskii, he overstated his case in an effort to give stronger resonance to his arguments against ideological abuses of science.[91]

Searching for ways to protect its tradition and cultural role, the Academy turned to a historical recounting of its own contributions to the advancement of scientific thought in Russia and to the progress of modern science in general. In 1921 Vernadskii, in a paper presented at a meeting of the Academy, noted with sorrow that Russia did not have a single organization involved in the study of the history of scientific and philosophical thought or of scientific creativity. His colleagues responded to his lamentation by founding the Commission of the History of Knowledge, with Vernadskii as its first chairman. The commission embarked on an ambitious plan to study the continuity of scientific thought in Russia, to illuminate the historical depth and breadth of Russian science, and to show that "the history of science from the eighteenth to the twentieth century cannot be either understood or adequately surveyed without a study of the history of Russian science."[92]

The Academy sought to counteract the current efforts by the Communist Academy to develop a Marxist history of science that emphasized the economic and technological roots of scientific theory and the organic unity of science and Soviet ideology. Most of all, the Academy wanted to show that *it* was the prime mover in the development of scientific thought in Russia, as well as the chief Russian contributor to the pool of world science. The founding of the St. Petersburg Academy of Sciences, according to Vernadskii, was "one of the most important events in the history of European thought in the first half of the eighteenth century."[93] The mathematician V. A. Steklov took time to write a book on M. V. Lomonosov, the father of Russian science and the main contributor to making the St. Petersburg Academy of Sciences a national institution and the propelling force in the evolution of Russian science.[94] An inordinate number of studies were dedicated to recounting the achievements of the past giants of Russian science, led by N. I. Lobachevskii, P. L. Chebyshev, D. I. Mendeleev, A. M. Butlerov, and I. M. Sechenov. While Marxist theorists cast their eyes toward a future emergence of socialist science, the scientific community worked arduously to demonstrate the real historical roots of science in Russia.[95] The Commission of the History of Knowledge did not limit its publications to studies in the history of Russian science. In 1927, for example, it published a special collection of essays commemorating the two-hundredth anniversary of Newton's death.

KEPS: A BRIDGE BETWEEN THE ACADEMY
AND THE GOVERNMENT

While the government was reluctant to abolish the Academy, it was more than eager to dislodge it from its position of preeminent influence in the national system of scientific institutions. To achieve this goal, it pursued many lines of action. In budgetary allocations to the Academy, it favored the natural sciences, a practice that reduced the humanities and the social sciences to a most precarious existence. Nor was the financing of the natural sciences above discriminatory practice, for the government clearly favored the disciplines deeply involved in applied research. The government inaugurated the policy of establishing new scientific institutes not attached to the Academy and of expanding selected old centers. Thus, for example, in 1927 the Geological Committee, which employed 1,146 engineers and technicians, had a staff equal in size to that of the Academy.[96]

The government clearly followed a line of functional decentralization of scientific research and placed the Academy on a par with other scientific institutions. A few years after the revolution, the Optical Institute, the Roentgenological and Radiological Institute, the Institute of Physico-Chemical Analysis, and the Central Chemical Laboratory were established. These became not only model research centers with pronounced practical orientations but the main filters for new theoretical ideas brought forth by the revolutionary discoveries in Western laboratories. Most new research centers were organized to serve as direct links between theoretical science and industry. A large number of them were offshoots of the individual research units of the Commission for the Study of the Natural Productive Forces (KEPS), founded in 1915 to strengthen the participation of the Academy in national defense.[97]

The academicians quickly realized that they had to demonstrate their willingness and readiness to follow a course that placed heavy emphasis on applied science. Government authorities, in return, showed a readiness to tolerate—and even to encourage—an infusion of theoretical concerns into practically oriented research undertakings. Moreover, the government's emphasis on practical research was not merely a special expression of the Marxist view of science; it was also a confirmation of ideas shared by an influential group of academicians, the founders and members of KEPS. Led by V. I. Vernadskii, these academicians openly admitted that the Academy of Sciences—and the Russian scientific community in general—was grossly negligent in its duty to contribute to the bolstering of national defense. Vernadskii noted with sadness that despite its rich mineral resources the country had reliable information on deposits of not more than thirty elements, and that not all these were exploited.[98]

In January 1917, less than two months before the February Revolution, the chemist L. A. Chugaev noted that, with minor exceptions, "scientific research received neither sympathy nor help from the government, the society in general or private individuals." In Russia, he said, science was tolerated but not encouraged.[99] The work conducted in laboratories was limited primarily to educational purposes. At the same time, the chemist N. D. Zelinskii put his finger on the real dilemma of Russian science when he stated that the impressive achievements of Russian scholars in scientific theory were matched by a calamitous lack of interest in scientific problems of an applied nature.[100]

Maksim Gorky had the same problem in mind when he stated in 1917, just before the October Revolution, that "regardless of the inexhaustible quantities of natural riches in and on our land, we gain limited benefits from them and from our labor. Industrial and enlightened countries look at Russia as another Africa—as a colony—where it is possible to obtain all kinds of raw material at low prices, because our indolence and ignorance prevent us from transforming them into industrial goods."[101] Gorky added that only "the creative power of science can make our country rich and can stamp out one of the most sordid, wicked, and shameful sides of our life." One of the major reasons for Russia's poor showing in World War I, wrote Vernadskii, must be sought in the unwillingness of both Russian society and the Russian government to support a sustained and comprehensive study of "Russian nature." At no time, he said, were government-sponsored studies part of larger research plans; nor were they carried out with requisite vigor.[102]

Zelinskii, Vernadskii, and their peers may not have approved of the Marxist philosophy of science or of the new government's determination to link the notions of planned science and planned society. But they did feel considerable sympathy for Lenin's inclination to make the Academy a center for the study of natural resources and to elevate KEPS to a position of strategic importance.[103] This, however, did not preclude their devotion to the Academy as a national center of theoretical exploration. Lenin and the new government went so far as to endorse KEPS's current plan for new research units. In January 1918, soon after the October Revolution, the government made it known that it was in favor of involving the Academy in a large-scale and systematic study of natural resources.[104] A month later, KEPS issued a special memorandum that underscored the need for an expanded participation of the Academy in the study of natural resources and their economic potentials, but that also noted the duty of the government to protect the freedom of scientists and the institutions of "the scientific creativity of the Russian people."[105]

A. P. Karpinskii, president of the Academy, assured Lunacharskii that the Academy was determined to continue, expand, and diversify the work of KEPS. He noted, however, that the study of natural resources—as well as research in general—was made exceedingly difficult by the sparse ranks of qualified scientists. Indeed, Karpinskii recommended that one of the first and most practical tasks of KEPS and the Academy was to speed up the ongoing survey of national scientific manpower and scientific institutions. He lamented the crippling effects of a strong tendency among Bolshevik activists to view professional specialization in science as a particular manifestation of antidemocratic attitudes, and he pleaded with the authorities to continue the two commissions established by the Provisional Government that were assigned the task of preparing legislation safeguarding free inquiry and academic autonomy. Karpinskii also used the opportunity to refer to the October Revolution as a "misfortune" of Russian history that broke the continuity of national growth, without which there is no solid base for the "creative life."[106]

On April 12, 1918, A. V. Lunacharskii recommended that the government undertake the obligation of financing the Academy's "important and unpostponable" task of searching for the most rational ways of organizing industry and utilizing the natural resources.[107] The government also expected the Academy to prepare plans for a more efficient regional distribution of industry.[108] The program passed by the Eighth Party Congress in March 1919 stated explicitly that the Communist party considered work on the development of the productive forces of the country the most pressing duty of scientific institutions.[109]

As a sure sign of its determination to encourage practical research, the government allocated a disproportionately large share of the Academy's budget to KEPS,[110] and a special order issued by Lenin assured the commission of adequate supplies of printing paper.[111] Lunacharskii treated KEPS as a model organization dedicated to coordinating the efforts of Russian scientific manpower and to guiding scientific theory to the vast and challenging areas of practical research. To support his ideas, Lunacharskii relied on a citation from Vernadskii, the founder of KEPS: "Now, when the publication of . . . *Materials for the Study of the Natural Productive Forces of Russia* has begun and when parts of the symposium on *The Productive Forces of Russia* are being published or are being readied for publication, and when we have accumulated concrete experience, it is possible to seek solutions for new and more complex problems deeply affecting our lives."[112]

During World War I Vernadskii stressed the need for intensive study of radiation, methodical search for radioactive elements, and national production of optical instruments. In 1918–1920 the government provided funds for carrying out these national requirements. KEPS's insistence on the pressing need for the creation of a network of research institutes freed from educational obligations became the central idea in the science policy of the new government. A product of their own initiative, KEPS gave the academicians an effective mechanism for making up for the "sins" of the past. It employed the methods and techniques of science to solve pressing economic problems and to establish closer ties between the government and the Academy. KEPS set up special units to search for deposits of precious minerals and to study the Kursk magnetic anomaly, the economics of the textile industry, and a wide range of other economic problems. It combined interest in theoretical exploration with applied research of direct economic value.

For example, the Institute of Physico-Chemical Analysis, a subsidiary of KEPS, worked intensively on the practical aspects of the chemistry of alloys; at the same time, the institute became the national center for modern exploration into the theory and methodology of inorganic chemistry. In his extensive and original elaborations of J. W. Gibbs's phase-rule theory, N. S. Kurnakov, the founder and moving spirit of the institute, made an involved and challenging effort to establish analogies between transformations in chemical equilibrium systems and topological transformations of space.[113] Physicochemical analysis, according to a contemporary document, was both "a scientific discipline dominated by a practical orientation" and an important source of ideas that contributed to "the solution of the fundamental questions of chemical philosophy."[114]

In 1927, on the occasion of the celebration of the tenth anniversary of KEPS under the Soviet regime, A. E. Fersman credited the organization with pioneering work in interdisciplinary research and structured group endeavor in science. KEPS provided a functional model for research organizations equipped to render professional service to the country without giving ideological support to the government, and it made the first effort to gather and systematize information on scientific manpower in the Soviet Union. The air of practicalism fostered by KEPS provided a stimulus for changing the name of the Archeological Commission to the Academy of the History of Material Culture; the latter, in turn, lost no time in organizing the Institute of Technological Archeology, which was concerned with the evolution of the tools of production since the beginning of human history.[115]

KEPS was the institutional center for most expeditionary assignments undertaken by the Academy.[116] Despite staggering difficulties, an expeditionary group was sent to the Ukhtinsk district in search of new deposits of petroleum. In 1919 the first research teams were stationed in the areas of the Kursk magnetic anomaly, and in 1920 an expeditionary group went to the Kola Peninsula in search of mineral deposits of various descriptions. The "Mongolian Expedition," organized a few years later, undertook a systematic and thorough survey of mineral resources in an area 2,000 kilometers in length. The Iakut Expedition, set up in 1925 for a five-year period, was heralded as the most ambitious and versatile exploratory project undertaken by the Academy in its entire history.[117]

INSTITUTIONAL INDEPENDENCE

While the government was ready to establish a working relationship with the ideologically alien scientists on a professional level, it opposed all current efforts favoring a wider base of academic autonomy. The Russian scientific community had a long and distinguished history of efforts to create a national organization for the protection of the rights and privileges of scholars. In the early 1890s V. I. Vernadskii and the geologist A. P. Pavlov prepared a plan for a Russian equivalent of the British Association for the Advancement of Science, an organization devoted to both the diffusion of scientific knowledge and the protection of the professional interests of the scientific community.[118] Again during the 1905 revolutionary wave, the members of the Academy took an active part in both the criticism of the educational policy of the government and the preparation of a charter for a national organization of professors. Both efforts demanded wider domains of associative autonomy than the government was willing to grant, and both were abortive. In 1916 the government approved the charter of the Association of Russian Naturalists and Physicians, but wartime conditions and the October Revolution prevented this body from moving into action.

In 1917, in the wake of the February Revolution, a group of scientists, led by the most prominent members of the Academy of Sciences and by Maksim Gorky, founded the Association for the Advancement and Diffusion of Science. The most influential and determined supporter of the new organization, Gorky was moved by a firm conviction that scientists, more than any other professional group, needed a base of unity that would create the conditions for accelerated advancement of their

interests in Russia.[119] The neurophysiologist I. P. Pavlov, the plant physiologist K. A. Timiriazev, and the writer I. A. Bunin were among the
most influential supporters of the new society.[120] The mathematician
A. A. Markov was unanimously elected the honorary chairman of the
founding session.[121] The widely publicized association undertook two
clearly formulated tasks: to develop a broad network of channels for the
dissemination of scientific knowledge and to assure the scientific community of a wider and more active part in shaping the national science
policy. It also called for the establishment of a national Institute of Positive Sciences in honor of February 27, 1917, the day of the overthrow
of tsarist rule. The institute was envisaged as a network of laboratories,
museums, libraries, and lecture halls. Its basic aim was to give nationally
selected young scientists an opportunity to devote themselves to
research under the most favorable conditions.

In inaugurating the new association, Gorky dreamed of a future "City
of Science" in which "the men of exact knowledge," living in "an atmosphere of freedom and independence," would create "a love for reason"
and an admiration for its "power and beauty."[122] "There is no future,"
he wrote, "without science in democracy." Aware of the cultural values
of the "Protestant ethic"—although he showed no overt signs of having
read Max Weber—Gorky noted that, in order to create sound conditions
for the advancement of science, Russia needed not only free expression
of ideas and the unobstructed cultivation of rational faculties but also
the elevation of work to the level of a paramount cultural value. In his
praise for the liberating effects of the February Revolution, he found it
important to note that the time had come to stop treating work as a
"curse" and to grasp the full scope of its unsurpassable value.[123]

While the scientists, led by the members of the Academy, were gen-
uinely pleased to have Gorky as a spokesman for the needs of the
scientific community, Bolshevik leaders showed little enthusiasm for his
emphasis on "political" democracy rather than on "social" democracy,
or for his laudatory reference to W. F. Ostwald, the man who, in Lenin's
view, represented the guiding force in the revival of idealism in physics.[124] After the October Revolution, the association moved slowly in its
search for recognition—and for survival. Soon it became clear that the
new conditions did not favor organizations based on associative autonomy. The government's early promises of help quickly gave way to
mounting bureaucratic strangulations, oblique expressions of ideological
criticism, and discriminatory decisions. In 1918 the association ceased

to exist, leaving behind only residual reminders of a long-fought battle for academic autonomy.

Despite the setbacks in the battle for autonomy, the Academy of Sciences continued to serve as the main forum for scholars who favored administrative decentralization and scholarly independence. Beginning in 1919, A. E. Fersman, an influential member of the Academy, made no secret of his firm conviction that a successful pursuit of science required decentralized scientific institutions, free alliances of research centers engaged in similar work, personal initiative in the choice of research topics, and local support of scientific inquiry.[125]

In the early 1920s, a group of academicians proposed a plan calling for the formation of "unions of scientific institutions" that would bring together scientists working in the same and related fields. Unions, they emphasized, would be sufficiently flexible to combine the dual function of scientists: to contribute to the practical needs of society, as articulated by government authorities, and to attend to the theoretical needs of individual sciences, as dictated by the internal logic of the growth of scientific thought. The common interests of these organizations would be attended by a central Union of Scholars, a body essentially independent of the government and scrupulously mindful of the fundamental principles of academic autonomy.[126] The drafters of the project for the new body stated explicitly that its function was "to protect the autonomy of scholarly institutions and the institutions of higher education."[127] The academic project, which called for limiting the role of government to supplying financial resources for research, received little sympathy from the authorities and quickly ceased to be a topic of public discussion. Nevertheless, it left a legacy that helped to convince the government not to dispense with all the attributes of administrative autonomy in the national system of scientific institutions and, particularly, to recognize—though not with unimpeachable consistency—the realistic wisdom of the philosophy that measures the social usefulness of science by the richness, depth, and continual progress of basic theory.

The struggle for academic autonomy took yet another form: a public discussion of the role of "individual expression"—or "individuality"—in scientific creativity. This topic received dramatic emphasis in V. I. Vernadskii's *Essays and Addresses* (1922), consisting of papers originally published before the October Revolution. Consistently and vigorously, Vernadskii pointed out the grave danger of government curbs on critical thought and unorthodoxy, the true vehicles of scientific creativity.[128] A political ideology such as Marxism might be based on a scientific world view, but scientific world view and scientific creativity were two different categories of phenomena. The former, a product of "social psychology," did not depend on laboratory verification and was often openly opposed to established facts and theories of science. Scientific creativity, the only source of scientific progress, could be validated only by laboratory experiments and objective experience. Vernadskii's analysis contained a warning that the current emphasis on the paramount role of the scientific world view as a cultural basis of society was not by itself a sufficient base for the development of science—and that this emphasis might easily be turned into an enemy of science, particularly if the scientific world view became an umbrella for pseudoscientific ideas.

In a contribution to the symposium on creativity, published in Moscow in 1922 by a group of scholars and artists, S. F. Ol'denburg chose to stress two ideas of paramount relevance to current developments in the Soviet Union.[129] First, as a creative force, individuality in science and individuality in art are essentially the same phenomenon. In both, progress is made by a resolution of conflict between various creative individualities rather than by suppressing individual digressions from arbitrarily selected norms. In both, progress is made by a constant search for consensus of professional judgment, unrestrained by external influences. Second, in forming the institutional base of science, the intrinsic interests of science require no less emphasis than the political interests of the state. Ol'denburg obviously resented the summary abolition of such time-honored institutions as the Free Economic Society and the Pirogov Society, on the ground that they did not meet the political test of the new authorities.

MECHANISMS OF GOVERNMENT CONTROL

While the Academy led the forces in search of academic autonomy, the Bolshevik leaders moved slowly and at first unsteadily in making science and the scientific community vital areas of government control. On August 16, 1918, the Supreme Council of the National Economy (VSNKh), in close consultation with the leaders of the Academy of Sciences, established the Scientific and Technical Section (NTO), the first agency entrusted with centralized control over scientific research, with relating the development of science to the most acute needs of industrial and agricultural technology, and with creating new research units engaged in applied science.[130] According to Robert Lewis, the section was intended to have "wide-ranging functions" in controlling research and development and in providing "advisory services on matters concerning science and technology."[131]

During its formative years, the section was headed by the chemist V. N. Ipat'ev, a member of the Academy. As it developed subsequently, the NTO regulated the work of research bodies not under the Commissariat of Education and exercised authority primarily in applied research. At the outset, it was in full charge of fifteen research bodies, including the very active Central Aerodynamic Institute and such burgeoning organizations as the Institute of Applied Physical Chemistry, the Scientific Automobile Laboratory, and the Institute of Pure Chemical Reagents, encouraging them to establish contractual relations with industrial enterprises and to operate as self-contained financial units. In addition to guiding the work of institutes subordinated to it, the NTO organized and financed special research projects conducted by individual scientists or teams of scientists outside its jurisdiction. For example, it sponsored A. F. Ioffe's research projects on the X-ray analysis of materials. Representing the interests of the VSNKh, the section organized several expeditionary teams to study regional industrial resources. For example, it sponsored the Northern Scientific-Industrial Expedition (*Sevekspeditsiia*) in 1920, headed by academicians A. P. Karpinskii and A. E. Fersman. The expedition undertook the most extensive work on the Kola Peninsula, the Pechora territory, and the western reaches of the Soviet Arctic. The work of this expedition laid the foundations for the industrialization of northern Russia.[132]

The Scientific Commission, consisting of carefully selected faculty members of the leading Moscow institutions of higher education, was the braintrust of the section. Divided into seven (and later eight) sectors, the commission covered a wide field of specialties from "physics, electrotechnology, and geophysics" to "agricultural economy and biology." Its job was to appraise research projects prepared by various technical groups in response to specific requests by the VSNKh. It sought new ways of expanding the role of government in creating an effective institutional base for science, and it examined government plans for future developments in technology. Generally unrestricted in its work, the commission provided a forum for a continuing exchange of ideas. In the absence of an adequate inflow of Western scientific literature, the proceedings of the commission became important channels for the diffusion of up-to-date developments in the theory of relativity and in the laboratories of Rutherford, Bohr, and other pioneers of modern physics.[133] The commission was part of an effort by the new authorities to gain technical help from an ideologically alien professoriat.

The NTO was in charge of the Bureau of Foreign Science and Technology, attached to the Economic Delegation of the RSFSR in Berlin. The bureau was the central state agency that purchased Western scientific materials and organized and financed the publication of Russian scholarly books abroad, making up for the shortage of printing materials at home. German presses published a series of volumes prepared by the State Optical Institute and the Russian Physical and Chemical Society. Among the books published in Germany was a Russian translation of Einstein's effort to present a "popular" account of the special and general theories of relativity. In the introductory note, specially written for this translation, Einstein noted that "the troubled world makes it so much more imperative to rely on artistic and scientific work for forging friendly relations between peoples who speak different languages."[134]

As viewed by its Soviet interpreters, the NTO was "the first central government agency" in the world entrusted with the task of coordinating the work of "virtually all" national institutions engaged in applied research.[135] The new body was based on three principles that became the backbone of the science policy of the new regime: administrative centralization of experimental science, close relations between science and the national economy, and national planning of research activities.[136] The NTO underwent continual changes, mainly adaptations to wavering policies of the government and personal proclivities of major administrators. In 1926 it was expanded and renamed the Scientific and Technical Administration, which then became embroiled in far-flung controversies on the best method of coordinating science and technology.[137]

In December 1921, after several years of ambiguous policies and vacillating administrative moves, the government established the Main Administration of Scientific and Scientific-Artistic Institutions (Glavnauka) at the People's Commissariat of Education.[138] The new agency brought the Academy of Sciences, the research facilities of the institutions of higher education, libraries, museums, and learned societies into a unified administrative system. One of its more important functions was to establish acceptable models for the internal organization of learned societies and scientific associations. In 1924 it was the chief government supervisor of 150 such bodies.[139] Although the Glavnauka guided all the institutions involved in "pure science," its main task was to bring the world of theoretical science closer to the technical needs of the economy and to the revolutionary spirit of socialist construction.[140] It underwent a series of transformations, all aimed at strengthening its hold on the national effort in science. But the rise of parallel agencies—such as the State Scholarship Council (GUS)—at the same commissariat only added to the prevailing confusion in the search for a central authority in the administration and development of scientific institutions.

This confused situation was compounded in 1922 when the government established the Provisional Committee on Science to serve as a clearinghouse for "all scientific and material needs of scholarly institutions and to take the necessary steps toward meeting them." Actually, the new committee dealt with general policies related to organizational and procedural matters: organization of expeditionary research, travel to Western Europe, publication of general reports on the achievements of Soviet science and technology, scheduling of national congresses of scientists, and preparation of international scientific meetings.[141] The establishment of the committee showed clearly that the government and the scientific community did not see eye to eye in interpreting the grave predicament of science caused by the ravages of the Civil War. In initiating the creation of the committee, the scientists, led by the Academy, wanted a central agency to solve two critical problems: the rapid deterioration of research equipment and resources and the substandard living conditions of persons devoted to scholarship. The government, however, pressed the committee to give the highest priority to work aimed at improving relations between science and production. Amorphous and unwieldy, the committee ceased to meet in 1924. This happened at the same time that V. A. Steklov recommended to the Council of People's Commissars that the committee be given a "permanent"—rather than a "provisional"—status and that it be assigned the task of dealing with the problems of science "common to all republics" and with the international relations of the Soviet scientific community.[142]

In its short history, the Provisional Committee produced no startling results; however, it served as a clear indicator of the government's serious commitment to the idea of a unified and centralized administration of the national effort in science. At the same time, the job of the Glavnauka and its unsteady predecessors was made relatively easy by the Civil War and the economic plight of the country, which drastically reduced the activity of scientific institutions. The gradual consolidation of central controls did not lead to immediate improvements in financing science. The physicist A. K. Timiriazev, in a report on the development of science during the first five years of the Soviet system, noted that although prior to 1921 the government advanced adequate sums of money for research, no scientific equipment was available for purchase; after 1921, he added, such equipment was available but the government was stingy in advancing requisite funds.[143]

The Glavnauka was one part of the dual composition of the Academic Center, the key agency of the Commissariat of Education working in the domain of science policy. The second part consisted of the State Scholarship Council.[144] The Glavnauka was in charge of "academic institutions"—all institutions involved in research—which at the time were dominated by "bourgeois" scholars. The State Scholarship Council was concerned with the fundamental principles of three distinct areas involving science: the politics of science, the place of science in public education, and the relationship of science to technology. It was a Marxist body in that its main job was to give Soviet science a distinct Marxist orientation and organization. Its immediate duty was to guide the Glavnauka and to search for methods to involve "bourgeois" scientists in socialist economic projects. With regard to the Academy of Sciences, it performed two functions: it searched for a place for the Academy in the new network of scientific institutions, and it sought to forge a compromise between the traditional notions of academic autonomy and the newly evolved government controls over national involvement in science. Political and ideological uncertainties, magnified by the absence of historical models of a Marxist organization of scientific institutions, made the Academic Center a transitory and rather ineffectual body. It quickly ceased to exist.

In the spring months of 1918, the government tried not only to lay the foundations for centralized national control of scientific institutions but to "reorganize" the Academy of Sciences. In April 1918 V. T. Ter-Oganezov submitted to the State Commission on Education a general plan for the creation of a new supreme body in scholarship, to be known as the Association of Sciences. The projected organization was clearly intended to replace the Academy of Sciences, from which it differed in two respects: it was intended to be a forum of distinguished scholars, delegated by scientific institutions and societies, rather than a self-perpetuating body; and it was intended to represent the full spectrum of scholarly disciplines. Although the members of the commission agreed unanimously that the Academy should be subjected to extensive reorganization, the Ter-Oganezov plan, for reasons not made public, was not discussed again.[145] Lenin's advice that the Academy should be handled with utmost care and patience might have been the deciding factor.

In their effort to integrate the scientific community into the new polity, the Bolshevik authorities did not rely solely on the administrative agencies of the government. They also made extensive use of mass organizations created and controlled by the Communist party and its expanding network of auxiliary groups. Among these groups, the Section of Scientific Workers, a branch of the Union of Education Workers, occupied a key position. Founded in 1922, the section convened the First All-Russian Congress of Scientific Workers in 1923, which ratified its formal organization and elected the members of various administrative boards. The new body attended to "professional, material, and legal interests of scientific workers" and contributed to the shaping of the national science policy.[146]

From the outset, however, the section showed much more interest in marshaling support for government policies than in generating plans or actions for improving the well-being of its members. It drew proposals for a civil-service classification of scientific workers and uniform academic degrees. It concentrated on publicizing the official views on the organization and administration of scientific institutions and on providing liaison service between the government and the leading scholars, particularly the members of the Academy of Sciences. In time, it became more bellicose and doctrinaire. After 1926 it conducted numerous campaigns against scientists labeled "enemies" of the new system. Although it counted several academicians—including S. F. Ol'denburg, N. Ia. Marr, and A. E. Fersman—among the members of its Central Council, it made the Academy of Sciences an object of many direct and indirect attacks and investigations by specially appointed committees. It treated the Academy as a bastion of ideological "neutrality" and "pure science." It urged members of the Academy to make dialectical materialism their only and all-pervasive philosophy of science, to coordinate academic research with the needs of the national economy, and to work toward achieving the full integration of the Academy into the political fabric of the Soviet system. At the end of the decade, when its militancy reached a climax, it relied on the coercive tools of organized authority to accelerate the "dissolution" of the "old intelligentsia."

The Section of Scientific Workers was the Soviet regime's answer to the demands of the scientific community for constitutional guarantees for associative autonomy. The various models of this autonomy proposed or supported by the Academy of Sciences had one common goal: protecting the pursuit of science from excessive external controls. The basic function of the Section of Scientific Workers, however, was to strengthen and consolidate the multiple mechanisms of government control over the scientific community. It referred to the Academy as the most glaring example of unacceptable ivory-tower detachment from the realities of life and as a bastion of opposition to dialectical materialism as the method of science. The section was a transitional organization: once the government established unchallengeable authority in the scientific community, it quietly disappeared from the scene.

The ambivalent status of the Academy in the total configuration of political authority did not hamper the gradual expansion of its research operations and institutional base. In 1917 the personnel of the Academy consisted of 41 academicians and a total staff of 220 people, half of the latter group classified as scientists.[147] During the following years the Academy lost some members to emigration and to an unusually high rate of mortality;[148] however, the number of people engaged in research soon reached new heights, mainly through the rapid expansion of KEPS. In 1925 the Academy employed 873 people, half of whom were classified as "scientific workers."

At the same time, the Academy undertook measures to adapt itself to the new political conditions. For the first time it made some of its meetings open to the interested public. Selected scientists not associated with the Academy were granted the right to an "advisory vote" in the meetings of individual departments. Scattered signs indicated that the government was ready to make the Academy a vital part of future plans for a national organization of institutions involved in scientific research. Despite ominous administrative detours and political threats, optimism grew steadily after April 1918, when Lenin called on the Academy to set up an unspecified number of commissions to participate in the preparation of a general plan for "the reorganization of industry" and general "economic improvement."[149] Lenin's message showed that the Bolshevik authorities had early thoughts on integrating the national network of scientific institutions into a system of centralized economic planning and of making the Academy a strategic component in this machinery. In the same month, N. P. Gorbunov, secretary of the Council of People's Commissars and the chief liaison between the central government and the Academy, made it known that the authorities were willing to help the Academy in "all possible ways."[150]

A clear definition of the legal status of the Academy did not come until 1925, when the institution celebrated its two-hundredth anniversary. In addition to publishing extensive reports recounting the glories of past achievements, the Academy held a series of festive meetings that included an impressive number of scientists from abroad—among them the German physicist Max Planck, the English geneticist William Bateson, the Italian mathematician Tullio Levi-Civita, and the Indian physicist C. V. Raman.[151] Government leaders greeted the celebration with glowing pronouncements about the affinity of the new regime with the best interests of science and about the vital place of the Academy in shaping a socialist culture. M. I. Kalinin, representing the government at the main ceremony, expressed unlimited optimism in the future of the Academy as a pillar of socialist society.[152] A. V. Lunacharskii, the commissar of education, noted that it would be impossible to recount all the great achievements of the Academy and expressed his faith in an even greater future for the institution. He found it important to point out the key role of the Academy in the preparation of the new Cyrillic orthography and calendar reform. The Academy, he said, assisted the government in negotiating with neighboring countries, prepared detailed ethnographic maps of Bessarabia, and helped many Soviet nationalities in acquiring their first alphabets.[153] Leon Trotsky, speaking at a special gathering, contributed to the celebration by offering a Marxist analysis of the scientific contributions of D. I. Mendeleev, the main hero of Russia's involvement in science. Mendeleev, he said, showed the superiority of science over religion and all other modes of inquiry; he also advanced a philosophy of science that recognized the organic interdependence of pure theory and applied knowledge.[154] L. B. Kamenev, in a talk before the Moscow Workers Council, emphasized the kinship of socialism with the science of such "destroyers of the divine order" as Giordano Bruno, Galileo, Denis Diderot, Charles Darwin, and Robert Mayer.[155]

All the speakers presented science as a liberating force and the most reliable source of designs for a social life governed by reason and justice. They expressed faith in the Academy as the guiding force in the national commitment to the substance and the ideals of science. Upon his return to England, Bateson wrote in *Nature* that all the speeches delivered on the occasion of the two-hundredth anniversary celebration created the impression "that the revolutionary government is perfectly sincere in its determination to promote and foster science on a very large scale." "Signs were not wanting," he wrote, "that science, especially perhaps in its applications, is regarded by the present governors of Russia as the best of all propaganda." He noted that "it was interesting to hear the faith that the advancement of science is a first duty of the State proclaimed by professional politicians."[156]

Individual members of the Academy took advantage of the festive mood generated by the commemorative sessions to reaffirm their institution's deep engagement in scholarly work consonant with the pressing needs of the new society. Ol'denburg made it known that the Academy had broadened its activities and had become the main link in a centralized system of national research institutions.[157] There were, however, academicians who did not ignore the need for uncompromising defense of "pure science." P. P. Lazarev readily conceded that the future progress of the country depended on advances in technology; but he was quick to add that technology could not be separated from "pure science" and that the emphasis on "pure science" was the surest path to the recovery of the country's shattered economy.[158] V. L. Omel'ianskii, the much-respected microbiologist, took it upon himself to give clear and unequivocal expression to the importance of "pure theory":

> Science is not created in practical laboratories, regardless of how splendidly they may be equipped, or by drives for fleeting practical achievements. . . . It is created in the calmness of scientific laboratories, in the disinterested, persistent, and often selfless quest for absolute verity— for the "beautiful truth" that Descartes elevated above every other truth. Here are created lasting values; here great possibilities are transformed into full realities; and here are born grand ideas, often making a powerful impact on the entire world and influencing every sphere of human life.[159]

More as a tactical move to placate the government than as a realistic assessment of the status of the Academy in the total network of scholarly bodies, Ol'denburg felt obliged to state—in his annual progress report for 1926—that the members of the Academy had participated in "most of the scientific work in the Union" and that there existed "real and living ties between the Academy, other scientific bodies, institutions of higher education, and technical centers." He noted the rapidly increasing participation of academicians in editing professional journals and serial publications sponsored by scholarly organizations outside the Academy, as well as the "close ties" of members of the Academy to activities of the central planning authorities (GOSPLAN).[160] As concrete proof of the growing commitment to "practical" research, the Academy added the Special Committee for the Study of Union and Autonomous Republics to the previously established Commission for the Study of the Iakut Autonomous Republic and the Commission for the Study of Tribal Population.

The government found the 1925 celebration a propitious occasion at which to give a more precise definition of the place of the Academy in the Soviet polity. It ordered that the name of the institution be changed from the Russian Academy of Sciences to the Academy of Sciences of the USSR. The ties connecting the Academy with the central government were raised to a higher plane: instead of its subordination to a section of the Commissariat of Education, it was now placed under the direct jurisdiction of the Council of People's Commissars. To consolidate its position, the government gave the Academy a new charter in 1927, replacing the charter issued in 1836.[161] With the introduction of a presidium and a system of administrative bureaus, the Academy's internal order took the shape of a typical Soviet institution. The new charter grouped the Academy's research units into two departments: the Department of Physical and Mathematical Sciences and the Department of the Humanities. The new departmental organization left no room for research of a sociological, juridical, political, or economic nature; these instead became a monopoly of the Communist Academy and allied institutions. The most fundamental change, however, took place not in the internal organization of the Academy but in the range of its autonomy: the academicians lost the exclusive right to select new members. The new charter ruled that henceforth learned societies operating outside the Academy, as well as various mass organizations—such as local units of the Communist party and the trade unions—and individual scholars, be invited to take part in nominating candidates for vacant academic positions. This measure broke the backbone of academic autonomy and accelerated the process of bringing the Academy into the institutional mainstream of Soviet life. It gave government and party outposts an effective mechanism for reaching deep into the inner sanctum of academic life.

GENERAL ORIENTATION

During the 1920s the Communist authorities made two historical decisions that laid the foundations of the national science policy and profoundly affected the future of the Academy of Sciences. The first decision put an end to the Proletkul't campaign to launch the new republic on a vigorous search for a proletarian science that would be a negation of traditional—that is, bourgeois—scientific thought. A. A. Bogdanov, the codifier of the Proletkul't ideology, had argued vehemently and quixotically that the cause of socialism could not be advanced without socialist science, which he envisioned as proletarian "by its very nature, that is, by its original point of view and method of analysis and presentation."[162] He did not suggest that "proletarian science" should have its own axioms and laws; rather he placed primary emphasis on the unique principles of socialist integration of scientific knowledge and arrangement of scientific abstractions.[163] Because of his preoccupation with discontinuities in the evolution of scientific thought, Bogdanov was generally regarded as prophet of a sui generis "proletarian science"—and as the chief ideologue of the movement that saw no role for "bourgeois" scientists in building the Soviet model of socialist society.

Bogdanov went so far as to point out the main lines of the future development of socialist science: namely, the elaboration and refinement of his own "general theory of organization," or "tektology"—a term he borrowed from Ernst Haeckel.[164] Bogdanov defined tektology as a science of the universal aspects of nature and society best revealed in the laws of organizational uniformities. He was ready to admit that the beginnings for such a science were provided by Einstein's conceptualization of relativity as a universal view of nature based on the unity of physics and geometry, by "the objective monistic view of nature" built into the modern theory of energy as the basic and unifying substratum of the universe, and by the work of Yugoslav mathematician M. Petrovich on "the theory of analogies," concerned with "similarities of structures" in pivotal components of nature and human life.[165]

Bogdanov's vision of proletarian science as a radical departure from bourgeois science had few adherents. Although the Communist Academy was firmly committed to the search for a distinctive Soviet science—a Soviet theory of relativity, a Soviet version of the foundations of mathematics, a Soviet theory of organic evolution and sociology—and for a unique Soviet organization of scientific activity, the majority of its members rejected Bogdanov's ecstatic clamor for a proletarian science. The Eighth Party Congress and Lenin's writings at this time dealt a mortal blow to Bogdanov's vision. They stressed the need for extensive reliance on native "bourgeois" scientists and on scientific advances in the West.

Lenin stated explicitly that socialism could not be built without the help of bourgeois science. Kendall Bailes has noted that "although Lenin was preoccupied until 1922 with the Civil War and the consolidation of power, he frequently intervened to protect the interests of old specialists and generally took their side in work disputes with nonspecialists."[166] All this did not mean that the Leninist policy kept universities and other research centers isolated from the heat of the growing class struggle. Lenin treated the involvement of bourgeois scientists in socialist construction as a variety of "class warfare" that combined elements of "persuasion" and "coercion": it included cooperation with, and extensive "reeducational" work among, scientists loyal to the government, on the one hand, and uninterrupted purges of scientists classified as saboteurs, ideological opponents, or champions of political neutrality, on the other.[167] Whether relying on persuasion or on coercion, on cooperation or on purges, the leaders of the new political system were advised by Lenin never to forget that the future of socialism in the Soviet Union depended on "a successful blending of the victorious proletarian revolution and bourgeois culture—bourgeois science and technology."[168] Lenin chastised the "pseudoradical" party comrades who contended that it was possible to surpass the bourgeois system without "learning from bourgeois specialists" and without working along with them for a long period of time.[169]

The second decision of the authorities that affected the Academy dealt with the organization of scientific institutions. In the early 1920s a group of Marxist theorists and leading engineers advocated the transfer of research institutes to industrial establishments or the building of industrial establishments around individual research institutes. They wanted a total organizational unity of the two. This idea was a specific response to the avalanche of criticism directed against the "ivory-tower" atmosphere in the Academy of Sciences. Although government authorities never stopped emphasizing the need for a practical mobilization of all activities in science, they were careful not to submerge scientific bodies into the institutional maze of the national economy. The final word came in 1925, when F. E. Dzerzhinskii, chairman of the VSNKh, stated forthrightly that a full institutional coalescence of research centers and industrial plants would narrow the horizons of scientific inquiry; industry, he said, could operate on a local or regional level, but science must work on a national level.[170] Dzerzhinskii worked on two fronts: he halted efforts to transfer applied-science centers to industry, and he inaugurated a long line of official efforts to establish close ties between science and industry without abrogating their institutional separateness and without abandoning A. V. Lunacharskii's advice that the critical economic situation in the country demanded that no research be carried out outside a "collectively organized" and "planned" endeavor.[171]

The Academy of Sciences benefited from both decisions. The first decision recognized its "bourgeois" members as useful citizens and helped the Academy to overcome a major hurdle in its critical struggle for survival, for the Proletkul't movement had viewed the Academy as the institutional center of all moral aberrations and intellectual degradations of bourgeois science and had demanded its dismemberment. The second decision helped the Academy in its search for a new and viable *raison d'être*. The Academy thus moved into the wide domains of practical research without surrendering its institutional identity.

In the meantime, the Academy had digressed from the paths of tradition. Gradually and haltingly, the center of academic gravity had shifted from the humanities to the natural sciences. It had also embarked on a policy of establishing institutes that facilitated a concentration of talent in the key fields and created a broader base for cooperative work in science. The Mathematical and Physical Institute, the Physiological Institute, the Chemical Institute, and the Institute of Soil Science were new creations and indicators of future developments in the organizational makeup of the Academy. By organizing huge institutes, the Academy both carried out the institutional dictates of the new science policy and set up models of organized research for other centers of scholarship. The new trend was clearly noted by M. Ia. Lapirov-Skoblo and his team, who conducted a systematic inquiry into the evolution of scientific institutions during the first ten years of the Soviet system. "The expired decade," wrote Lapirov-Skoblo, "was a watershed dividing two different periods in the history of scientific work in our country. We have entered the phase of scientific institutes. In these institutes we see a decisive transition from isolated individual work to a broad collective endeavor. They have built the foundations for new scientific schools and for tightly woven collectivities of talented young workers guided by the leading scientists."[172]

At the same time, expeditionary work conducted under the sponsorship of the Academy underwent rapid expansion: from 1917 to 1928 the Academy had sponsored 300 expeditionary teams.[173] In tsarist days, most expeditionary work concentrated on linguistic and ethnographic studies of numerous tribal societies inhabiting the country and on broad geographical descriptions in the manner of natural history. In the early

Soviet period the entire spectrum of scholarly interests was covered, but the emphasis was clearly on specialized and technically rigorous exploration of a geological, mineralogical, and paleontological nature.

The Academy became the primary advocate of new types of scientific congresses. After observing the work of national and international congresses abroad, the academicians became advocates of the so-called American type of meetings: the concentration on narrow scientific areas that called for a precisely structured and limited number of papers concentrating on questions of fundamental significance as seen by the leading specialists in individual fields.[174] After 1925 members of the Academy occupied a privileged position in obtaining permission—and finances—to attend international congresses of scientists, making them a valuable and extensively relied upon source of recommendations in organizing communication lines in the Soviet scientific community.

The first ten years of the Soviet system were dominated by a mixture of excitement and anxiety in the pursuit of science and by widespread experimentation and uncertainty in the molding of scientific institutions. It was a period in which the traditional forces first helped give Soviet science institutional continuity and historical individuality and then gradually retreated before the overpowering onslaught of the combined revolutions in science and politics—revolutions in the scientific way of looking at the universe and in the fundamental structure of social existence. The Academy of Sciences became the bulwark of tradition in the new fabric of social relations and cultural values, a position it could not hold for long. At the end of the 1920s the time had come to make it an organic part of the new polity—to make it an institution guided by the dictates of Marxist philosophy and Soviet ethos and dedicated to upholding tradition only inasmuch as that answered the needs and the spirit of the new society. In one sense, during the first ten years under Soviet rule the Academy was involved in a grueling struggle to regain the growth momentum lost at the beginning of World War I: it was not until 1928 that its publication output reached the prewar level.

III
THE FORGING OF A SOVIET
INSTITUTION (1929–1940)

THE SOVIETIZATION OF THE ACADEMY

At the end of the 1920s—soon after the celebration of the tenth anniversary of Bolshevik rule—the Soviet Union entered a period of revolutionary change known as "socialist construction" or "cultural revolution." This was the time of "all-out collectivization" of agriculture and the inauguration of the five-year plans for the development of the major sectors of the national economy. It was also the time of concerted efforts to make Soviet culture an embodiment of Marxist-Leninist cultural values. The rapidly changing structure of society reflected not only the new alignment of social classes but also the new configuration of professional groups. Relying on militant groups of Marxist loyalists, the authorities pressed for stepped-up action to remove from sensitive positions all individuals not trusted to go along with the new policies. The Shakhty trial in 1928 signaled the advent of a sharp rise in official actions aimed at eradicating the last strongholds of "bourgeois" resistance to Bolshevik politics and ideology. Persons engaged in science were among the main targets of the housecleaning campaign, and the Academy of Sciences was no exception. The time had come for this institution to shed all residues of traditionalism that prevented it from becoming an organic part of the socialist polity. The swiftly moving developments transformed the Academy into a true Soviet institution; with lightning speed, the Academy experienced radical changes in the composition of its personnel and in the style of its work.[1]

In early 1928 the Academy invited nominations for forty-two new academicians, a number that would double the current roster of full

members. As nominations poured in from various professional institutions and mass organizations (party and trade-union units and their
numerous affiliates), the press—particularly the *Leningradskaia Pravda*—
supplied a running commentary on favored candidates, mainly individuals with distinguished records in the much-publicized engineering and
industrial projects.

The press also carried a long string of articles by trusted scholars who
concentrated on the need for an extensive redesigning of the internal
organization of the Academy. N. I. Vavilov, the famed geneticist, called
for a new charter broadening the role of the Academy in socialist construction.[2] S. Iu. Semkovskii, a Marxist philosopher of science, stated
explicitly that the Academy had only two alternatives: either "to adapt
itself to the changes wrought by the revolution" or to go out of existence.[3]
It was strange, wrote M. V. Serebriakov, that ten years after the October
Revolution, only ten members of the Academy's staff—and not one a
scholar—were Communists, and that the Academy showed no hesitation in employing people dismissed by other organizations for political
reasons.[4] D. B. Riazanov, impressed with the model provided by the
Institut de France, a product of the French Revolution, went so far as
to suggest that the future belonged not to the Academy as a scientific
center but to an association of specialized scientific institutes.[5] He suggested that not only the Academy of Sciences but also the Communist
Academy, as "universal academic societies, must be supplanted by specialized research institutes aggregated into scientific associations." He
pleaded for a closer study of the Kaiser Wilhelm Gesellschaft in Germany, founded on the eve of World War I, as a model for new organizational measures in the pursuit of science.[6]

The deluge of acrimonious comment made it clear that the survival
of the Academy depended on its readiness to become a true Soviet
institution and, as a first step in that direction, on its willingness to elect
a solid core of new academicians imbued with the spirit of the new
society. The Communist press asserted that only the election of a representative group of pro-Soviet scholars to the positions of full membership could make the Academy a useful instrument of socialist
construction.[7]

To help place the Academy on the right track, the government sent
a special commission to Leningrad to "assist" in selecting the nominees
for the coveted positions. After numerous public debates, the seven
selection committees, representing the major areas of scientific activity,

came up with a list of no more and no less than forty-two nominees, the exact number of new positions. The list comprised, for the most part, three categories of nominees: scholars with widely recognized achievements in their specialized disciplines, such as the mathematicians S. N. Bernshtein, N. N. Luzin, and I. M. Vinogradov, the geologist V. A. Obruchev, the physicist L. I. Mandel'shtam, the aerodynamics pioneer S. A. Chaplygin, and the medieval historian D. M. Petrushevskii; leading Marxist theorists, led by N. I. Bukharin and A. M. Deborin; and specialists with distinguished records in the reorganization of the national economy, typified by I. M. Gubkin. One-fifth of the nominees were Communists, and another fifth were persons with established records of loyal participation in government-sponsored activities.

` A group of senior academicians opposed both the railroading techniques used by various selection commissions and the role of the party press in trying to break the last strongholds of academic autonomy. I. P. Pavlov and V. I. Vernadskii, the most courageous and influential academicians, criticized both the high-handed techniques of choosing the nominees and the unsatisfactory levels of achievement of individual candidates.[8] Above all, they resented plans to make dialectical materialism the official philosophy of academic science. Vernadskii categorized Deborin's interpretation of dialectical materialism as nineteenth-century idealism with no basis in science.[9] Pavlov criticized the disregard of scientific accomplishment in the selection of candidates for academic positions. He expressed particular dissatisfaction with the election of D. K. Zabolotnyi, an epidemiologist who produced noted studies on the plague and cholera epidemics.[10] Pavlov was displeased with Zabolotnyi's increasing detachment from basic research as a result of his extensive involvement in "practical" activities sponsored by the government.

At its annual meeting, on January 12, 1929, the Academy's General Assembly voted in favor of electing thirty-nine candidates; it turned down three nominees, all Marxist scholars: A. M. Deborin, N. M. Lukin, and V. M. Friche. The leaders of the campaign to transform the Academy into a Soviet institution were quick to interpret the negative vote as part of an antirevolutionary strategy of the "reactionary" forces anchored in the Academy. A RANION resolution claimed that the Academy's General Assembly turned down the nominations of Deborin, Lukin, and Friche solely on the ground that they were Marxist theorists.[11] An assembly of instructors and students of the Institute of Red Professoriat called the action of the General Assembly "an anti-Soviet political demonstration" and joined the forces that demanded a thorough reorganization of the Academy. A. V. Lunacharskii castigated uncooperative academicians as "supercilious mandarins in science" and the main force working against the Sovietization of the Academy.[12] While the voices of dissatisfaction were still growing in volume and bitterness, the government ordered the Academy to conduct new elections. The General Assembly met on February 13, 1929, and, aided by the votes of the new members, this time elected the three men to full membership. But this action did not bring an end to the Academy's troubles. Indeed, it was at this time that mass organizations intensified their harassment of the Academy by demanding a broad recasting of its administrative procedures, staffing methods, and philosophical-ideological orientation.

The All-Union Association of Workers in Science and Technology for Cooperation in Socialist Construction (VARNITSO) served as the pivotal organization in the ubiquitous campaign to eradicate the real and imaginary centers of resistance to the Sovietization of scientific activities and higher education.[13] Founded in 1928, in anticipation of the First Five-Year Plan, VARNITSO was organized and guided directly by the central authorities of the Communist party. It conducted an extensive and punitive campaign to accelerate the splintering of the "old" intelligentsia as a first step toward creating a new intelligentsia loyal to the Soviet system. According to the biochemist A. N. Bakh, it united the "left wing" of the Soviet intelligentsia in a relentless war against "counterrevolutionary forces."[14] Furthermore, it organized special sessions at congresses of scientists to discuss ways of making the work of scholars, teachers, and engineers part of socialist construction and to criticize digressions from orthodox thought and expected behavior. VARNITSO articulated the ideology of the new intelligentsia, which placed primary emphasis on forging closer ties between science and industry, planning scientific work, making education polytechnical, and giving unqualified recognition to dialectical materialism as the universal method of science. In universities and other institutions of higher education, it inaugurated a policy of periodic evaluation of professional personnel, a policy that resulted in dismissals of individuals considered dangerous enemies of socialist construction.

VARNITSO made a particularly strong issue of the urgent need for a major reorganization of the Academy. A. N. Bakh, a revolutionary of nineteenth-century vintage and now a Marxist in conviction if not in party membership, was the chief coordinator of VARNITSO criticism. He concentrated on the theme of the social utility of science and on the elaboration of ties between the Academy and industry.[15] In a public statement made immediately after the Deborin affair, VARNITSO labeled the Academy one of the last strongholds of reactionary forces and called for its full reorganization.[16]

In the early spring of 1929, a special government commission, headed by Iu. P. Figatner, a high official of the Workers and Peasants Inspection, descended on the Academy in an effort to carry out a thorough surveillance of the personnel and their overt and covert activities. In addition to three academicians—S. F. Ol'denburg, V. L. Komarov, and A. E. Fersman—the commission included representatives of the Academy trade-union organizations, several Leningrad industrial plants, and Leningrad party organizations. It took the commission several months to complete its work. In gathering material it depended heavily on information supplied by various mass organizations, and particularly on the voluminous files of the secret police. By passing fragments of information to the press at carefully planned intervals, it gave wide publicity to "flagrant violations" of the trust placed in the Academy by the Soviet state. According to one accusation, the Academy served as a haven for all sorts of ideologically incorrigible persons, including an unusually large representation of the high aristocracy. It concealed secret archival materials left by the Social Revolutionaries and Kadets as well as rich materials on the history of revolutionary movements in nineteenth-century Russia. And it not only withheld secret documents related to the last days of the Romanov dynasty but also allowed some of them to find their way out of the country.[17]

The surveillance led to a wave of rapidly executed purges of undesirable persons at all levels of the academic hierarchy. The purged persons did not share identical fates. Many persons were not only expelled from the Academy but were forbidden to take other jobs. Others, typified by academicians S. F. Platonov, M. K. Liubavskii, and V. N. Peretts— all medieval historians—were exiled to undisclosed locations and never reappeared.[18] Academicians E. V. Tarle and N. P. Likhachev were exiled but eventually reappeared, reacquiring their old positions and resuming their duties. S. F. Ol'denburg, the respected permanent secretary of the Academy and an old Kadet, was relieved of all administrative duties but did not lose the title of academician. The swiftness with which the

purges were carried out clearly demonstrated that they were long in
preparation. Loren Graham has described the general climate and the
political strategy of the purges:

> While Figatner's claim that the Academy was a center of counterrev-
> olutionary activity was merely an ignominious and feeble rationalization
> for his actions, there is little doubt that the Academy contained large
> numbers of people who would have preferred another type of govern-
> ment. As one of the few isolated and nonpolitical institutions in Soviet
> Russia after the Revolution, the Academy had indeed attracted people
> who sought shelter. But the Academy was in no sense of the word, so
> far as one can determine from available information, including the
> evidence advanced by the Party, a locus of actual counterrevolutionary
> activity. Its worst sin was political aloofness, not political opposition.
> Yet in the Soviet Union in the year of the "great break," such aloofness
> was a major transgression in itself, incurring disgrace and loss of posi-
> tion, if not worse.[19]

The triumph of Marxism in the Academy made it much easier for
Communist mass organizations to continue with their harassment of
scholars suspected of political heterodoxy. For several years VARNITSO
activity grew in magnitude and fierceness. Special groups worked on
tracking down and breaking up various opposition factions made up of
the "old intelligentsia." They fought real and imaginary efforts of the
intelligentsia to organize a unified front against the established political
system. Their main targets were individuals and groups suspected of
"political neutrality," "apolitical" views, "rightist leanings," and outright
opposition to government policies that incorporated science into economic planning and threatened to lock the Academy into a limited world
of practical activities.[20] Often combining its efforts with those of the
Section of Scientific Workers and the Society of Militant Dialectical Materialists, VARNITSO, with a national membership of 15,000, worked in
many directions: it harassed Marxist deviationists, it sponsored meetings
at individual research units for purposes of ideological "self-criticism,"
it campaigned against philosophical distortions in college textbooks, and
it fought "pseudo-Marxists"—the opportunists who placed a Marxist
façade on all kinds of theoretical constructions incompatible with the
current attempts to make science a unique expression of the Stalinist
creed.

The ruthlessness with which VARNITSO conducted ideological warfare can best be described by noting a publicly recorded episode in its
violence-dominated history. In 1932 academician E. F. Karskii, a Russian

historian, died, and the Academy asked academician N. S. Derzhavin, an expert in Slavic studies, to deliver the customary encomium. Shortly thereafter the local VARNITSO authorities called Derzhavin on the carpet, accusing him of not having used the occasion to expose Karskii's reactionary ideological stance and alleged distortions of the national identity of the Belorussian people. VARNITSO directed Derzhavin to place an article in the *Herald of the Academy of Sciences* pointing out errors he had made in his commemorative speech and supplying specified corrections.[21] When the article appeared in the October 1932 issue of the *Herald*, VARNITSO called the incident closed, but not until it had been widely publicized so as to forewarn potential violators of the new ethical code of scholarship.

The 1929 elections and purges completed the process of the Bolshevization of the Academy; it only remained to work out the details of the future course of its development as the institutional hub of Soviet science. In the process of restructuring the Academy, five developments were particularly significant: the rapid expansion of the role of Marxist scholars in the administration of academic affairs; the synchronization of the work of academic personnel with the national five-year plans; the reorganization of academic research centers; the broadening of the geography of academic institutions; and the creation of a new system of postgraduate training. Each of these developments merits closer scrutiny.

MARXISTS IN THE ACADEMY

Approximately one-third of the academicians elected in 1929–1931 were Marxists. The main source of their power was not in their number but in the administrative positions they held. In 1930 a Marxist—V. P. Volgin—became the permanent secretary of the Academy, the main link between the Academy and the government. A student of "utopian socialism," Volgin added little to the branch of scholarship he chose to pursue. His influence and power derived instead from the wide range of positions he occupied in Marxist scholarly organizations. He headed the Communist Academy's Commission for the Publication of the Classics of Socialism; he was an organizer and the most active member of the Central Committee of the Section of Scientific Workers; and he was closely associated with the rise of the workers faculties, the Institute of the Red Professoriat in Moscow, and RANION.[22] In 1928, in a report to the Section of Scientific Workers, he made the "isolation" of the Academy

of Sciences a subject of scathing attack. He advocated the speedy elim-
ination of the last residues of "social-class" enemies in the Academy.

Among the Marxist academicians, G. L. Krzhizhanovskii presented
the most attractive curriculum vita. He came to the Academy with two
strong assets: he was an Old Bolshevik, and he could show a long list
of practical achievements in socialist construction; in particular, he was
the first director of the Commission for the Electrification of Russia and
of the State Planning Commission. In 1929 he was appointed vice president of the Academy, with the task of eliminating internal opposition
to official moves to bolster the involvement of the Academy in the technological projects of the First Five-Year Plan. Since A. P. Karpinskii,
president of the Academy, was in his eighties and had withdrawn from
all administrative activities, Krzhizhanovskii stood an excellent chance
of becoming the main figure in the drama of academic change. But he
did little to advance himself or to interfere with the Academy. Unprepared for scholarship, he limited his activities to advocating a broader
representation of applied science in the Academy and to propagandizing
the guiding principles of the new science policy.

According to a 1933 official report, 13.6 percent of the total staff of the
Academy were Communists; however, 37.5 percent of the "top-level
personnel," a category that included full and corresponding members,
were members or candidates of the Communist party.[23] Marxist academicians received credit for having been "a vital and constructive" force
in transforming the Academy. Some of them, however, were plagued
by personal problems and uncertainties, particularly in their relations
with powers outside the Academy. For example, N. I. Bukharin enjoyed
more prestige as a scholar among the non-Marxist members in the Academy than as a political figure among the governing authorities. The
election of Bukharin to membership in the Academy coincided with
intensified attacks in the press on his "right-deviationism." He was
accused of "political" and "theoretical" opposition to the Soviet government and, at one point, of close ideological proximity to Leon Trotsky.[24]

In November–December 1929, the Communist Academy held a series
of sessions criticizing Bukharin's sociology as an ensemble of revisionist
postulates, all proclaimed to be contrary to dialectical materialism.[25]
Worst of all, he was labeled the leader of a strategically located group
of Marxist deviationists engaged in a pernicious war against the established order in the Soviet Union. Stalinist moves to bring unity to the

divided house of Marxist theorists led to an attack on A. M. Deborin, the leader of "dialecticians," a group accused of having been unappreciative of Lenin's contributions to Marxist thought. In 1931 D. B. Riazanov, the deposed head of the Marx-Engels Institute, was humbled by a public announcement of his expulsion from the ranks of academicians as well as from the Communist party; he was found guilty of close ties with emigrated Mensheviks.[26] These aberrations, however, did not impede the impetus toward a strategic Bolshevization of the Academy. The contribution of men like Deborin was more in helping to annul the "extraterritoriality" of the Academy than in engineering an imminent victory for the Marxist view of science.

Nevertheless, it should not be overlooked that for every Marxist academician of doubtful behavior there were several Marxist academicians whose record was unmarred by theoretical, ideological, or other deviations from Stalinist norms. In their quest for effective Sovietization of the Academy, the authorities relied less on Marxist scholars than on representatives from the swelling ranks of "technical science" with proven loyalty to the Soviet system; in 1933 the Academy's Technical Group consisted of eighteen members, the largest single group in the Academy. The government also relied on a growing number of representatives from theoretical disciplines—such as A. N. Bakh in biochemistry, N. I. Vavilov in plant genetics, and S. I. Vavilov in physics—who had amply demonstrated their loyalty to Bolshevik ideology and a readiness to work for the government and for socialist construction.

In the meantime, the party organization gained a solid footing in the Academy. In 1929 the Academy had only one party group; in 1932 eighteen primary organizations, each with an average of twenty-one members, spread their activities throughout the entire structure of the Academy—holding patriotic rallies, sponsoring ideological lectures, keeping an eye on the lines of communication, seeing to it that the right persons were nominated for elective positions, cooperating with trade-union organizations, and sponsoring seminars on dialectical and historical materialism. One of their key—but silent—functions was to help the authorities forestall the creation of informal groups and relations untutored by the state and considered potential nuclei of opposition to the system. Close to daily work, they supplied a constant flow of information on popular reaction to specific national policies related to scientific activity. Because of the new policy of admitting aspirants for higher degrees, graduate students formed the backbone of party strength in the Academy. In 1933 graduate students and nonacademic administrators made up two-thirds of total party membership.[27]

PLANNED SCIENCE AND SOCIALIST CONSTRUCTION

The new atmosphere in the Academy favored making science a clearly defined practical activity, thereby harnessing it for direct application in socialist construction. The growth of the Technical Group (subsequently renamed the Technical Department) into the largest single component of the Academy showed clearly that the government favored applied science.[28] A. N. Bakh's pronouncement that science was "integral and indivisible in its nature" and "utilitarian in its origin" expressed the reigning thoughts that transformed the spirit and the body of the Academy.[29] The integration of the Academy's activities into the five-year economic plans became the main instrument of the new emphasis on practical research.[30] Practically oriented research was expected to perform a dual function: to make science the moving force in the ongoing economic transformation and cultural revolution and to help authorities combat "the individualism of scientists," which often took on "political coloration."[31]

The tendency to make social utility the ultimate criterion in judging the worth of scientific theories expressed itself with particular force in the growing interest in the philosophy of Francis Bacon, who had maintained that science was the infallible source of social well-being and the most reliable index of cultural progress. By rescuing science from the clutches of Scholastic metaphysics and by bringing it down to earth, Bacon met the most pressing requirement of scientific practicalism. A Marxist writer noted that Bacon was neither a "dialectical materialist" nor a "mechanical materialist," but that he was an "analytical materialist"—the first systematic advocate of inductive methodology and concrete analysis of experience. Bacon's philosophy of science, this writer maintained, was the first major stride on the path to a full triumph of dialectical materialism. Under his influence, the narrow and obscure trails emerging from the darkness of Scholasticism became a broad avenue leading to the "sunny glade" of modern natural science.[32] Marxist writers disagreed about the details of the relations between Bacon and the modern philosophy of science; however, all were willing to name him the forefather of modern materialism as both a philosophical and a sociological category. All agreed that he was the founder of modern studies of the social utility of science, the true prophet of the growing role of science in the development of technology, and the first advocate of extensive government participation in the development of scientific knowledge.

A. V. Lunacharskii did not live to complete his most-favored literary undertaking: a biography of Bacon for the popular series "The Lives of Famous Men." In a published fragment, he stated that "the personality of Bacon is of great value to us," for he provided a model of an active warrior against "all previous ideas" that stood in the way of establishing man's control over nature and in favor of enhancing the role of the individual in "social authority." History, according to Lunacharskii, was made neither by pessimists, represented by Hamlet, nor by social recluses, represented by Prospero, but by activists typified by Bacon.[33]

Other writers hailed Bacon as an astute observer of the dialectical nature of "matter in motion" and as an investigator of the positive and negative sides of ancient materialistic philosophies.[34] V. F. Asmus, one of the more perceptive and erudite Soviet philosophers—and one who also managed to stay outside the controversies of the day—began his notable *Marx and Bourgeois Historicism* with an essay on Bacon. *Under the Banner of Marxism*, the leading journal in Marxist philosophy, devoted several approving articles to the two pillars of Baconian philosophy: inductionism and utilitarianism. In the meantime, Bacon's selected works appeared—or reappeared—in Russian translations and were lauded by reviewers in the leading journals. All studies agreed that Bacon was the first philosopher to view science as a social institution and as a true synthesis of the most sublime intellectual achievements.

Bacon was indeed the father of a utilitarian or pragmatic philosophy of science. His utilitarianism was, however, very broad: while Bacon stressed the wealth of practical consequences that would flow from the pursuit of science, he went beyond mere utility when considering the criteria for choosing research topics, the logic of inquiry, and the methodology of practically oriented science. Bacon, like Spinoza and Descartes, thought that scientists did particularly well when, in their research for practical wisdom, they were neither blinded by excessive practicalism nor limited by the specific goals of individual research projects.[35]

In two respects, government authorities and academicians were in basic agreement: both accepted and espoused the Baconian utilitarian view of science, and both viewed advance in theory as the most reliable index of progress. They also agreed that the entire spectrum of scientific endeavor could contribute to the progress of the national economy. Beneath this unity, however, lay a deep cleavage. When the authorities emphasized the practical aspect of science—the active role of science in socialist construction—they adduced the urgent need to plan research thoroughly and integrate it into the national annual and five-year plans of economic and cultural development. Academicians, however, considered planning procedures a method of subjecting their professional involvement in science to a complete control by government authorities. The intellectual power and social utility of science, they argued, were contained in the internal logic of scientific development, not in mechanical responses to externally generated and coordinated practical needs. As S. F. Ol'denburg, citing the famous historian S. M. Solov'ev, said in 1918, society at large benefited most from science when it abstained from interfering with its internal dynamics—the true area in which its progress was forged.

Planning scientific activity was a key vehicle of the new utilitarianism in science. Planning research on a national scale was a new experience that posed great difficulties and called for careful and systematic preparation. N. I. Bukharin indicated the enormity of the problem when he stated that planning must consider no less than five cardinal problems: (1) technological impact of scientific breakthroughs; (2) urgent technological demands; (3) current technical and economic plans; (4) long-range developmental perspectives; and (5) "the revolutionizing reverse influence of science on technology" and the potentialities of "grand theory."[36] The best way to establish a sound relationship between the external demands of technology and the internal pressures of the natural logic of scientific growth, Bukharin believed, was to submerge the planning of today within the vision of tomorrow and to combine the responses of science to the needs of technology with the responses of technology to anticipated advances in science. Bukharin called for an all-out war on two fronts: against "the old academicism" and its distaste for practical work; and against technologicism and its distaste for "grand theory." His plan was too ambitious and too complex to be translated into reality, but the basic principles it expressed continued for a long time to dominate discussions centered on the methods and philosophical subtleties of planning the national effort in science.

On April 28, 1931, the General Assembly of the Academy of Sciences approved the first annual plan of research to be carried on by its numerous institutes and laboratories, and a year later it made its research operations a working part of the Second Five-Year Plan.[37] Until 1936 the annual plans of the Academy were mechanical summations of plans submitted by individual members of the research staff. After 1936 the Academy tried to prepare integrated plans, focused on the basic targets of the five-year plans. The new type of planning was designed to make the work of individual scholars controllable parts of a collective effort. Only group research, it was now emphasized, could eliminate the last vestiges of bourgeois mentality in science and, at the same time, bolster the socialist foundations of the Academy. In shaping the practical orientation of the plan for 1939, the Academy's technical, biological, and chemical departments held numerous conferences with representatives of the central agencies of various "labor organizations" of "scientific workers."[38]

The leading scientists showed no willingness to abandon the claim that the real practical strength of individual branches of scientific knowledge lay in the depths of their theory; the ideologues, however, placed primary emphasis on research related directly to the economic and cultural needs of the day. While the leading scientists resolutely adhered to the idea that science is a prime mover of technological development, the ideologues argued that the needs of technology should be the prime mover of scientific research. The government was careful not to deny the role of theory as the motive force of scientific knowledge; however, it favored theoretical exploration related to the practical necessities of the day.

While a typical academician found it inopportune to voice publicly his opposition to planning, here again Pavlov and Vernadskii were exceptions. In refusing to prepare a plan for his work in 1933, Pavlov stated explicitly that by abiding to the letter of a plan, he would deprive himself of many unplanned and unanticipated leads in experimental research.[39] Vernadskii resented the planning of research activities because it demanded from each scholar a specified commitment to participate in the cultural revolution of the day, necessitating that he become an active defender of dialectical materialism as the official philosophy of science and as a "universal methodology."[40]

Vernadskii also led a small group of scholars who bitterly criticized the decision of the Academy to replace KEPS with the new Council for the Study of the Productive Forces of the USSR. The latter was entrusted with the task of directing the practical orientation of research carried out by all the units of the Academy rather than research carried out only by its own components. They argued that the new move discouraged theoretical exploration, stultified scholarly initiative, and violated the internal impulse of scientific development.[41] The new authorities, for their part, viewed the resistance to official designs for planned effort in science as a unique expression of academic "individualism" with easily detectable "political coloration," a form of resistance to the Soviet regime.[42] As a response to government dichotomization of "pure" versus "applied" science, some academicians thought of splitting the Academy into two huge institutes: the Lomonosov Institute, which would unite the research units engaged in basic theory, and the Mendeleev Institute, intended to coordinate and direct practically oriented research.[43] This plan did not receive sufficient official support and quickly faded away.

In implementing the new philosophy of planned utilitarianism, the government heeded the claims of influential academicians that the most abstruse branches of modern science were potentially the most promising sources of technological progress. It came as no surprise that the first five-year plan of scientific research prepared by the Academy gave the highest priority to the study of the microstructure of matter as staked out by quantum mechanics and its offshoots. This policy quickly led to efforts to expand the research base of modern physics. The grand challenges and enormous potentials of theoretical nuclear physics attracted many brilliant young scientists, and the discipline appealed to the government as a new source of economic and military strength. In the early 1930s the country had enough nuclear physicists to hold national conferences and to attract leading Western scholars—such as Paul Dirac and Frédéric Joliot-Curie—to participate in their deliberations.

In 1934 the Institute of Mathematics and Physics, founded in 1921, split into two institutes—the V. A. Steklov Institute of Mathematics and the P. N. Lebedev Institute of Physics—each expanding rapidly and becoming a true barometer of the strength of Soviet science. It was at this time that P. A. Cherenkov's experimental discovery of the so-called Cherenkov effect and D. D. Ivanenko's advancement of the proton-neutron model of the atomic nucleus signaled the beginning of a steady stream of Soviet contributions to the most advanced branches of modern physics. At this same time, A. N. Kolmogorov constructed his classic set-theoretical axiomatics of probability theory; this took place even as Marxist writers were attacking probability theory as a branch of mathematics applicable more to the whims of the capitalist economy than to the clearly hewn regularities of the socialist economy.[44] In 1930 N. N. Luzin, the founder of an internationally noted school of Moscow mathematicians, published his *Leçons sur les ensembles analytiques*, which helped to build a bridge between set theory and mathematical logic. In the same year, V. A. Fok produced a mathematical apparatus that, in the view of P. A. M. Dirac, was "for many purposes the most convenient representation for describing states of the harmonic oscillation," a "cornerstone" of the theory of radiation in quantum mechanics.[45] In 1935 L. V. Kantorovich, at the age of twenty-five, produced a series of papers that contributed a new and fundamental orientation in functional analysis: the theory of semiordered spaces.

As soon as the new idea of scientific utilitarianism became a reality, it began to invite all kinds of abuses. The ideologues of the Stalinist "Great Break" gave it a prominent place in patriotic appeals and antiintellectual slurs. The new purveyors of socialist chastity, usually in search of self-aggrandizement at the expense of both government and science, made it an integral part of the rhetoric of socialist construction. The pressure on practical orientation in science helped T. D. Lysenko to mount a ferocious attack on genetics as a science far removed from the immediate needs of Soviet agriculture. This attack diverted chemists from tackling the theoretical and methodological challenges of quantum mechanics and nuclear physics. And it prevented mathematicians from building upon suggestive national creations in mathematical logic and foundations of mathematics.

The planning of science was a complex process operating on many levels—from planning the work of individual scholars to planning national scientific development. On the national level, the integration of scientific research into the annual and five-year economic plans was one matter; the creation of an institutional mechanism for carrying out planning activities was another matter. Despite the growing pressure during the 1930s for making science part of national economic plans, the country did not have a central coordinating agency—or a set of central coordinating agencies.[46] The Scientific and Technical Section and the Glavnauka had gone through a succession of stages until they had ceased to exist. The proliferation of sectoral research units, operating under the jurisdiction of individual departments of the government, added to the administrative fragmentation of scientific activities. S. I. Vavilov opposed the frequently heard suggestions to make the Academy the administrative center of science—a "commissariat of science." But this did not mean, he said, that the country was not much in need of a national coordinating body in science. Robert Lewis has noted that

> the application of planning to scientific research was probably the feature of the organization of science in the Soviet Union in the inter-war years which attracted the greatest interest elsewhere, [and that] the Soviet case was to be frequently cited in the discussion and controversy that took place in Britain in the 1930's on the possibility of directing the power of science to improve the lives of all members of the society, which culminated in the publication of J. D. Bernal's *The Social Function of Science.*[47]

While the Academy multiplied the institutional mechanisms that brought it closer to the economic pulse of the country, and while it experimented with planning techniques, the Technical Group—the strongest expression of the new philosophy of practicalism—grew by leaps and bounds. In 1935 it became a special department of the Academy: it was elevated to the same administrative status as the Department of the Mathematical and Natural Sciences and the Department of the Social Sciences. It included five "groups" of research units concentrating on energy, technical mechanics, technical chemistry, technical physics, and mining.[48] The rapid rise of technical research units marked a serious defeat for the traditionalist wing of the Academy—the advocates of basic research as the main wheel of technological progress. The engineers—typified by A. V. Vinter, E. V. Britske, I. P. Bardin, and B. E. Vedeneev—who rose to high positions in the academic hierarchy, became the model academicians. Unconcerned with the grand theory of modern science, they were able to avoid entering the murky waters contaminated by ideological pressures.

Opposing this development were the "fundamentalists," who contended that science and technology could advance at an accelerated pace only if the line separating them was clearly marked and mutually respected. In 1936 L. D. Landau claimed that to force scientists to deal with technological questions was to force them to enter an area beyond their competence. He said that the first duty of Soviet physicists was to raise the level of their discipline and, in order to achieve this goal, to improve experimental procedures.[49] The real and critical problem, he asserted, was that the country had only a limited number of physicists fully familiar with the latest theoretical and methodological developments in their discipline. He implied that physicists would be much more useful in the rapid industrialization of the country if they stuck to their own work—the advancement of the theoretical and experimental base of their science. Although the opposition of the "fundamentalists" to the "technicians" quickly ceased to be an open affair, it remained a force of considerable importance in the inner dynamics of the Academy, ready to flare up at a propitious moment.

The "practical" orientation of the Academy went beyond intensive planning of all phases of research activities, in order to integrate the operations of the Academy into the national five-year plans. So-called external sessions, sponsored by the Academy, were another form of the new utilitarianism: teams of academicians visited selected industrial cities to conduct seminars on the common problems of science and production with managers and engineers. Mass meetings with factory workers were yet another form. The first such meeting was held in Leningrad on November 25, 1931, when a group of academicians spoke to an assembly of 4,000 workers. All reports had a common theme: the production resources of the Leningrad province. N. I. Vavilov spoke on "northern agriculture."[50]

Contracts signed by individual people's commissariats for specific types of research were still another form of involvement in socialist construction, often completely unrelated to the Academy's role as a center of fundamental research. In 1932 the Academy signed a "general contract" with the People's Commissariat of Heavy Industry that covered a wide spectrum of research activities to be undertaken by its institutes. Among many other commitments, the Academy undertook to prepare a comprehensive survey of research topics related to heavy industry, to intensify the quest for new sources of energy, to study the geochemistry of rare metals, and to hold two national conferences on scientific topics related to heavy industry.[51] Contracts of this kind were often too ambitious and too unwieldy to be realized within a specified time (normally one year), and more often than not they were too general to be controlled with any degree of efficiency. Sometimes they were modified versions of projects already "in operation." General contracts were often used to dramatize the socialist spirit of the Academy rather than to produce concrete results.

In 1932 60 percent of the Academy's revenue came from special contracts, a clear indicator of the commitment to "practically" oriented research. Pressure generated by leading academicians, however, who pointed out the potential danger of "wasting" highly trained talent on practical problems that could be handled by lower-level research bodies, led to a reversal in fiscal policies: in 1937 70 percent of the revenue came in the form of budgetary allocations from central and local governments.[52]

The Academy did not hesitate to document abundantly its views on the role of science in the economic development of the country. In 1935 it published a book-length report addressed to the Seventh All-Union Congress of Soviets, one ostensibly designed to present significant details of changes in the Academy's "research activities," "organizational structure," and "ideological profile," all for the purpose of realizing a more effective participation in socialist construction. The real purpose of the book, however, was to present the Academy's claim that frequently the most abstract and remote theoretical concerns in science led to the most practical and substantial results, and that without daring— and "impractical"—theoretical exploration, science would stagnate. The report on physics, for example, noted the Academy's primary concern with such fundamental questions as the structure of atomic nuclei, the nature of elementary particles, the scattering of neutrons, the dynamics of light quanta, the composition of cosmic rays, and the penetration of electrical particles into crystals.[53] While recording the many avenues of the Academy's participation in socialist construction, the report also served to educate the government about the true meaning of practical orientation in scientific research.

The government's intent was not to force the Academy to abandon "pure science"; however, it wanted the Academy to embark on more practical activities as well—on activities that were measurable by current indices of economic growth. From then on, the Academy's institutes were continuously confronted with the tedious task of balancing pure research, dictated by the rapid growth of theoretical and experimental physics, with practical research, dictated by the national economic plans. At no time during the 1930s did the Academy abandon the idea that basic theory was the heartbeat of science and that the pure science of today should be judged by the practical results of tomorrow.[54] The leading members of the Academy argued consistently that they should limit their practical research to problems depending on high theory.

On March 14–20, 1936, the Academy of Sciences sponsored a special session at which the leading physicists reported on the state of their discipline to a select group of scientists, party workers, and engineers.[55] The meeting represented the first public show of the triumphs and dilemmas of intensive efforts to make the Academy an integral part of the national system of economic and cultural planning. It inaugurated a long series of public appraisals used by the central government as an elaborate technique to keep the Academy in line with official utilitarianism and ideological dictates.

At the March meeting, A. F. Ioffe gave the main report, which attracted the most attention and criticism.[56] The criticism came from two sides. Physicists, led by L. D. Landau, criticized the low theoretical level and "primitive" experimental designs of the Physico-Technical Institute, headed by Ioffe and heralded as the cradle of modern physics in the Soviet Union.[57] Physics, in Landau's view, could not become a powerful tool of technical knowledge unless it continued to advance its theory and experimental designs. Landau received strong support from I. E. Tamm, a corresponding member of the Academy. In a special report, Tamm stated that atomic physics promised to unlock untold quantities of energy, but that it was not at the time sufficiently developed on a theoretical plane to answer the practical questions of the possible uses of nuclear energy.[58] Implicit in his argument was the idea that in order to render the maximum service to society, physicists must be allowed to concentrate on advancing scientific theory unhampered by considerations of the economic needs of the day. Party workers and their allies among the physicists pursued the opposite line in criticizing Ioffe's report: they accused physical institutes of a disproportionate engagement in theoretical problems and a gross neglect of the needs of industry.

Unlike some of its successors, the conference was not tightly controlled by external authorities, and the criticism it generated was mild in tone. The press focused on achievements in theoretical physics as well as deficiencies of research practices and organizational styles.[59] Strangely enough, more criticism came from physicists than from philosophers. The purpose of the session was not to settle ideological issues but to clarify the ramifications of the mounting official pressure for closer ties between the Academy and the national economy. Its job was to translate Baconian utilitarianism into the science policy of the five-year plans.

During the remainder of the Stalin period, the Academy was an object of annoying and distracting pressure, from the party and from trade-union locals, to widen the base of its participation in building the industrial foundation of Soviet socialism. From time to time the government made known its unflinching concern with the nature and scope of the Academy's involvement in matters of a technological nature. It was equally eager, however, to show its awareness of the role of the Academy as the national center for theoretical exploration. In 1938 the government criticized the Technical Department for its reluctance to tackle the concrete questions of the national economy.[60] In the same year a *Pravda* editorial stated that "true Soviet science" had no use either for "abstractions separated from life" or for "narrow practicalism."[61]

ORGANIZATIONAL ADJUSTMENTS

The transformed Academy required a change in its style of work—in the institutionalization of its research activities. The new situation favored institutes as basic units in the overall division of scholarly endeavor. The emphasis on the institutes was a recognition of the advantage of "group work" in science and of the large-scale concentrations of talent, laboratory facilities, specialized libraries, and publication outlets. Heeding A. F. Ioffe's plea for a careful study of *all* the numerous areas opened by quantum mechanics, the government showed readiness not only to make the newly founded Institute of Physics one of the largest and most active research units in the Academy but also to create a new institute—the Institute for Physical Problems—for P. L. Kapitsa, who previously headed the Mond Laboratory at Cambridge University. Kapitsa, in turn, attracted many young and established physicists, such as L. D. Landau, to the Academy. Rapid developments in new physics provided the stimulus for founding the first modern research institutes outside Moscow and Leningrad. Among these new bodies, the most noted were the Ukrainian Physico-Technical Institute in Khar'kov and the Siberian Physico-Technical Institute in Tomsk.[62] The Physico-Technical Institute in Leningrad, founded and guided by A. F. Ioffe, served as the model for the organization of these and similar centers—all contributing to making quantum mechanics and related disciplines the focal point of the national dedication to science.

The new organizational policy called for two types of units intended to avoid the usual divisions and bring various specialists, institutes, and laboratories together for interdisciplinary research. The new charter (1930) encouraged the formation of "groups" made up of experts from various fields, working specifically in areas where cooperation between several disciplines would be advantageous. These groups were a specific institutional adaptation to both the ongoing revolutions in physics and mathematics and the Stalinist pressure for collective effort in science. They were highly flexible, engaged primarily in resolving problems related to planning and organizational procedures. The establishment of groups was part of a larger design to neutralize the alleged tendency of academicians to work in isolation or to build personal schools. Groups gave research associates a more influential place in the Academy than they had had in the past. Although they were a passing phenomenon, groups left behind a legacy that emphasized interdisciplinary exploration and cooperative research as the mainstay of the Academy.[63]

Associations, another new institutional component of the Academy, brought together entire institutes, laboratories, or other research bodies. In this case, too, the aim was to facilitate interdisciplinary research, but on a much broader scale. Associations found a particularly attractive home in the Department of Mathematical and Natural Sciences, one of the two general administrative branches of the Academy (the second was the Department of Social Sciences). The Chemical Association, for example, consisted of the Laboratory of General Chemistry, the Institute of Physico-Chemical Analysis, the Platinum Institute, and the Laboratory of Organic Synthesis. From the very beginning, the associations were exceedingly fluid and difficult to administer. Although they had a skeletal administrative apparatus, they were functional rather than structural components of the Academy; they provided for new working relationships between the established organizational components. In the government's view, the associations made the Academy more flexible and more adaptable to the needs of the many activities that went under the general name of socialist construction.

Various "committees," "councils," and "commissions" not only addressed themselves to specific problems of a practical nature but also served as instruments for cooperation with talent from outside the Academy. Commissions, in particular, were dynamic, interdisciplinary, and— unlike groups—concretely defined research units, emphasizing either theoretical exploration or practical aims. Some of them were set up for the explicit purpose of preparing organizational changes in the Academy necessitated by the opening of new research vistas and theoretical orientations. Some of the commissions were subordinated directly to the

General Assembly of the Academy; others were anchored in single or combined institutes. The Commission on the Origin of Domestic Animals was an affiliate of the Institute of Language and Thought, the Institute of Anthropology and Ethnography, the Zoological Institute, and the Institute of Genetics.[64]

ACADEMIC OUTPOSTS

The restructuring of the Academy went hand in hand with extensive efforts to establish research outposts in selected communities scattered throughout the country.[65] This action had several distinct motives. It spoke to the rapidly expanding study of natural resources and to the creation of industrial complexes in previously underdeveloped regions. Part of the strategy that favored a decentralization of scientific activity, it was a positive response to pressure, from KEPS, for regional scientific institutions, and to deficiencies built into the country's economic and nationality policies. The chief articulator of this philosophy was V. I. Vernadskii, who criticized the tsarist policy that deprived many national groups of institutional mechanisms for direct involvement in science. The Academy finally realized that the establishment of permanent research centers at the periphery would ease the burden of sending expeditionary teams from the center to do the work that could be done by local experts. From 1919 to 1928 the Academy of Sciences alone sent over 300 research teams to various parts of the country. The projected outposts promised to strengthen the role of the Academy in both the national commitment to science and the coordination of scientific activity. Geographical deconcentration and administrative centralization of scientific activities were now established as the cornerstone of the Soviet science policy. These outposts were envisaged as the main links in cooperative relations between academic, university, and sectoral research units.

In 1930 the Academy established the Council for the Study of Productive Forces (SOPS), made up of several previously existing bodies, including KEPS.[66] While KEPS concentrated on organizing expeditionary research teams, the new council undertook the ambitious task of coordinating the entire effort of the Academy in the study of natural resources. One of its first duties was to organize stationary research outposts at strategic economic and cultural locations. In 1931 a special commission was given the task of preparing a general plan for local research units. In the same year, the Central Executive Committee of the USSR approved the Academy's recommendation for the establishment of "complex institutes" in regions where large-scale economic construction projects were either in progress or at the planning stage.

After careful consideration of local cultural conditions and economic potentialities, the Academy adopted a plan for creating two kinds of outposts: "bases" would be established in the areas with underdeveloped networks of scientific institutions, and secondary economic potentialities and "branches" would be set up in more advanced communities. Bases were intended to be small and undifferentiated: their basic task was to assist local organizations in planning scientific activities and in organizing the study of natural resources. Branches, in contrast, were from the beginning composite and functionally differentiated bodies: they consisted of clearly defined institutes, laboratories, and research sectors. In 1931–1932 the Academy established the first branches, led by the Transcaucasian Branch, the Far Eastern Branch in Vladivostok, and the Central Asian Branch in Alma-Ata, Kazakhstan.[67] In 1935 the Transcaucasian Branch split into the Azerbaidzhanian, Armenian, and Georgian branches, located in Baku, Erevan, and Tbilisi, respectively. The first bases were established on the Kola Peninsula and in Arkhangelsk and Tadzhikistan.

While the base was a single, undifferentiated institute, the branch was a system of specialized institutes. In branches the disciplines not covered by institutes were attended to by smaller units identified as sectors or sections.[68] The work of both branches and bases was initiated and supervised by corresponding research units of the Academy of Sciences. Of the two categories of research units, branches were much more dynamic and subject to rapid expansion and organizational mutations. They supplied republican governments with scientific answers to pressing economic questions and organized research in local culture, language, and history. The Georgian Branch, according to a report from 1936, concentrated on the study of ferromanganese and subtropical plants, as well as on the archeology of the Caucasian region.[69] The Far Eastern Branch, one of the more versatile new academic outposts, had special research sectors for work in hydrology and geobotany as well as a microbiological laboratory.

During the second half of the 1930s most branches began to establish research units concerned with the theoretical aspects of individual sciences, without challenging the primary emphasis on the local economy and culture. The Georgian Branch attracted national attention for its work on the dynamic theory of seismic stability. The astronomical observatory at Erevan University, established in 1934, trained a formidable group of young astronomers who, in turn, made it possible to establish the Biurakan Astrophysical Observatory and to make Armenia one of the leading republics in the study of the galaxies.[70] The Ural branch was particularly active in research, being deeply involved in the technology of the ferrous metal industry, for which purpose it built special chemical laboratories. In addition to research, all these units served as centers for developing local scientific talent.[71] After widening their base of operation, some leading branches became academies of sciences of individual republics and key links in a gigantic and versatile science empire.

TRAINING FOR HIGHER DEGREES

The new Academy became a vital center for postgraduate training leading to higher degrees. After 1926 the Academy experimented with providing research experience to small groups of "practicants" or "aspirants," as they were soon to be called.[72] At first only the Department of Physical and Natural Sciences—the forerunner of the Department of Mathematical and Natural Sciences—admitted young people to postgraduate studies, the task of training future scholars in the humanities and the social sciences having been assigned to the Communist Academy. The method of admitting "aspirants" to the academic research units did not please the authorities outside the Academy. Selected by individual academicians, the trainees did not always show particular attachment to the Marxist philosophy of science; indeed, in 1928 a representative group of aspirants staged a public protest against efforts to pressure them into taking courses in dialectical materialism.[73]

The Commission for the Training of Aspirants, established in October 1929, presented an approach to the selection of graduate students more in keeping with the spirit of the time: academicians lost the right to be the sole judges in selecting students to be trained under their supervision.[74] The new Committee for the Training of Scientific Cadres, dominated by Marxists and pro-Marxists (such as A. M. Deborin, V. P. Volgin, and N. Ia. Marr), acquired the sole authority to select aspirants.[75] The committee consisted of representatives of the Academy, the central government, the Communist Academy, the party group in the Academy, and previously admitted aspirants. Despite the new policy, leading scholars continued to play a strong part in choosing the aspirants for higher degrees working under their supervision.

At this time, official certification of prescribed competence in dialectical materialism became a requirement for candidacy for the higher degrees. Indeed, special seminars in Marxist philosophy admitted representatives of all age groups, academic strata, and professional specialties. A party decision of August 18, 1929, decreed that persons of working-class origin be favored in the selection of aspirants for higher degrees. A few months later, a *Pravda* correspondent complained that only 7–8 percent of the country's total of 24,000 "scientific workers" were party members, and that an overwhelming majority of these were in the humanities.[76] In 1928, according to one report, 67.5 percent of Communist scholars in the Russian republic were in the humanities and the social sciences, 15.4 percent were in medicine, 8.9 percent were in the natural sciences, and 8.2 percent were in engineering.[77] Responding to the new pressure, the Academy saw to it in 1929 that 60 percent of the new graduate students belonged either to the Communist party or to the Young Communist League and that 73 percent came from the families of workers and peasants. In 1934 the total number of graduate students at the Academy was 246; 60 percent came from peasant and worker families, and 70 percent were members of either the Communist party or the Young Communist League.[78]

During the 1930s the educational level of aspirants for higher degrees rose as a result of two strategic developments: high school and university training returned to traditional standards emphasizing a solid grounding in the basic disciplines, and social-class origin ceased to be an important criterion in selecting university students.[79] In 1935 graduate work leading to a doctorate was added to—and raised above—the requirements leading to a candidate's degree, and the requirement for proven competence in two foreign languages was enforced with more rigor. Furthermore, institutions granting higher degrees required all prospective graduate students to submit certificates of political loyalty.[80] The government retained the right to select the institutions engaged in postgraduate studies and to have the last word in certifying higher degrees. From then on, personnel and curricular changes required final approval by specifically designated government authorities.[81]

Despite notable successes in expanding and improving postgraduate training, scientific institutions grew more rapidly than the supply of competent research scholars. Fully aware of the paradoxical situation, S. I. Vavilov went so far as to recommend in 1935 that the number of research bodies be reduced in order to avoid a counterproductive scattering of sparse talent. In fact, he recommended both that there be fewer scientific institutions and that the size of individual institutions be smaller.[82] In his opinion, the need for more research scientists could be met by inaugurating a program of sending promising young scholars to Western laboratories for advanced training. In 1936 L. D. Landau pointed out "the catastrophic situation with regard to the cadres of physicists," illustrated by the "huge discrepancy" between the number of available physicists and the national need for them.[83] He also said that many persons classified as physicists did not have sufficient qualifications to be so labeled. It was under criticism of this kind that the government came around to recognizing research experience as a vital component of university education.[84] The process of equipping the universities with adequate laboratories and libraries, however, proved to be exceedingly slow.

The new system of training graduate students made it possible to undertake the first surveys of the magnitude, dynamics, and differential quality of advanced studies. On October 1, 1938, a total of 7,331 graduate students attended 228 institutions of higher education.[85] Statistical data showed clearly that graduate students working for higher degrees at research institutes were usually in a more advantageous educational environment than those working in institutions of higher education. In the first group of institutions, 60.4 percent of graduate students worked under advisers with doctoral degrees; in the second group, only 42.7 percent. Research institutes had much larger and more functional holdings of modern Western scientific literature. Surveys also showed that most graduate students were Russians (53.3 percent), followed by Jews (21.8 percent), Ukrainians (11.6 percent), Armenians (2.5 percent), and Belorussians (1.8 percent). Although the Academy of Sciences followed the national pattern of ethnic representation, it may be safely assumed that the percentages of Russian and Jewish students were somewhat higher.

Equipped with more advanced laboratories and larger library holdings, and staffed by personnel with a comparatively high level of professional achievement, the Academy attracted the most promising students, the future elite of Soviet science. Indeed, at this time the authorities agreed that the Academy should provide postgraduate training of "a higher level"—that it should admit only candidates who had shown great promise as graduate students at institutions of higher education and who had been successful in their scientific work.[86] The Academy henceforth became the main supplier of experts staffing its own institutions and laboratories and advancing to the highest rungs of the academic ladder. An important function of the Academy was to preserve and advance the strong national traditions in science—to assure the "scientific schools" anchored in the Academy of uninterrupted growth. The share of the Academy in the output of advanced degrees increased gradually in the 1930s: while in 1933 the Academy provided training for 2.74 percent of all graduate students, in 1940 its share was 6.06 percent.[87]

A succession of structural changes after 1929 made the Academy a true microcosm of the Soviet political system and the leading and guiding force of the labyrinthine institutional structure of Soviet science. In 1934 the Academy was moved to Moscow to be closer to GOSPLAN and other agencies of the central government engaged in restructuring the national economy and Soviet culture according to the principles of Marxism-Leninism as defined by Stalin. Explaining this memorable event in the history of the Academy, V. P. Volgin noted: "All of us know that the moving of the Academy to Moscow will inaugurate a new epoch in the life of our institution, for it will reinvigorate science and will mark a turning point in the relations between the top-level Soviet scientific institutes and government agencies engaged in economic and cultural reconstruction."[88] The moving of the Academy closer to the Council of People's Commissars and to GOSPLAN (and the establishment of the Sector of Science at the Central Committee of the Communist party) was part of a new design for unified and centralized national management of science. The charter of 1935 formalized the changes effected during the preceding six years and gave the Academy organizational stability that was to last for many years. Nevertheless, G. M. Krzhizhanovskii, in his annual report on the achievements of the Academy in 1936, stated that while the basic restructuring of the national economy had been achieved, no such claim could be made for the Academy.[89]

DIALECTICAL MATERIALISM AND
THE UNITY OF SCIENCE

After 1925 A. V. Lunacharskii used many occasions to note that the absence of a unified view of nature, society, and science limited the effectiveness of the Academy as a research center. Arguing in more general terms, N. I. Bukharin noted that the introduction of the dialectical-materialistic method into the rapidly widening span of sciences promised to intensify the cross-fertilization of various branches of knowledge and to give new strength to the methodological unity of science in general.[90] Marxist theorists attached much importance to the notion of the unity of science on yet another ground: such unity was a necessary precondition for the full alliance of science and ideology and for making the national effort in science conducive to precise and efficient planning. In 1929 dialectical materialism became the official philosophy of the Academy. E. Ia. Kol'man, the fast-rising artificer of Stalinist orthodoxy, gave a straightforward and unadorned description of the new situation:

> Only recently a discussion of the radical changes in philosophy had taken place. Now all of us understand clearly the main point of that discussion: under the dictatorship of the proletariat, neither philosophy nor any other discipline can exist in isolation from politics and Party leadership. Now we all know that any effort to view a theory or a science as an autonomous and self-contained discipline would mean objectively to oppose the general party orientation and the principles of the dictatorship of the proletariat.[91]

The victory of the new philosophy was slow in coming. In some fields the representatives of Marxist thought disagreed among themselves on the interpretation of the fundamental principles of modern science. In 1930 A. M. Deborin delivered a paper entitled "Lenin and the Crisis in Modern Physics," in which he hailed Einstein's theory of relativity as a grand synthesis of such "contradictory" notions as "gravitation and inertia," "space-time continuum and gravitation," and "electromagnetic field and gravitation."[92] In 1933 the physicist V. F. Mitkevich, another scholar close to Marxism, spoke on the same theme at a similar occasion. Unlike Deborin, however, he rejected the general theory of relativity on two grounds: it did not allow for the existence of ether, and it depended on excessive mathematical formalism.[93]

In most disciplines non-Marxist scholars ignored philosophy altogether, and research proceeded more or less in the same vein as before the 1929 takeover of the Academy. There was another, and rather unanticipated, development: the rebellion of eminent scientists, typified by V. I. Vernadskii, against dialectical materialism as a philosophy of science. In 1929 Vernadskii used philosophical arguments to bolster his campaign against the candidacy of A. M. Deborin for full membership in the Academy. Deborin was then the leader of the "dialectical" wing of Soviet Marxism, which became the ruling party in the Communist

Academy after the Second Congress of Marxist Institutions; the losing party was the "mechanist" wing, led by L. I. Aksel'rod and A. K. Timiriazev, who were accused of close identification with mechanical materialism.

But the Deborinist victory did not bring lasting peace to the Communist Academy. In a speech at the Moscow Institute of Red Professoriat on December 9, 1930, Stalin identified the views of the Deborin group as an expression of "Menshevizing idealism."[94] On January 25, 1931, the Central Committee of the Communist Party made the theoretical journal *Under the Banner of Marxism* "the militant organ of Marxism-Leninism," defending "the general party line" and fighting the two categories of philosophical deviationism: "the mechanical revision of Marxism, as the main danger at the present time, and the idealistic distortions of Marxism by the Deborin group."[95] Deborin quickly acknowledged his errors and apologized particularly for his "unsupportable" assertion that while Plekhanov was primarily a theorist, Lenin was first of all "a practical person, a revolutionary and a leader."[96] He also admitted his flirting with "formalism," a stance that threatened to take philosophy away from "the concrete conditions of phenomena."[97] "I must admit," he said at a specially convened session of the Communist Academy, "that my grossest error was in misunderstanding and insufficiently appreciating the Leninist phase in philosophy and in Marxism in general."[98] Nevertheless, Deborin did not fare well in the Communist Academy, because his Marxism was not strong and pure enough; and he did not fare well in the Academy of Sciences, because his Marxism was considered too rigid and antiquated.

Taking part in this debate strictly as an outsider, and before the codifiers of Leninist philosophy had made the decisive attack on both "mechanist" and "dialectical" theories, Vernadskii focused on the relative merits of the two orientations in regard to their relations to science. Unlike Stalinist theorists, he concluded that the philosophical stance of the mechanists was more realistic and more in tune with the spirit of twentieth-century science. The mechanist orientation, he said, was more satisfactory because it was further removed from Hegelian idealism and was closer to eighteenth-century materialism, which was based on the achievements and the logic of science and did not try to impose its authority on science.[99]

In a letter written in September 1930, Vernadskii told V. P. Volgin, permanent secretary and the protector of Marxist interests in the Academy, that a mandatory introduction of dialectical materialism into laboratory work would lead to "a decline of science in our country." To give his concern a more general basis, he asserted that "in science nothing has ever been discovered with the help of the philosophical method."[100] In 1931 Deborin accused Vernadskii not only of refusing to accept dialectical materialism as a philosophy of science but of advancing philosophical views in direct opposition to Marxist thought. That Vernadskii survived the attacks led by Deborin was proof of the willingness of the authorities to tolerate selected established scholars, particularly in the natural sciences, despite demonstrative refusals to make dialectical materialism part of their thinking.

The task of creating a "unified scientific method" based on dialectical materialism was passed on to the Communist Academy, which, however, was prevented by general unsettledness, staggering developments in physics, limited numbers of experts, and smoldering ideological conflict from meeting the challenges of modern scientific thought. The Marxist writers were not always united in separating "friends" from "adversaries" among non-Marxist scholars. S. A. Ianovskaia, the leading Marxist among mathematicians, identified V. F. Kagan, the country's leading expert on Lobachevskii's non-Euclidean geometry, as a scholar "very close to us" and "loyal to the Soviet system," despite residues of Henri Poincaré's conventionalism in his thinking.[101] A. A. Maksimov, however, showed no inclination to view Kagan in a positive light; instead, he saw him as a thinly veiled anti-Marxist and, by definition, as an enemy of the Soviet system.[102] Kagan's non-Marxism, he said, was only a cover for thorough and concerted anti-Marxism. At a time when the Academy of Sciences was undergoing internal reorganization and housecleaning, the Communist Academy was busy examining the views of its members on the philosophical foundations of natural science.

At the end of December 1930 and the beginning of January 1931, the Communist Academy held a conference that dealt with "the situation on the front of natural science." The conference concluded that a solid group of Communist Academy members did not understand "the general party line" related to "socialist construction" and "cultural revolution," and that some members had confused "the new works of bourgeois science" with dialectical materialism.[103]

A conference on biological theory, held in March 1931, revealed that the Communist Academy was deeply involved in a bitter feud among

the representatives of various theories of organic evolution—ranging
from extreme Lamarckians to the most inflexible supporters of classical
genetics.[104] The conference provided ample evidence that Marxist views
on modern biological theory were characterized by deep-seated disunity,
rampant personal conflict, and debilitating uncertainty on the scientific
front. A year later the Academy sponsored a symposium on the philosophy and history of mathematics, with the purpose of providing
concrete illustrations of the use of the dialectical method as an analytical
tool in the study of the social background of science.[105] The symposium
lacked methodological unity and theoretical depth, an obvious result of
the serious shortage of expert scholars combining an interest in Marxist
philosophy with competence in mathematics.

These and similar discussions inaugurated the Leninist phase in Marxist theory, which arose on the ruins of the "mechanical" and "dialectical"
orientations dominant in the 1920s.[106] The "dialecticians" were accused
of underplaying the role of Lenin in the history of Marxist thought; the
"mechanists" were charged with "misrepresenting" Lenin by trying to
substitute a mechanical model, based on such stationary attributes as
"equilibrium," for a dialectical model, based on the dynamics of "conflict." The "dialecticians" and "persons close to them" were accused of
being eager to make the grand theories of quantum mechanics and
special and general relativity fully congruent with dialectical materialism
while remaining silent about pronounced idealistic inclinations among
the leaders of modern physics. They were criticized also for their tendency to favor modern genetic theory over both Lamarckism and Darwinism, and for their efforts to make Kantianism and Machism the
philosophical pillars of biology.[107] The "mechanists," for their part, were
rebuked for their strong inclination to ignore the triumphs of modern
physics altogether and to create a sociology on the models of classical
physics.[108]

The attack on "mechanists" and "dialecticians" inaugurated the Leninist stage in the evolution of Marxist philosophy and led to the recognition of Stalin as an infallible interpreter of Leninism. It achieved—
or it tried to achieve—yet another goal: the crowning of the Communist
Academy as the main interpreter and codifier of the Leninist philosophy
of science. In philosophical matters the Academy of Sciences was now
expected to recognize the superior authority of the Association of Natural
Science, the natural-science branch of the Communist Academy.[109] Leninist philosophy became Stalin's tool in the struggle for ideological unity,

on the one hand, and for the unity of science and ideology, on the other. Leninist philosophy had many ramifications. It was presented as a modern Marxist theory of imperialism, of "the state and revolution," and of "proletarian dictatorship." But it was also presented as a modern Marxist interpretation of the twentieth-century revolution in natural science. M. B. Mitin, a leading codifier of the Leninist phase of Marxist theory, summed up the new attitude:

> Lenin's works in the field of dialectical materialism, particularly *Materialism and Empirio-Criticism*, are a new phenomenon. They present a generalized description of the achievements of natural science after Marx and Engels. They analyze the current state of natural science, particularly physics. And they identify the significance of the deep crisis in modern science from the point of view of materialistic dialectics.[110]

Henceforth, Soviet philosophy of science was centered on the epistemological postulates set forth by Lenin in *Materialism and Empirio-Criticism*, and it revolved around six basic premises. First, all scientific ideas, even the most abstract ones, are not constructions of the "pure mind" (as Einstein said) but are true—even though incomplete—reflections of a reality existing independently of the human mind. Second, scientific knowledge is relative, for it contains both lacunae and misrepresentations. The history of science is a history of the gradual reduction of these inadequacies; the relativity of scientific knowledge is a historical phenomenon. Marxism-Leninism has rejected epistemological relativism as a cover for philosophical agnosticism and even solipsism. Third, no theory can be proved or disproved by another theory: practical experience—praxis—alone is the ultimate judge of the validity of scientific wisdom.[111] Fourth, causality is the fundamental principle of scientific explanation. Science is the study of causes in nature and society; and as the study of causes it is a study of time—of genetic relations. Emphasis on causality is the basis of the pervasiveness of historicism in Marxist thought. Fifth, there is a limit to the reliance on mathematical methods in the explanation of both natural and social reality. The philosophers were mindful of Lenin's warning that "too much mathematics" is the surest path to idealism. "Too much mathematics" leads to "formalism," an escape hatch for scholars with an ivory-tower mentality. And sixth, "matter is inexhaustible": science cannot lose its *raison d'être*, for it can never solve all the riddles of the universe.

Leninist philosophy of science was not confined to the domain of epistemology. It also defined the basic principles of Leninist sociology of science as a special branch of the sociology of knowledge. The true guiding principle of science and the scientific community was *parti-inost'*—a consistent defense and advancement of the interests of the social class in power. Soviet scholars were constantly reminded that Leninist *partiinost'* was based on a full identification of scientific truth with the interests and the class consciousness of the proletariat. In practice this meant that scholars were expected to be ready to wage a war on, and to reject or modify, all theoretical ideas advanced by modern science that were interpreted as incompatible with Soviet ideology as articulated by Stalinist philosophers. The label "objectivism," designating attitudes opposed to *partiinost'*, now acquired a wide circulation among Stalinist philosophers. According to M. B. Mitin, an "objectivist" in science, as seen by Lenin, ignores the social-class vantage point in the analysis of phenomena, makes little use of a concrete analysis of "reality," overlooks inner contradictions in historical processes, and limits himself to "abstract logical schemes and theories" from which he tries to deduce the real world. He does not understand the unity of the logic and the history—of the theory and the class dynamics—of social phenomena.[112] In brief, an objectivist refuses to make science a weapon of class warfare. Since science could not be separated from class interests, to be labeled an objectivist meant to be accused of using the stance of ideological neutrality as a cover for pernicious attacks on proletarian values. To accuse a scientist of objectivist proclivities soon became the same as accusing him of unpatriotic behavior, and Stalinist philosophers hence became an arm of the secret police. To the police, digressions on the philosophical front were important clues for digressions on the political front.

The "new" philosophy was simple and straightforward. It shifted the emphasis away from the more abstruse and elusive ontological questions to the more earthbound and direct epistemological and ethical questions—from the ontological foundations of the universe to the epistemological attributes of science and the moral obligations of scientists. The new philosophy was dominated by ceremonial repetitions of philosophical imperatives and their use as criteria in judging the current developments in science, particularly in physics and biology. Thus, it became easier to spot "aberrations" in the texture of modern science and in the professional behavior of individual members of the scientific community. But above everything else, it became easier for scientists to enter the world of philosophy, a field that most of them found increasingly distasteful.

Despite mounting pressure for ideological conformity, expressed in epistemological and ethical terms, the voice of dissent was not completely silenced. Ia. I. Frenkel', a widely heralded physicist, and since 1929 a corresponding member of the Academy of Sciences, was one such rebel: he not only belonged to the group of philosophical skeptics but was willing to give his views public airing. In December 1931, at a national conference of experts in physical chemistry, after expressing his unwavering loyalty to the Soviet system, he launched a blistering attack on dialectical materialism as the official philosophy of the Academy. In what proved to be the last public attack on dialectical materialism in the age of Stalin, Frenkel' stated:

> There is no justification for the claim that dialectical materialism performs a guiding role in science. Our politics goes to the extreme in forcing the views of dialectical materialism on our scholars and on our youth. Socialism needs a [theoretical] base, and this is provided by historical materialism. Dialectical materialism, on the other hand, is an obstacle to the development of science. Neither Lenin nor Engels is an authority for physicists; Lenin's book [*Materialism and Empirio-Criticism*] is an affirmation of elementary truths which should evoke no arguments. Dialectical materialism is a reactionary philosophy and I hope that the government will soon be convinced that I am right. Some recent remarks by Bukharin seem to be going in that direction. There can be no proletarian mathematics or proletarian physics.[113]

I. E. Tamm, one of the most productive Soviet researchers in modern physics, took another line of attack. In an article published in *Under the Banner of Marxism* in 1933—the year in which he was elected a corresponding member of the Academy—he chose not to attack dialectical materialism but to criticize its interpreters in the Communist Academy. Having acknowledged "the great and incontestable success of dialectical materialism in the realm of the social sciences," Tamm made it clear that the situation in the natural sciences was not so favorable. Unlike Frenkel', however, he chose to place the blame on the philosophers rather than on dialectical materialism. His main argument was that the Marxist interpreters of modern physics could not meet the challenges of the ongoing revolution in science. "In the best cases," he wrote, "their knowledge did not go beyond the level of science at the end of the nineteenth and the beginning of the twentieth century." Even a college student, he contended, was able to detect the "scientific illiteracy" of the Marxist philosophers, who concealed their ignorance by "profuse and empty verbosity" and by "shifting the emphasis from fundamental to ephemeral questions."[114]

Frenkel' and Tamm voiced the views and sentiments of many colleagues. Nevertheless, their protests did not affect the course of history. It did not take long for the Academy of Sciences to enter the seething waters of dialectical materialism. Two occasions provided academicians with an excellent opportunity to take stock of the then-current search for unity between science and Marxist theory: the commemoration of the fiftieth anniversary of the death of Karl Marx in 1933 and the commemoration of the tenth anniversary of the death of V. I. Lenin in 1934. Both symposia produced huge volumes, made particularly attractive by the elegance of their technical composition and the quality of paper. Commenting on the first symposium, a spokesman for the Academy stated: "The reports at the Marx commemorative session have been important witnesses to the recognition of the general methodological importance of Marx's dialectic. They reveal the growing scope of the influence of Marxism on our scientific thought."[115]

The two commemorations produced one important result: they inaugurated the philosophical involvement of an influential group of academicians who were not members of the Communist party but who became supporters of dialectical materialism. The physicists A. F. Ioffe and S. I. Vavilov quickly emerged as the leaders of this group. They helped assure dialectical materialism of a firm footing in the Academy of Sciences, but they also became influential champions of the strategy to abate the conflict between Leninist epistemology and modern scientific theories, particularly in physics. The two conferences served as an outward manifestation of the triumph of dialectical materialism in the Academy, but they also demonstrated that the members of the Academy were far from united in their views on the Marxist interpretation of modern physics.

V. F. Mitkevich, for example, used the Marx commemoration to lament the disappearance of ether from the scientific scene and to criticize quantum mechanics and the theories of relativity for their heavy dependence on mathematical formalism.[116] Ioffe, for his part, emphasized the triumph of modern physics as a triumph of dialectical materialism; he chose to stress the modern physical theory of emission and absorption of light as a dialectical synthesis of wave and corpuscular theories.[117]

Ioffe received support from S. I. Vavilov, who asserted boldly that the theories of relativity and quantum mechanics—in addition to their other contributions—provided "powerful examples of the method of mathematical extrapolation."[118] The strength of new science was in the rapid and universal mathematization of methodological tools. Mitkevich emphasized the duty of physicists to suppress those theories of their science that were incompatible with classical Marxist theory; whereas Ioffe and Vavilov chose to follow Engels's advice that dialectical materialism be treated as a flexible philosophy, ready to absorb and adjust itself to advances in science.

The session commemorating the tenth anniversary of the death of Lenin served as an occasion for the Academy of Sciences not only to show its adherence to dialectical materialism as the official philosophy of science but to manifest its entrance into the Leninist phase of Marxist philosophy. The occasion provided an opportunity for the sinners of the past to redeem themselves and to rejoin the ranks of orthodox philosophers. Among the "sinners" were two leading Marxist theorists: N. I. Bukharin and A. M. Deborin, who in their previous writings were less than generous in praise of—and in references to—Lenin's work and ideas, and who were repeatedly accused by their Marxist comrades of serious misrepresentations of dialectical materialism. Point by point, Bukharin used the solemn occasion to bestow high praise on the pivotal points of the Leninist theory of knowledge; the crispness of his metaphors and the freshness of his philosophical detours added to the glamour of the occasion. Damned by the reigning powers and haunted by the police, Bukharin looked for a safe place in the Stalinist configuration of power. In his earlier writings, he recognized Lenin's inordinate talents as a man of practical politics; now, for the first time, he recognized his philosophical acumen. He said:

> Lenin's greatest contribution was in enriching the philosophy of Marxism in the wide sector of materialistic dialectics by utilizing the most diverse materials from the exact sciences and their multiple reflections in philosophy. With the power of a great thinker, he subjected the achievements of modern science to a critical analysis; he destroyed the prejudices and distortions of the newest bourgeois philosophy and, relying on the granite-like base of Marxism, raised materialistic dialectics to a qualitatively new and higher level.[119]

Bukharin knew that his contribution to Leninist philosophy was a contribution to the ideological unity of Stalinist power. If his aim was to enter Stalin's camp, he did not succeed. In 1937 Bukharin was imprisoned, and in 1938 he was executed an an "enemy of the people," a label applied to persons suspected of conspiring to stem the rising tide of Stalin's authoritarian rule.

A. M. Deborin found the Lenin symposium an invitation to give public expression to his readiness to apotheosize Leninist philosophy. In a very long essay he dealt with every major aspect of Lenin's philosophical legacy; his aim was to present Lenin's thought as a base for two types of operations: building a modern philosophy of science in full harmony with Soviet ideology; and conducting an unyielding war against the rising forces of revisionism. In fact, he leveled his harshest and most extensive criticism against Hendrick de Man, the author of *The Socialist Idea* and the *Psychology of Socialism*, who in his opinion advanced a revisionist philosophy by relying on the "pseudoscientific" theories of Karl Mannheim's sociology of knowledge, Freudian "biosociology," Spenglerian philosophical interpretation of "culture" and "civilization" as antithetical forces, and Max Weber's sociology of religion. De Man saw socialism where Deborin could see only the work of the most decadent forces on the modern intellectual scene.[120] Although Deborin survived the Stalinist purges, he did not regain the strong influence in the philosophical community he had held prior to 1930.

During the 1920s Marxist philosophers fought their most crucial battles among themselves. In the 1930s, with Stalin's Leninism firmly established, there was no room for internal squabbles; now philosophers could concentrate on attacking the scientists who exhibited detectable deviations from the new philosophical creed. Since the early 1930s the Communist Academy had already been deeply involved in such a campaign. It fought "idealism" in physics and mathematics, in biology and geochemistry. The Marx commemorative session brought the Academy of Sciences into philosophical confrontation with nonconforming scientists in and outside its fold. In an essay devoted to the second law of thermodynamics and "idealistic" interpretations of entropy as the inescapable path to the "thermal death" of the universe, a Marxist commentator could not resist the temptation to direct slurring remarks at such Soviet "followers" of Mach's philosophy of science—which was bitterly attacked by Lenin in *Materialism and Empirio-Criticism*—as L. D. Landau and M. P. Bronshtein, the "young school" of Soviet physics.[121]

The Lenin session unveiled another key development: the readiness of leading physicists to fight the pronounced tendency among the new breed of philosophers to turn dialectical materialism against the pivotal theories of modern physics. A. F. Ioffe warned against the rising tendency among philosophers to place the label of idealism on "Bohr, Schrödinger, Dirac, Heisenberg and Frenkel', and all others who strive to find an adequate explanation for the microscopic attributes of the atomic world."[122] Ioffe was careful to note that he did not consider all of Heisenberg's ideas "a holy truth and [pure] dialectical materialism," but his message was clear and forceful: dialectical materialism must not be transformed into a philosophy hampering the work of physicists by undue restrictions and paralyzing distortions.

In 1936 the government abolished the Communist Academy, an organization that never got off the ground and that made no tangible contribution to bringing dialectical materialism in tune with the revolutionary discoveries of modern science. With the silencing of long and often exciting debates over the "correct" interpretation of the theoretical base of Marxist thought, and with the reduction of Marxist philosophy of science to a few elementary postulates of Leninist theory of knowledge, it had ceased to be a dynamic source of creative responses to the general problems of the day. Its major institutes were transferred to the Academy of Sciences, without enriching it in any way. Among the transferred units was the Institute of Philosophy, which made the Academy of Sciences the center of philosophical studies in the country. The main task of the transferred philosophers was to render philosophy in its totality a tool of Stalinist ideology, to make Leninist epistemology the general "methodology" of science, and to conduct a constant campaign against idealistic "aberrations" in science at home and abroad.

The aggressiveness and ominous zeal of the newly acquired philosophers did not stop individual scientists from trying to make dialectical materialism an open philosophy of science—a system of thought adaptable to theoretical advances in science. Fighting against orthodoxy in Marxist thought, these scholars pointed out the total incompatibility of the "reactionary" views of science expressed by a growing breed of "new" philosophers—mainly graduates and associates of the Communist Academy—with the spirit of the "scientific foundations" of Soviet socialism. Ioffe, Fok, and Frenkel' led a group of physicists who resisted efforts by the philosophers to identify the key theories of Einstein, Bohr, and Heisenberg as specific expressions of "physical idealism" and to make "Soviet science" an intellectual and cultural category different from "bourgeois science" in both method and substance.[123] They received support from a dwindling group of philosophers, including S. Iu. Semkovskii, the author of a noted book on Einstein's theoretical thought, who as late as 1935 was willing to defend the view that the fundamental principles of modern physics confirmed the philosophical postulates of dialectical materialism.[124]

In biology the confrontation of scientists and philosophers did not follow the same course as in physics. Although there were dissenting voices, the physicists were generally united in protecting the guiding principles of modern theory against attacks from philosophical quarters. In biology, however, the emergence of T. D. Lysenko made the discipline a battlefield of two feuding groups of scientists—supporters and foes of genetics. In physics the disunity on the philosophical front made it easier for scientists to ignore ideological and philosophical preconceptions in interpreting the theoretical foundations of their science and to wage a quiet but persistent war against the Stalinist advocates of a "Soviet physics" built on the foundations of Leninist epistemology. In biology philosophers were much more united than in physics. Led by M. B. Mitin, they were unanimous in viewing the theory of modern genetics as totally alien to dialectical materialism and in treating Lysenko's theories as a true echo of the Marxist philosophy of science and of Soviet ideology. They made Lysenko's theories "Soviet biology," and they accepted Lysenko's distortions of dialectical materialism as true Leninism. No philosopher was willing to come to the defense of the beleaguered geneticist N. I. Vavilov, the main target of Lysenko's wrath. Nor was any philosopher willing to defend N. K. Kol'tsov, an early ancestor in the field of molecular biology, from vicious attacks by I. I. Prezent, Lysenko's philosophical aide. While in physics a united community of scientists was poised against a disunited camp of philosophers, in biology a disunited front of scientists faced a united front of philosophers, all dedicated to pruning science of theoretical thought regarded as alien to dialectical materialism.

In biology the iron hand of Lysenko made philosophical explanations clear, uniform, and unchallengeable. In most other sciences—including physics and mathematics—philosophy invited diverse, often contradictory, explanations. It was easy to declare dialectical materialism and Marxist-Leninist ideology the cornerstones of Soviet science. It was not so easy to arrive at a general consensus on the Marxist interpretation of individual scientific theories, particularly those of the post-Engels era.

The paralyzing confusion in the search for a Marxist interpretation of modern science was compounded by the unfamiliarity of Marxist philosophers with the theoretical and methodological subtleties of modern science and by the philosophical amateurism of Soviet scientists. Most scientists found it convenient to avoid all issues that involved philosophy and ideology. Some tried to bring dialectical materialism in tune with the basic theories of quantum mechanics, the theories of relativity, genetics, and other landmarks of modern science: they saw the strength of dialectical materialism in its adaptability to all advances in science. Another group declared war on many aspects of modern scientific theory regarded as incompatible with dialectical materialism; the most ardent philosophers of the 1930s belonged to this group. All Marxist philosophers argued that science could not advance without consistent and deep philosophical involvement. Academician L. I. Mandel'shtam, representing the physicists, argued that science and philosophy should be treated as separate and fully independent systems of thought or modes of inquiry.[125]

V. I. Vernadskii, an eloquent spokesman for the scientific community, thought that philosophy could be of great help to science only as long as different philosophical orientations were allowed to coexist. He welcomed the current revival of various vitalistic orientations on the ground that they added new dimension to modern philosophy, particularly in the area where it came in direct contact with science.[126] Vernadskii did not have much use for the basic postulates of vitalism per se, but he firmly believed that the wider the range of philosophical criticism, the stronger the pressure on science to sharpen its methods, logic, and verification procedures. Marxist critics contended that Vernadskii's excursions into the domain of vitalist thought were part of a deliberately designed "eclectic union of diverse ideas held together by the thread of antimaterialism."[127] Vernadskii, it was now argued, studied nature for the purpose of formulating "empirical generalizations" purporting to show "the value of religious modes of inquiry and the truthfulness of vitalism." His theoretical views were treated as part of a general vitalistic interpretation of the universe and an effort to present religion as a "creative force."[128] At least one critic considered Vernadskii's theories a major barrier on the path of socialist reconstruction of science and technology.

The differences between Vernadskii's philosophical views and the Stalinist rendition of dialectical materialism were fundamental and irreconcilable: Stalinists accepted the revolutionary advances in modern science—particularly those in physics and biology—only inasmuch as they were confirmations of the philosophical claims of dialectical materialism; Vernadskii, in contrast, thought that modern developments in science were so far ahead of all existing philosophies that there was an acute need for new philosophical ideas and insights.[129] The new science did not bestow the rights of unchallengeable authority on an existing philosophy; instead, it showed the need for fresh ventures in philosophical elucidation. Vernadskii found it increasingly difficult to publish his heterodox philosophical ideas. His correspondence and unpublished manuscripts show that philosophical issues continued to occupy his attention and that the basic ingredients of his philosophical orientation were too open and too fluid to be incorporated into doctrinaire thought. His philosophical views reflected the dynamics and perturbations in the logic and method of modern science and the changing cultural panorama of the scientific world view.

Divergent philosophical positions coexisted for a relatively long time not because the Academy encouraged the free spirit of philosophical speculation but because the guardians of Marxist theory did not evolve a precise and generally accepted dialectical-materialistic interpretation of the revolutionary advances in science. Stalin and his advisors, busy with more pressing problems, such as masterminding the ongoing political purges, did not find the task of fully integrating the theoretical foundations of modern science with the ideological dictates of Marxist-Leninist legacy so pressing that it required immediate attention. The conspicuous inability of the authorities to provide unified guidance produced one favorable result: it left wide areas of scientific thought generally undisturbed by ideological constraints and, therefore, relatively free of undue external interference.

The debates of the 1930s demonstrated clearly that Marxist theorists regarded the philosophy of science exclusively as a specific elaboration and refinement of the ideological superstructure of science. They also showed that the sciences were not equally exposed to philosophical and ideological interference. Crystallography was typical of the sciences that maintained almost complete isolation from philosophical issues and steered clear of heavy ideological commitments. Neurophysiology, however, was among the sciences that became deeply enmeshed in intense campaigns to harmonize their theoretical views with the leading postulates of dialectical and historical materialism. Other disciplines lost their philosophical-ideological *raison d'être* completely and ceased to exist as legitimate branches of scientific knowledge. For example, historical materialism robbed sociology and social psychology of their subject matter and reduced them to derogatory labels for "bourgeois aberrations" in social theory. In addition, the growing attack on the scientific legitimacy of genetics shifted its main operation to ideological grounds.

Marxist philosophers made a serious effort to gain a Western audience for their general theoretical interpretations of the main currents of scientific thought. The Second International Congress of the History of Science and Technology, held in London in 1931, presented them with an ideal opportunity to air their views before an international audience. Boris Gessen's paper on the determining role of the technical needs of the economy in shaping Newton's scientific concerns attracted most attention.[130] From the philosophical point of view, the papers by B. M. Zavadovskii on the physical aspects of organic evolution and Ernest Kol'man on "dynamic and statistical regularity in physics and biology" were particularly illuminating. These two papers were part of a common effort to present dialectical materialism as a moderate philosophy of science opposed to the excesses of the mechanist orientation, at one extreme, and various kinds of idealism, at the other. Zavadovskii argued against physical reductionism and vitalism in modern biology and suggested a compromise between the Lamarckian tradition and the new genetics.[131] Kol'man, following the same line of reasoning, claimed that the Marxist philosophy of science sided neither with the old-fashioned "mechanistic fatalism" nor with the modern versions of "indeterminism" in physics and biology, which, he thought, invited "formalism" and "idealism." In contrast, he advocated the dialectical unity of contradictory theoretical explanations of natural phenomena, including the dialectical unity of necessity and chance—of Laplace's "dynamic" determinism and Heisenberg's "statistical" determinism.[132] Western listeners found little solace in the distracting terminology used by the Soviet scholars; but at least some leading representatives of Western science felt that their own views did not diverge in principle from those of Zavadovskii and Kol'man. Most, however, had little use for dialectics as a method of explaining the processes of nature and human thought; but they did agree with the claim of their Soviet colleagues that mechanistic and idealistic theories were the prime enemies of modern science.[133]

The case of Kol'man, in particular, showed that the Soviet guests were much tamer in their London speeches than in their writings at home. In his London speech, Kol'man found gaping idealistic aberrations in the work of only one Western physicist: Arthur Stanley Eddington. In his writings at home, he found the entire leadership of modern physics—from Einstein and Bohr to Heisenberg and Schrödinger—guilty of gross idealistic deviations that necessitated a complete revamping of quantum mechanics and the theories of relativity. He also criticized the young physicists of the "Leningrad school"—including Ia. I. Frenkel', G. Gamow, L. D. Landau, and D. D. Ivanenko—who made no effort to rise above the "idealistic" and "Machian" biases of their Western mentors.[134] In his London paper, he saw the future of science in the search for a dialectical unity of contradictory theories and concepts occupying a dominant position in contemporary scientific thought. Unlike its capitalist counterpart, socialist science, he thought, was in the position to undertake a *planned* effort to bring contradictory theories and concepts into a unified system of scientific knowledge and to protect science from metaphysical intrusions.[135] In his writings at home, he was a chief spokesman for so-called Soviet science, dominated by unique philosophical propositions, methodologies, institutional settings, and styles of work—and unencumbered by irreconcilable theories and concepts. Soon after his return to Moscow, Kol'man published an article in which he chastised Gessen for having been tardy in separating himself from "Menshevizing idealism."[136]

Gessen, however, was rewarded for his London paper: in 1933 he was elected a corresponding member of the Academy of Sciences. Nevertheless, his theory of the history of science, cast in simple and direct lines of historical materialism, or economic determinism, did not find many supporters in his native land. There were two reasons for this. First, his paper was too rigid even for the more orthodox Marxists. For example, he did not ask such complicated but necessary questions as, Why did Newtonianism triumph over Cartesianism? Second, he fell victim to the Stalinist purges and quickly ceased to be a citable source. Gessen's study of Newton's *Principia* was not only the first but also the last historical study of science conceived within an orthodox framework of Marxist social theory.

While ignoring Gessen, B. M. Kedrov, a young historian and philosopher of science (particularly of chemistry), went in the opposite direction. In 1940 he advanced a new and clearly less orthodox Marxist approach to the history of science: he argued that before a historian could undertake a successful study of the causal ties of a science to the national economy, he must study the internally propelled growth of that

science, as well as the influence of intellectual developments in neighboring disciplines and in philosophy.[137] Preoccupied with the more
pressing ideological questions, Stalinist philosophers paid no attention
to Kedrov's intellectualistic view of the development of science. While
continuing to shy away from rewriting the history of science in the spirit
of economic-technological determinism, they did not abandon the general faith in the economic causes of theoretical work in science.

The inordinate and intense activity on the ideological front during the
1930s did not achieve its main goal: the unity of philosophy and natural
science. But the search for this unity grew in intensity and unbending
determination. The portentous tone of this dedication was stated clearly
by V. L. Komarov at the Eighteenth Party Congress held in 1939:

> In the natural sciences, where the richness of factual material staggers
> the intellect of the scholar, only a clearly defined ideological approach
> can illumine the research material by a guiding idea and lead to correct
> conclusions. Such a guiding idea is presented in the work of Marx,
> Engels, Lenin and Stalin. Therefore, our experts in natural science and
> technology must be in command of dialectical materialism—the philo-
> sophical method of Marxism. Soviet philosophers, in turn, must be
> deeply concerned with the problems of natural science. Our Academy
> of Sciences, which unites the representatives of philosophical thought
> with the representatives of progressive natural science, must become
> the materialistic center of world science. It must work consistently on
> applying the great philosophical ideas of dialectical materialism to the
> modern problems of natural science.[138]

HISTORICAL MATERIALISM: THEORY OF HISTORY
AND THE SOCIAL SCIENCES

During the 1920s the Academy of Sciences stood apart from all endeavors to advance historical materialism as a general Marxist theory of society. This job was instead one of the major concerns of the Communist
Academy. The latter body, however, could not devise a generally
accepted interpretation of the theoretical and methodological foundations of historical materialism. One group, led by N. I. Bukharin and
related to A. A. Bogdanov, treated historical materialism as Marxist
sociology—as a substantive science concerned with the general structure
and dynamics of human society. In developing his theory, Bukharin
tried to answer the modern critics of Marx; and in answering the critics
he borrowed from them, as Stephen Cohen has pointed out.[139] Another

group, led by A. M. Deborin, and heavily represented in the Communist Academy, treated historical materialism as a "scientific method"—as an application of "dialectical method" to the study of social phenomena. According to this school, sociology was a clearly demarcated science that possessed its own subject matter but relied on historical materialism as its main source of methodological tools. It rejected Bukharin's treatment of sociology and historical materialism as synonyms. The third group, led by N. Karev, argued that historical materialism and sociology were radically different bodies of knowledge, but also that sociology—in contrast to historical materialism—was a pseudoscience, irretrievably lost in the morass of idealistic and mechanistic philosophy. Karev was perfectly content to see sociology taken off the roster of legitimate scientific disciplines. Sociology was not a synonym of but an antidote to historical materialism.

Karev's views won out, not because of his authority in Marxist circles but because his position was favored by the turn of historical events during the years of the consolidation of Stalin's power. During the 1930s the term "sociology" came to be used extensively as a derogatory label for non-Marxist social theories, even though the article on "historical materialism" in the first edition of the *Great Soviet Encyclopedia* stated that "the general methodological importance" of historical materialism lay in its providing the social sciences with a common "sociological" point of departure—a common recognition of the economic "base" as the ultimate determining force of all social phenomena.[140]

F. V. Konstantinov, a prominent figure in the generation of Stalinist philosophers, was the last—in the Stalin era—to define historical materialism as Marxist sociology. In an article published in 1936 in *Under the Banner of Marxism,* he asserted that one of the great contributions of Marx and Engels was the founding of Marxist sociology as a science of social development. He could see no logic in rejecting the identification of historical materialism with sociology on the ground that the latter was "vulgarized" by bourgeois scholars and Marxist revisionists.[141] Instead, Konstantinov reminded his readers that Lenin, in "Who Are 'the Friends of the People'?" and in *Materialism and Empirio-Criticism,* called historical materialism the only scientific sociology. He even found a favorable reference to sociology by Stalin in 1928. Konstantinov's plea fell on deaf ears, however, and before long all references to sociology became derogatory in nature. From that time until the mid-1950s, sociology was most commonly referred to as a tool of "Western imperialism."

The posthumous attack on M. N. Pokrovskii, one of the founders and the first president of the Communist Academy, and a member of the Academy of Sciences from 1929 to 1932, the year of his death, marked the official end of sociology as a legitimate discipline. The attack was long in coming; it was the culminating point in a sequence of skirmishes between Pokrovskii and his numerous adversaries, a dispute that had begun to receive public airing in the 1920s.[142] In 1936 the central authorities of the Communist party stressed the urgent need for freeing historical scholarship from the "anti-Marxist, anti-Leninist, and antiscientific" views of the Pokrovskii school. In 1938 the Central Committee of the Communist party ordered the organization of a national propaganda campaign in favor of the newly published *Short History of the Communist Party*. The order made special reference to the need for a total war on the Pokrovskii school. [143] The organization and the execution of the attack on the ghost of Pokrovskii were entrusted to the Institute of History, which responded in 1939 by publishing a two-volume collection of essays condemning this newly discovered class enemy and his school for all kinds of devious efforts to force historical materialism into "the Procrustean bed of a sociological scheme."[144]

Once called "the leading Marxist historian" by A. V. Lunacharskii, Pokrovskii was now attacked for having exaggerated the role of "mercantile capitalism" in the evolution of Russian society and for having "vulgarized" economic determinism.[145] He was accused of denying the objective nature of history and of treating the work of bourgeois historians as totally useless to Marxist scholars. He was labeled a peddler of antihistoricism—the greatest condemnation of a Marxist. By deed and by theoretical inference, he identified himself with N. I. Bukharin and Leon Trotsky, the alleged archenemies of Leninism-Stalinism. The growing chorus of emotionally charged criticisms presented Pokrovskii as a synthesizer of many theoretical and methodological sins. He was depicted as a bizarre and tragic example of what happens to Marxist scholars unwilling to toe the line of Marxist orthodoxy as interpreted by Stalinist ideologists.

The brutal attack on Pokrovskii was double edged. On the one hand, it marked the climax of a long-brewing resistance among the community of historians to the intellectual authoritarianism of Pokrovskii and his school. On the other hand, it was one of many moves Stalin made to bring unity to the social sciences by making them subservient to the general theory of historical materialism. In all this, the historians had little choice: they were disencumbered of the sociological bias of the Pokrovskii school, which was never so rigid as to be seriously harmful, only to be pushed headlong into the ironclad structure of Stalinist ideological dictates, which were completely closed to critical scrutiny by the scholarly community.[146] The result of the attack on Pokrovskii was that a challengeable and constantly eroding authority in historical scholarship was replaced by an unchallengeable authority bent on dominating every phase of academic life.

The attack on Pokrovskii was the first large-scale involvement of the academic community in an enterprise forced on it by outside authorities—in this case by the high command of the Communist party. It was a major step in the realization of the Stalinist goal of achieving unity between science (in all its branches) and ideology as the basic prerequisite for the unity of Soviet society. The most alarming feature of the anti-Pokrovskii crusade was the large number of historians who spared no arguments in damning their fallen comrade. The campaign made history a servant of ideology: from then on, historical scholarship was guided by the idea that all deviations from the official theory of history were deeds beneficial only to "the enemies of the people," a label reserved for the critics of Stalinism. The triumph of Stalinism represented the end of critical inquiry into the theory and methodology of historical and sociological scholarship.

The campaign to make the Stalinist breed of historical materialism the grand theory of social structure and social change removed the need for a further search for general theoretical postulates in ethnography, political science, social psychology, and economics. The grand theory, now given a fully crystallized form, tolerated no challenge; it had the authority of Scholastic philosophy, backed by the coercive power of the state. Marxist officialdom was willing to admit that historical materialism did not rest on absolute principles, impervious to the vicissitudes of the human condition; but, at the same time, it reserved for itself alone the right to make theoretical changes. If theoretical interests continued to draw the attention of social scientists, they were limited to clarifications and amplifications of Stalinist thought. In the broad issues of social analysis, scholars were given not only a general and authoritative outlay of theory but also the questions to be asked and the answers to be given. For serious scholars this was the beginning of a long period of retreat into obscure research areas far removed from the pathways of Stalinist theory.

While the Stalinist codification of historical materialism crippled historical scholarship, it completely wiped out sociological theory. The new creed ruled that Soviet society—as a socialist society—was dominated by complete consonance between the "means of production" and the "social relations in production" and between the "base" and the "superstructure." Soviet society did not need sociology to study its structure because the latter was stated clearly and authoritatively in the official interpretation of historical materialism. Nor did it need sociology to study deviations from structural norms, for such were not recognized. Social deviance that did exist had roots in residual survivals of "precapitalist mentality" among certain classes—such as the peasants—rather than in the structural makeup of Soviet society. Crime and other kinds of deviance only illumined the structure of pre-Soviet society. Stalinist thinking allowed no room for a general science of society unrestrained by ideological imperatives.

During the 1920s it was customary for Marxist theorists to examine the ideas of such stalwarts of modern sociological thought as Georg Simmel, Émile Durkheim, Max Weber, and Karl Mannheim—all as part of an effort to arrive at a general theory of Marxist sociology. During the 1930s interest in the work of these sociologists gave way almost entirely to another interest: pre-Marxist sociology as embodied in the work of the forerunners of modern socialist thought. The readers of *Under the Banner of Marxism* could now satisfy their interest in sociology by reading articles on the social theories of the likes of Adam Ferguson, Charles Fourier, and Robert Owen. They could also read papers clarifying and amplifying such topics presented by the official interpretation of historical materialism as "the relationship of 'base' to 'superstructure' in human society," "the role of the individual in history," "the forms of social consciousness," and "the place of the intelligentsia in socialist class structure."

PRESSURES, PURGES, AND ADAPTATIONS

The Academy did not escape the mammoth political purges of 1936–1939, which swept the country like a hurricane and left untold thousands of victims and a bewildering institutional dislocation in its wake. The first sign of the forthcoming purges in the Academy came in 1935 when N. P. Gorbunov, a Communist since 1917 and at one time Lenin's assistant, was appointed permanent secretary of the Academy. To bypass the formal procedures for such an exalted appointment, he was made an academician on direct orders by the government. His scientific fame rested primarily on his participation in the first successful ascent of the highest peak among the Pamir Mountains, which he promptly named Stalin Peak (and was subsequently renamed the Peak of Communism). Upon assuming the new position, Gorbunov stated his major task by announcing publicly that the Academy had still a long way to go to catch up with "general development of the country."[147] But he disappeared in 1937, just as the purges had begun to gather momentum; many years later it was revealed that he too was a victim of Stalinist terror.[148]

Since Gorbunov was not replaced, the time-honored position of permanent secretary—modeled on the corresponding position in the Paris Academy of Sciences—went out of existence. For the time being, the Presidium, in which scholars with proven loyalty to the regime formed a clear majority, became the government stronghold in the Academy. The purges left deep scars on every component of the academic body. As if to justify the intensity and the sweep of the purges, V. L. Komarov, the new president of the Academy, stated in 1937—on the occasion of the celebration of the twentieth anniversary of the October Revolution— that during this entire period the Academy was "the breeding ground of counterrevolution."[149] While the academic purges of 1929–1931 concentrated on non-Marxists, those of the late 1930s were limited mainly to active or former members of the Communist party. The list of the full members of the Academy who were executed or repressed in a way that led to their de facto removal from the roster of academicians during the late 1930s included—in addition to Gorbunov and N. I. Bukharin—the social historian N. M. Lukin, the agricultural economist N. M. Tulaikov, and the literary critic I. K. Luppol.[150]

Bukharin was a victim of the political and ideological housecleaning staged for the purpose of strengthening the absolutism of Stalin's rule. Among the charges leveled against him were allegations of misdeeds directly affecting the Bolshevik spirit of handling institutional matters of science. A specific accusation stated that, as the head of the research sector of the Commissariat of Heavy Industry, he divided all sciences into "primary" and "secondary"—the former category including theoretical activities and the latter covering applied research. Bolshevik critics contended that the treatment of applied-science research units as "secondary" bodies went against the Bolshevik policy of placing primary

emphasis on practically oriented research—or at least against the Bolshevik style of harmonizing theoretical and practical research. Bukharin
was also blamed for using the journal *Socialist Reconstruction and Science,*
which he edited, to publish articles critical of the science policy of the
Soviet government. The official accusation noted that Bukharin was the
leader of subversive activities in the Academy; at the May 1937 session
of the Academy's General Assembly, N. P. Gorbunov made the open
charge that Bukharin was directly responsible for making the Department of Social Sciences—particularly the institutes of history and ethnography—a haven for "Trotskyites and right deviationists."[151] None of
the accusers, however, made an effort to substantiate the charges against
Bukharin.

Mark B. Adams has shown that, in and outside of the Academy,
biologists who avoided ideological involvement of any kind fared much
better than their colleagues who actively supported Marxist theory and
participated in the execution of government policy. A. N. Severtsov,
who never spoke in support of the Soviet government and Marxism,
had no apparent trouble with the authorities. I. I. Shmal'gauzen was
neither a Marxist nor a supporter of the official policy that made dialectical materialism the exclusive philosophy of the Academy; yet at this
time he managed to escape harassment by the authorities and the Com-
munist-led mass organizations. Indeed, in 1935 he was elected to full
membership in the Academy. In contrast, N. I. Vavilov and S. G. Levit,
closely identified with the government, died in prison.[152] I. I. Agol, a
dedicated Marxist biologist, disappeared in the Stalinist purges.

Many victims of the Stalinist purges were sentenced to death and
were executed. Others simply vanished and were never seen again. Still
others eventually reappeared—some of them years after the end of the
Stalin era. But there were also victims who were neither executed nor
sent to labor camps, but who were subjected to harassment of colossal
proportions. Party authorities went so far as to charge that the "Olym-
pian detachment" of scientists from the crucial social and economic questions of the day made it easier for various "Trotskyite-Zinovievite
bandits" to transform the Academy into their stronghold.[153]

Academician Luzin was singled out for a particularly bitter attack.
Luzin was the founder of the widely heralded Moscow school of mathematics that distinguished itself by general work in set theory, topology,
probability theory, and the foundations of mathematics.[154] In 1936, at a

time when the purges had moved into high gear, he became the model for the epithet *"luzinshchina,"* a type of unpatriotic behavior. He "sinned" by publishing his most important scholarly works abroad and by expressing rather uncomplimentary views on the future development of science in the Soviet Union. To add "academic" weight to the charges, the accusers contended that Luzin recommended mathematical textbooks of low quality for use in university classes, that he approved inferior dissertations for higher degrees, and that he was guilty of making the discoveries of his students appear as his own.[155] *Pravda* told its readers that the case of Luzin showed clearly that the "enemy" did not withdraw from the battlefield but, on the contrary, had resorted to more skillful and devious ways of fighting the established order.

Luzin survived the verbal lashing and the indignity of exclusion from the *Great Soviet Encyclopedia*, but, from that time on, Soviet scientists found it ill advised to publish their papers abroad or to participate in the work of Western laboratories and other research centers. At a special meeting of Moscow University mathematicians, A. V. Kolmogorov, an internationally reputed scholar, apologized publicly for his past publications in foreign journals and promised to write a Russian textbook in probability theory.[156] At the same time, Western scholars, discouraged by the Stalinist purges and by growing efforts to isolate the Academy from the world abroad, ceased to visit Soviet research centers and scientific conferences. Bohr, Dirac, and many other eminent scientists who visited the Soviet Union during the 1920s and early 1930s were no longer invited to the country, nor were they especially eager to go there.

As the names of purge victims reached the scientific community, individual defenders of Marxist orthodoxy tried to establish a causal relationship between the "political crimes" committed by condemned persons and their allegiance to alien philosophies of science. In 1938 M. P. Bronshtein, a promising young physicist working on the general theory of relativity, was executed as "an enemy of the people."[157] This did not surprise A. K. Timiriazev, an old-line physicist deeply disturbed by the theoretical claims of quantum mechanics and the theories of relativity, who lost no time in identifying Bronshtein's political disloyalty and "idealistic" leanings in physics as two sides of the same coin.[158] At the same time, the philosopher E. Ia. Kol'man came forth with a general claim that the Soviet advocates of "physical idealism" were closely linked with "anti-Soviet views and activities."[159]

Despite these and similar pronouncements, philosophical orientation in science was not an issue of decisive importance. There were individuals, typified by A. F. Ioffe and V. A. Fok, who were targets of vicious attacks by the more bellicose Stalinist philosophers, but who escaped recriminatory police action. After 1935 the philosophers who considered themselves defenders of Stalinist orthodoxy treated Fok as a leader of the idealistic wing among Soviet physicists. Nevertheless, in recognition of his scientific work, the Academy's General Assembly elected him a full member in 1939. But there were physicists, typified by L. D. Landau and P. I. Lukirskii, who steered clear of philosophical involvement and yet did not escape imprisonment. It was generally true that the scientific views of condemned persons became a matter of serious and defamatory questioning; it was equally true, however, that this came only after these persons were attacked on political grounds. At this time, many key questions on the relationship of dialectical materialism to science were still unsettled. Stalin was willing to delay the settlement of "philosophical" questions, particularly when they did not adversely affect the key issues of ideology; he was not willing to postpone the settlement of key "political" questions, particularly those related to the consolidation of his monolithic power.

The purges reached every stratum of the Academy's hierarchy, from the full members to the lowly research assistants. The condemned persons represented every age group, every social class, and every field of scholarship. The purges caused untold suffering and took many innocent lives, but they also produced many heroes—brave scholars who were willing to seek help for victims and who defended their own scientific views that were incompatible with Marxist orthodoxy. V. I. Vernadskii, N. I. Vavilov, D. N. Prianishnikov, and A. F. Ioffe, the four pillars of the Academy, spent endless hours pleading with police authorities to release their research associates from prison. P. L. Kapitsa combined pleas with threats in an effort to persuade the authorities to free the imprisoned L. D. Landau, the rapidly rising star of Soviet physics.[160] Many years later—in the post-Stalin era—Landau wrote that Kapitsa's efforts on his behalf required "superb courage, great humanity, and crystalline integrity."

When M. P. Bronshtein was arrested in 1937, the physicists S. I. Vavilov and V. A. Fok appealed to the procurator of the USSR for clemency; they appended a laudatory evaluation of the prisoner's contributions to physics.[161] N. K. Kol'tsov, when condemned by the Presidium of the Academy and the local party organization for the "idealistic" bent of his views on genetic theory, answered that he did not intend to retreat from his theoretical views and was ready to continue "to fight for science."[162] The American geneticist L. C. Dunn has written that Kol'tsov's paper on "the molecules of heredity," published in 1939, occupied an important place among the studies in the late 1930s that marked "the beginning of the period of transition from classical, or formal, genetics to molecular genetics as it exists today." Kol'tsov, in Dunn's view, pointed to "the converging research of physical chemists and of geneticists, working apart from each other, which centered on the idea of a long polypeptide with millions of isomeric parts."[163]

The climate created by the reign of terror produced not only victims and heroes but also opportunists who were ready to take full advantage of the new situation. T. D. Lysenko was the classic example of this type. A hero of the party press for many years and president of the Lenin Academy of Agricultural Sciences since 1938, he specialized in attacking modern genetics and extolling so-called Michurinist biology, based on the principle that man could improve the varieties of domestic plants and animals simply by improving the environment in which they lived. He emphasized the futility of genetics as a source of practical knowledge required by socialist agriculture, and with equal equanimity he defended the Lamarckian principle of the inheritance of acquired characteristics. He knew that he could succeed only by hiding experimental evidence for his sweeping "agrotechnical" claims, and he sought recognition and glory by presenting his orientation as a foolproof application and confirmation of dialectical materialism.

Conducting a carefully designed propaganda campaign in which he received strong support from Stalinist philosophers, Lysenko hammered away at three general ideas, all appealing to the ideological sensitivities of the political leaders. In the first place, he emphasized the materialism of his theoretical orientation—primarily by underlining the "idealism" of his enemies, the masters of modern genetics. Among its many other shortcomings, in his view modern genetics denied causality as a key explanatory principle in the natural sciences; instead of the unpredictability of mutations as a source of organic change (as stated by genetics), Lysenko stressed the predictability of directed evolution—evolution planned and engineered by man.

In the second place, he stressed the direct utilitarian orientation of his scientific concerns. His work, he said, was attuned to the dictates of the five-year plans; genetics, in contrast, suffered from too many theoretical ambiguities and pseudoscientific assumptions to be in a position to tackle questions of a practical nature. Although he did not present the necessary evidence, Lysenko made widely publicized claims that his work had made Soviet harvests much more abundant. He exploited the gullibility of the propaganda machine behind Stalin's economic policies, which was much more inclined to air news on concrete accomplishments in the economy than to examine their scientific bases and certify their accuracy.

In the third place, Lysenko played up the nationalism of his scientific orientation. Genetics, he claimed, was a science alien to Russia, for it was rejected by the leaders of Russian biology. His theory, however, had deep roots in the Russian scientific tradition: it combined K. A. Timiriazev's interpretation of Darwin's theory, V. V. Dokuchaev's and V. R. Vil'iams's soil science, and I. V. Michurin's hybridization techniques. He went along with Michurin's translation of "the unity of organisms and the environment" into the unity of Soviet organisms and the Soviet environment. Western breeders, wrote Michurin, could not be of much help to Soviet breeders for the simple reason that their methods applied to climatic conditions "completely different" from those in the Soviet Union.[164] The absence of references to the Western progenitors of his "agrobiological" theory was clearly the most telling feature of Lysenko's published papers.

The ecstatic publicity given to Lysenko's scientific claims reached a high pitch in October 1939 at a conference on the place of genetics in modern biology, sponsored by the journal *Under the Banner of Marxism*. Lysenko and N. I. Vavilov were the main speakers, each clinging to firmly crystallized ideas. The gallery was stacked with Lysenko's adherents, who viewed the conflict between Lysenko and his enemies as a political confrontation of large magnitude. The philosophers, led by M. B. Mitin, played the major role in producing a picture of Lysenko as an exemplary member of the scientific community. A correspondent of *Under the Banner of Marxism* summed up the proceedings of the conference by stating that Vavilov—"a leading scholar"—did not meet the expectations of the audience: he did not give a deep and critical analysis of the central questions of the dispute; he did not describe the present situation in biology; and he did not offer "a resolute self-criticism." Instead, he gave a speech that was suffused with "piety for foreign science" and "arrogance toward the innovators in Soviet science."[165]

Lysenko was presented as a pristine expression of Soviet patriotism and scientific materialism. His views were the views of the Soviet state and the Soviet people; they were also the true voice of the Russian tradition in science. Mitin took pains to present himself as a person of broad vision and judicious thought; he chastised both men for excesses in formulating their ideas and planning their action. But this was only a ploy, for he used the opportunity to proclaim Lysenko as representative of the true pulse of Soviet science. In Mitin's view, Vavilov represented the scholars clinging to "antiquated ideas" and resisting the efforts "to build new traditions, new norms and new organizations."[166]

The new-found glory helped Lysenko to pressure the Academy of Sciences to appoint special commissions to investigate the Institute of Evolutionary Morphology and the Institute of Experimental Biology, which treated genetics as a key source of information relevant to the theory of biological evolution.[167] A newly elected full member of the Academy of Sciences, Lysenko could at last concentrate on exposing the "unpatriotic" work of Soviet geneticists and on attacking the metaphysical delusions of the masters of Western genetics. He worked to make genetics an illegal discipline that should be erased as a research field and as part of the school curriculum. Commenting on Lysenko's election, V. L. Komarov, president of the Academy, stated:

> Growing in many directions, modern biology [in the Academy of Sciences] concentrates on giving Darwin's legacy more depth and more extensive application. Well known in our country are the names of the biological wing of our Academy—T. D. Lysenko and N. V. Tsitsin. These are people's scientists, dedicated to advancing the glorious traditions of K. A. Timiriazev and I. V. Michurin. They come from the people and are closely linked with the people. Their entire endeavor is devoted to the struggle for improving the yields of socialist fields.[168]

Lysenko concentrated his heaviest attacks on N. I. Vavilov, a master and chief defender of modern genetics. With the help of his philosophical advisors, the party press, and scholarly journals under his direct control, he presented Vavilov as a synthesis of attributes standing in direct opposition to Soviet patriotism. In addition to the "impracticality" and "idealism" of his scientific activity, Vavilov was now found guilty of maintaining close ties with Western scholarship, publishing scholarly papers in Western journals, participating in too many international scientific undertakings, and advocating the sending of young Soviet scientists to Western laboratories for additional research experience and exposure to modern techniques of experimental inquiry. Lysenko and his top advisors provided journalists with endless innuendos about Vavilov's bourgeois origin, Western mode of behavior, and ephemeral involvement in socialist construction. At the same time, Lysenko was more than willing to feed the popular press with information on his own personal and professional attributes that made him a true representative of Soviet-bred scientists. Having started his scholarly career during the Soviet period, he avoided and scorned all contact with the outside world, believed in the possibility of developing Soviet science as a unique historical and cultural phenomenon, and heeded Stalin's admonition to scientists not to become "slaves" of calcified ideas and vested academic interests, the main obstacles on the road to full socialism.[169]

Lysenko exploited every opportunity to present himself as a true personification of Stalin's notion of "Soviet intelligentsia," a social stratum combining a mastery of modern technical knowledge with ideological purity and a critical attitude toward tradition. On May 17, 1938, Stalin announced that Soviet-bred scholars had brought an end to the alienation of the scientific community from the society at large. "Old" science, according to Stalin, was dominated by "a priesthood of science," who, in search of a monopoly on knowledge, kept young scientists away from positions of authority and influence. "Old" science, as now interpreted, was tradition-bound and incapable of leading to previously unheralded treasures of secular knowledge. New scientists, as depicted by Stalin, were ready to ally themselves with older scientists, but only on the former's own terms. Their major job was to discard traditions shorn of pragmatic value. Party press and Communist discussion groups were quick to recognize Lysenko as an exemplar of the "new man" in Soviet science and to view Vavilov as a tradition-bound scholar with no living ties to the new generation of scientists.[170]

Lysenko's propaganda machinery, whose voice reached every corner of the country, created an image of Vavilov as a pseudoscientist and an unpatriotic citizen. Once the pride of Soviet science and the scientific community, Vavilov now became a target of daily surveillance by the secret police. In August 1940, on a research trip to the western Ukraine, he disappeared in the hands of the police. In July 1941 he was tried on charges of belonging to an international conservative organization operating from England and of deliberately sabotaging Soviet agricultural policies. The death sentence imposed by the court was later commuted to life imprisonment. He died in 1943, and the police record shows pneumonia as the cause of death. His life ended in Saratov, the city where in 1920 he read his paper presenting the law of homologous series in the variation of heredity, a contribution that catapulted him to the forefront of modern pioneers in genetics.

Lysenko provided the most striking example of a scholar who made "patriotism" a profession and who acquired extraordinary academic powers through support from authorities external to the Academy. He operated on two fronts: he joined Stalinist propagandists in describing the attributes of an ideal Soviet scientist, and he worked with unswerving determination to present himself as the purest personification of such an ideal. The manipulated press acted quickly in making him a paragon of scholarly devotion and achievement. This, however, did not make him a typical representative of the members of the Academy. After the elections in 1939, which brought the number of full members to 136, there were six additional categories of academicians, judged according to personal adjustment to the ideological and practical demands of government authorities.

The first and largest group consisted of persons who neither identified themselves with dialectical materialism nor expressed philosophical views contradicting Marxist theory. The growing Stalinist oppression made the professional existence of these scholars increasingly more precarious and subject to external criticism, particularly on issues of a philosophical nature. In general, however, the professional ideologues treated them as exponents of "natural-science materialism" and a force that did not encourage divisiveness on the philosophical front. At this time, to be a quiet non-Marxist was much more tolerable than to be a Marxist revisionist. In the field of biology, these academicians faced the threat of Lysenkoism, but Lysenko's influence was not yet strong enough to suppress heterodox ideas. The biologist I. I. Shmal'gauzen was a typical member of this first group: Lysenkoist criticism of his theoretical positions did not interfere with his deep engagement in the study of the distinct roles of "natural selection" and "mutation" in organic evolution or with his important role in the formulation of the theoretical edifice of "synthetic evolution."

The second category of academicians was made up of individuals who did not hesitate to endorse dialectical materialism as a philosophy of science but who knew little about Marxist philosophy and found no place for it in their scientific work. A. N. Kolmogorov, a typical representative of this group, did not hesitate to contribute to Marxist journals and symposia, always on topics in mathematics. He wrote the article on mathematics for the *Great Soviet Encyclopedia*, in which he affirmed, but did not explore, the interdependence of modern mathematics and dialectical materialism.[171] In none of these writings did he show even the most elementary familiarity with Marxist theory. His professional papers, which made him one of the most eminent mathematicians of his generation, not only ignored dialectical materialism but, in some respects, worked against it. At a time when Marxist writers were busy attacking probability theory as a mathematical adaptation to the chaotic nature of capitalist economy, he worked on giving this theory an axiomatic basis. And, at a time when Marxist writers condemned intuitionism, a general orientation in the foundations of mathematics advanced by L. E. J. Brouwer and H. Wyel and alien to dialectical materialism, he acquired international reputation by his effort to add new strength and breadth to it. His popular papers endorsed Marxism, but only in a most generalized form; his professional papers disregarded Marxism altogether.

The third group consisted of persons who became academicians as rewards for work on large construction projects and who played key roles in government efforts to make applied science the mainstay of the Academy. They were the top layer of the rapidly rising technical intelligentsia, the human power behind socialist construction in the age of Stalin. Not embarrassed by their total ignorance of Marxist theory, they were first and foremost academic politicians who articulated and propagandized a complex network of functional links between the Academy and the national five-year plans. Honored for the deeds of the past, they had little to offer to the Academy's deep commitment to fundamental science. Typical representatives of this group were G. L. Krzhizhanovskii and I. P. Bardin.

The fourth group included a relatively small number of established scholars who tried to establish working relations between dialectical materialism and science. These scholars were mainly physicists, workers in a discipline that spearheaded the modern revolution in science and occupied the leading position in Lenin's effort to formulate a modern Marxist philosophy of science. They took on the difficult task of protecting their discipline from mounting attacks by professional philosophers inclined to see an idealistic aberration in every great theory of modern physics. They resisted the philosophers who made physics the main battlefield of opposed ideologies and who confused the philosophical views of Western physicists with their own scientific ideas. Their major aim was to protect scientists from harassment by zealous philosophers and to reverse the trend of growing animosity toward the leaders of Western science. Represented most consistently by A. F. Ioffe and V. A. Fok, this group was exposed to unremitting verbal abuses by the more belligerent philosophers. These distinguished scholars received strong support from the growing ranks of young scientists who made the 1930s an era of lively and inordinately productive activity, particularly in physics and related sciences. This group became the true core of the scientific community and the first line of defense against uninvited political interference.

The fifth group consisted of Stalinist philosophers and their close allies. Ideologically, this group was united in extolling Leninism as "the new phase" in the history of Marxist theory. Professionally, it was united by a determination to keep the area of contact between science and philosophy as wide as possible. But its philosophy was narrow—consisting of a few epistemological principles spelled out in *Materialism and Empirio-Criticism*—and its knowledge of the revolutionary advances in modern physics was even narrower. As mediators between ideology and science, these philosophers became the defenders of conservatism in the Academy. The leader of the group was M. B. Mitin, who in 1939 lined up Soviet philosophy behind Lysenkoism and brought the era of Marxist tolerance of diverse views in biology to an end. A thinly veiled criticism of selected theoretical positions of both "formal genetics"—or classical genetics—and Lysenko's Michurinism led him to conclude that, in essence, the first orientation was deeply steeped in "mechanical materialism," and the second was closely allied with "dialectical materialism."[172] While the first orientation, as he saw it, represented regressive and antiquated scientific ideas, the second represented a "progressive step in the evolution of modern biology." Mitin and his associates helped to accelerate the process of alienating scientists from philosophy; they made philosophy both a bastion of Stalinist ideology and a disruptive force in the evolution of true scientific thought.

Behind Mitin and a small group of other academician-philosophers was a swarm of lesser philosophers—all brought in by the absorption of the Communist Academy—who were particularly boisterous in defending their orthodoxy, and who took pride in leveling all kinds of criticism against the leading physicists. Their pet targets were the epistemological "relativism" built into Einstein's theories, the "agnosticism" of Heisenberg's indeterminacy principle, and the logical and methodological "absurdity" of Bohr's principle of complementarity. It was because of the destructive work of these philosophers that V. A. Fok asserted in 1938 that the future of physics in the Soviet Union depended on the discipline's complete separation from philosophy.[173]

The sixth group was actually a dyad: it was made up of the neurophysiologist I. P. Pavlov and the geochemist V. I. Vernadskii, both of whom voiced open opposition to certain features of the national science policy and entertained philosophical views alien to Marxist thought. They resented tying the professional work of individual scholars to the national five-year and annual economic plans. They criticized the policy that, in their opinion, made scholarly achievement only one of the criteria in ranking the candidates for academic positions. And they made no effort to avoid public manifestations of their dissatisfaction with the Bolshevik takeover of the Academy.

In 1929 extensive preparations were made under the aegis of the Academy to celebrate Pavlov's eightieth birthday. The government ordered that all efforts be made "to create the most favorable conditions" for research in Pavlov's laboratories.[174] The presidiums of the Leningrad Regional Executive Committee and the Leningrad Regional Council were lavish in their praise of Pavlov's "gigantic work for the benefit of the national economy and the creation of a new culture—a culture of free and conscious labor, depending on the best achievements of human knowledge."[175] Nevertheless, Pavlov, obviously distraught by actions at the Academy, announced that he had decided not to attend any celebration in his honor. Not only did the government take no umbrage at Pavlov's provocative behavior but, on the contrary, it intensified efforts to make the theories of the great neurophysiologist a constituent part of dialectical materialism.

Vernadskii, in particular, disagreed publicly with the rising intellectual imperialism of dialectical materialism as the official—and the only—philosophy of the Academy. Not in Marxist thought, anchored in philosophical materialism, but in philosophies of idealistic and positivistic leanings did Vernadskii find preferred sources of scientific stimulation. Indeed, he made no secret of his feeling about the futility of efforts to perpetuate the nineteenth-century alliance of philosophical materialism and science. In philosophical commentaries of Henri Poincaré, Henri Bergson, Alfred Whitehead, and Bertrand Russell he detected fertile hints for ideas built into, and elaborated by, the revolutionary turns in modern science. He was much closer to Ernst Mach's epistemological relativism than to Lenin's epistemological absolutism.

The government made no effort to penalize Pavlov or Vernadskii; on the contrary, it showered them with privileges not shared by other academicians. For example, both men were given the unrestricted right to travel abroad to attend scholarly meetings and to visit research centers, and, in Vernadskii's case, to visit family members who had emigrated. By acting the way it did, the government was mindful of the international prestige these two men brought to Soviet science. Pavlov's work received worldwide recognition as a turning point in the evolution of modern neurophysiology and behavioral psychology. Vernadskii earned international repute through his pioneering work in two new disciplines: geochemistry and biogeochemistry. He must be counted among the founders of the systematic scientific study of the biosphere. The Communist leaders appreciated Vernadskii's consistent dedication to the practically oriented study of the country's natural resources. He was one of the rare professors favorably referred to by Lenin in *Materialism and Empirio-Criticism*. The academic community looked at Pavlov as the zenith of the national achievement in science; it looked at Vernadskii as the purest expression and the most consistent champion of the principles built into the moral code of science. The government, moreover, was aware of the advanced age of the two men: when Pavlov died in 1936 at the age of eighty-seven, Vernadskii was seventy-three years old.

These six groups were neither fixed nor all-inclusive. The boundaries separating them were not always clear. At the end of the 1930s, a particularly striking development signified the arrival of a new breed of academicians: scholars who chose to join the Communist party, even though they showed no familiarity with Marxist theory and had made no effort to make their scholarly work part of the relentless campaign to make dialectical materialism the "general methodology" of science. This group included, among others, the geographer A. A. Grigor'ev, the ecologist V. N. Sukachev, and the zoologist E. N. Pavlovskii. Some academicians made this step out of sheer patriotism; others found this to be an effective way of overcoming constant anxiety and personal insecurity; and still others acted out of a desire for professional success.

In one important respect, the members of the Academy—not including the elite of the Technical Department, where the lines were not clearly drawn—fell into two groups, depending on whether their scholarly achievement was certified by the scientific community or by the Communist party. The latter group consisted of individuals who received accolades from the Communist press—which treated them as model socialist scientists—but who received limited recognition from their peers, certainly not enough to warrant their treatment as the true leaders of Soviet science. O. Iu. Shmidt, V. R. Vil'iams, and N. Ia. Marr were typical members of this group.

Shmidt, a mathematician by training, was one of a handful of scientists who joined the Communist ranks in 1917. For the rest of his life, he was showered with special favors for this deed—usually favors of his own choosing. Taking advantage of his privileged situation, he sought activities full of glamour and excitement. Although he never mastered Marxism, he was in charge of the natural-science branch of the Communist Academy and the chief editor of the *Great Soviet Encyclopedia* and *Nature*, a popular scientific journal. In the early 1930s the new regime in the Communist Academy, engaged in an effort to unite the ranks of Marxist philosophers by burying the last remnants of "mechanical" and "dialectical" orientations, censured Shmidt's journal for minor and unintentional philosophical digressions.[176] Unlike a vast majority of deviators, however, he was quickly exonerated and reaffirmed as a hero of socialist science.

Shmidt organized an exploratory expedition to the Pamir Mountains and was in charge of the alpine section, which provided the most exhilarating experience in mountain climbing. Subsequently, he participated in six Arctic expeditions and had a newly discovered island named after him. In 1932 he realized M. V. Lomonosov's dream of making an uninterrupted northern sea voyage from Arkhangelsk to the Pacific Ocean. In 1938 Shmidt founded the Institute of Theoretical Geophysics, even though geophysics, geology, and physics were not his areas of competence. Although more exciting activities took him away from sustained work in mathematics, he managed in 1931 to hold a seminar on group theory at Moscow University, apparently an event of sufficient moment to warrant his being proclaimed a founder of a new school in that branch of mathematics. This passing and rather tenuous effort brought other favorable results: in 1933 he was elected a corresponding member and in 1935 a full member of the Academy of Sciences. Shmidt's various "scholarly" activities brought little benefit to science but an abundance of favorable publicity to himself. While his scientific work received little attention in the scientific community, the Communist press made him a symbol of the most sublime achievements in making science the propelling force of the new civilization. In the middle of the Stalinist purges, which affected the academic structure in its entirety, a party journal wrote:

> Soviet science has presented the world with examples of exceptional love for work and of full scientific enthusiasm inspired by unbounded loyalty to the socialist land. A model of such Bolshevik enthusiasm is provided by O. Iu. Shmidt's polar expedition, which combined careful preparation and sober scientific work with the heroic deed of four courageous sons of the Soviet land who spent a winter on drifting ice. This extraordinary devotion to the motherland and love for science— this harmony of most thorough preparation for and Bolshevik handling of a scientific undertaking—must become the style of work in all our scientific institutions and of their workers.[177]

N. Ia. Marr was a member of the Academy from 1912 to 1934, the year of his death. During the 1920s he was the only full member of the Academy to join the Marxist ranks. In 1930 he became a member of the Communist party and quickly rose on the ladder of Communist organizations. In 1931 he was made a member of the All-Russian Central Executive Committee and was given the Lenin Medal for achievement in science. His Marxist romance began in 1919 when, in return for his profusely manifested loyalty to the Soviet system, he was awarded sufficient funds to establish the Academy of the History of Material Culture, an institution that combined research in archeology with linguistic studies. Because of certain rudimentary similarities between his views and Marxist theory, Communist scholars were quickly accustomed to treating his theory as Marxist linguistics. Marxists were attracted to his Japhetic theory, which sought a common base for all the Caucasian languages, for Caucasian and Semitic languages, and for all languages of the world—the ideas of common origins and evolutionary unilinearity having been firmly built into Marr's thought and into Stalin's internationalism of the 1930s. Japhetic linguistics, or Japhetidology, started as a unique variety of Caucasian linguistics and ended as a general theory of "the origin, development, and future of language in relation to the stages of development of human society and consciousness."[178] Marxists also liked Marr's emphasis on the role of social classes in the development of language.[179] M. N. Pokrovskii, president of the Communist Academy, wrote in 1927: "As stated correctly by a Leningrad comrade, Marr's theory must recognize Marxism as its general philosophical and sociological base and Marxism must recognize the Japhetic theory as its special linguistic division."[180] Soon after Pokrovskii's overture, Marr was made a member of the Communist Academy and the head of its "subsection of materialistic linguistics."[181] An official pronouncement made it known that the time had come to accelerate the development of linguistics by combining Japhetidology with Marxist theory.

Marr's theory became the central theme of the Marr school in linguistics, which in turn aspired to become the official—and the only—Marxist linguistics. Although the latter effort encountered much resistance from Marxist as well as non-Marxist scholars, the Marr school succeeded in suppressing—but not eradicating—the most formidable enclaves of organized opposition.

After Marr's death—and the long string of commemorative meetings that contributed to portraying him as a builder of modern Marxist linguistic theory—the Marr school became more firmly entrenched and more ruthless in its war against real or potential enemies. The Stalinist claim that the theoretical unity of science was the basis for the ideological unity of society proved to be of great help to the Marr empire. The school was helped also by its inner unity and by the lack of unity in and among opposing groups. During the 1930s the leading philosophers took it upon themselves to demonstrate the purity and thoroughness of Marr's Marxist affiliation and to make his theories the backbone of Marxist linguistics. A combination of philosophers and linguists with vested interests in perpetuating Marr's theory made the Academy the center for a linguistic theory built on shaky foundations and unsupportable claims. The school grew because it received help from outside the scientific community. While Marr tried to adjust his theoretical views in linguistics to Marxism, his school concentrated on making these views the official Soviet linguistics and on eliminating all opposition. In pursuit of this goal, the Marr school relied heavily on the Marxist press, which portrayed Marr as a scientist of the highest distinction and as a peerless leader in revolutionary scholarship.

V. R. Vil'iams was a soil scientist whose work attracted little attention in the scientific community.[182] The reasons for this disregard were

Vil'iams's primary concentration on textbook writing and his closer ties with "practical knowledge" and "common sense" than with theoretical and experimental developments in science. Other reasons were the turnabouts in his views and the eclectic cast of his theory. At first he was concerned with the physical aspects of soil and the role of physical processes in soil formation, but later he shifted his emphasis to the role of microorganisms in the formation and fertility of soil. At first he sought a general theory of the evolution of soil; subsequently his main interests were in agrotechnology—in the techniques of soil conservation and improvement. He defended the so-called grassland system of rebuilding agricultural land; in the periodic planting of cereal grasses he saw the most reliable method of reestablishing the "structural equilibrium" of soil and ensuring its continued fertility.

On the occasion of the tenth anniversary of the October Revolution, specialists in soil science published a series of booklets on the history and major interests of their discipline. These studies made almost no reference to Vil'iams's work, and of the few that did refer to it, some were critical. One expert stated that Vil'iams's theory of natural humic acids—his major claim to fame—must not be taken seriously because it was not supported by experimental data and because the methodological procedures that led to its formulation were not stated.[183] In 1927 K. D. Glinka, a leader in the field, wrote a name-studded progress report on Soviet work in soil science during the first ten years of the new regime, but he made no reference to Vil'iams. Clearly, Vil'iams was not accepted by the scientific community as a noteworthy contributor to soil science.

In search of heroes in agricultural technology and economics, the party press stumbled upon Vil'iams's tireless and ambitious work on improving the quality of cultivated land. This publicity catapulted Vil'iams to a position of national visibility and to the ranks of "revolutionary scientists." He responded to favorable publicity by giving more breadth to his writing—in 1927 he published *General Geography and the Foundations of Soil Science*—and by making his writing simple, optimistic, and deeply imbued with socialist ideals. He joined the Communist party in 1928, at a time of widespread purges of disloyal intelligentsia. A staunch supporter of Stalin's collectivization moves in agriculture, he asserted that the larger the farms, the more efficient the application of modern techniques of soil conservation. In 1931, at the age of sixty-eight, he was elected a full member of the Academy of Sciences. To Stalin and Stalinist ideologues, Vil'iams represented the purest and noblest dedication to the ideology of socialist science and to the practical needs of collectivized agriculture. To the scientific community, he appeared as an overambitious scholar with a flair for sweeping generalization but with little patience for tedious work of an empirical nature.

The rapid rise of Lysenko and Lysenkoism after 1935 helped to make V. R. Vil'iams a leader of the "Soviet orientation" in soil science and agricultural technology. Like Lysenko, Vil'iams combined a belief in the unlimited potentialities of directed organic evolution with the view of biology as a discipline fully independent of the methods of physics and chemistry. He quickly emerged as an unofficial leader of the ongoing campaign to abandon the use of chemical fertilizers in Soviet agriculture. At a time when Vil'iams's authority appeared to be unchallengeable, thanks primarily to his close alliance with Lysenko, D. N. Prianishnikov, a full member of the Academy since 1929, issued a stern warning that the exclusive reliance on the grassland method of soil improvement was contrary to both the facts of science and the best interests of the national economy.[184] Prianishnikov considered the grassland method only one of several techniques for improving the quality of agricultural land; he worked with great determination to persuade the authorities to abandon the policy, which threatened to end the use of chemical fertilizers in Soviet agriculture. For more than a decade, Prianishnikov's voice received little attention in government circles. Often maligned by Vil'iams's followers and the party press, he became a highly esteemed leader in the smoldering struggle for the professional integrity of the scientific community.

The personal triumphs of Vil'iams—as well as those of Shmidt and Marr—show that in addition to the national heroes in science recognized by the scientific community, there were also "heroes" in science created by political authorities. In some cases—as, for example, in the case of I. P. Pavlov during the waning years of the decade—the same heroes appeared on both rosters; but in a vast majority of cases, the two lists were completely different. The heroes of the scientific community received quiet but unequivocal recognition in scientific publications; the heroes of Communist propaganda were celebrated in the popular press with lavish accounts of their accomplishments. They received more space in the *Great Soviet Encyclopedia*, more medals and other symbolic rewards for professional accomplishment, and more accolades from Stalinist philosophers. In the great emphasis on science as a noble occupation and as a primary source of social progress, they served as models to be emulated by young Russians.

INSTITUTIONAL CONSOLIDATION AND DILEMMAS OF GROWTH

Purged of real and imaginary enemies of Stalinism and isolated from Western scholarship, the Academy allowed little room for freethinkers and iconoclasts—at least in the regions of general scientific theory and philosophical views. But despite all the setbacks and dislocations of scholarly talent, the Academy retained sufficient strength to shape itself into a hardy institution dedicated to the advancement of most branches of science. Only in the most isolated cases did the government waver in its determination to provide funds for the expansion and moderniza- tion of research facilities and for the rapid growth of the ranks of com- petent scholars. The growth of the Academy was accomplished in three ways: the expansion of existing facilities, the creation of new research centers, and the absorption of many scholarly bodies previously outside its jurisdiction.[185]

The Institute of Experimental Biology, the Mendeleev Chemical Soci- ety, the institutes of the defunct Communist Academy, the Academy of the History of Material Culture, the famous Pulkovo Observatory, and many other research centers became integral parts of the Academy. They brought unique and in some cases rich traditions in science, and they helped substantially in making the Academy an institution engaged in scholarly work of many kinds. In 1939, following the election of new academicians and corresponding members, V. L. Komarov took note of the rapid growth of human and institutional resources of the Academy:

> In 1917, the Academy consisted of 45 academicians and 109 scientific workers. In 1939, its staff consisted of 136 academicians and 335 cor- responding members; its total scientific staff included 3,500 people. Instead of five laboratories and one institute, which existed before the revolution, now the Academy has 58 institutes and several independent laboratories; instead of 8 museums, now there are 20; and, finally, a large network of branches and bases of the Academy has been estab- lished in various regions and national republics.[186]

Komarov was certain to acknowledge that the Academy's institutional network made up only a small part of the total system of Soviet scientific institutions.

In order to make the administration of the growing empire of research bodies more efficient and better adaptable to the dynamics of rapidly expanding scientific tasks, the authorities decided to abolish "groups"— seen as unwieldy research complexes—and to divide the Academy into eight departments, each covering a major aggregate of related disciplines.[187] The new units were designed to provide institutional mechanisms for involving the leading scholars in the management of each department, at the same time preserving—and, in fact, strengthening— monocratic management.

Each department consisted of three distinct types of organizational components. The general assembly of the department served the scholars more in providing a forum for the sharing of one another's research experience than in wielding administrative authority. The bureau of the department, meanwhile, was a vital link in the chain of command that led through hierarchical links to the appropriate authorities in the Council of People's Commissars. Finally, the party organization was primarily engaged in public supervision of the distribution of cadres, the execution of specific components of science policy, and the organization of seminars in Marxist theory. Party units kept a watchful eye on the dynamics of informal groups and on working discipline. Quietly but persistently, they played the decisive role in selecting candidates for elective administrative positions.

Among research units making up a department, institutes occupied key positions. The most active centers of research in specialized fields, they were the main arteries of the Academy's dedication to science. After prolonged experimentation with the more fluid forms of research units, the authorities opted for the relatively stable institutes as the most desirable units with which to build the inner strength of the Academy. The institutes provided efficient mechanisms for controlling finances, calculating economic indices of the national investment in science, and meeting the rapidly changing organizational needs of modern science. They played an important role in finding a middle ground between government demands for practical research and the Academy's dedication to fundamental theory. They also faced the intricate task of harmonizing the collective aspect of modern research with the interests, background, and propensities of individual scholars.

Despite their extensive involvement in practical research, some institutes were dominated by the theoretical interests of their leading figures. While the I. P. Pavlov Institute of Physiology was ruled by the ideas and research methods of the great neurophysiologist, the Institute of Evolutionary Morphology was dominated by the efforts of A. N. Severtsov and I. I. Shmal'gauzen to work out the details of comparative morphology as the basis of a general theory of evolution.[188] The Institute of Physical Problems was immersed in Kapitsa's deep involvement in the theory and experimental study of the superfluidity of helium-2. The Institute of Experimental Biology was ruled by N. K. Kol'tsov's concern with human genetics and the molecular base of the structure of life. The Institute of Genetics placed primary emphasis on N. I. Vavilov's grand plan for producing a universal map of local concentrations of the gene pools of domestic plants and their ancestors; when Lysenko took over the institute it became a center of Michurinian evolutionism, which allowed no place for modern genetic theory. When Orbeli became the head of the Pavlov Institute of Physiology, after Pavlov's death in 1936, he immediately gave a broader evolutionary basis to the study of the higher nervous activity, an area of personal concentration. He tried to enrich Darwinian theory by adding evolutionary physiology to the already well-established evolutionary morphology.[189]

In a typical institute, however, no single individual occupied the position of preeminent influence on the choice of research problems and on the articulation of theoretical views. The Institute of Mathematics, for example, was dominated by three powerful individuals: S. N. Bernshtein, who was concerned with the modern approaches to the theory of probability; N. N. Luzin, who was engaged in the descriptive set theory; and I. M. Vinogradov, who worked in the theory of numbers. The Physico-Technical Institute was dominated by one person, A. F. Ioffe, its founder and a leading pioneer of modern physics in the Soviet Union. His personal involvement in research on semiconductors did not deter him from making his institute the general national center of atomic studies.[190] He not only organized and administered a model institute but also served as the most eloquent and successful advocate of the imperative need for a symmetrical development of all the leading branches of modern microphysics in the Soviet Union.

The Institute of Biochemistry represented yet another category of organizational style: it was dominated by one man, but not a strong man; nor was it divided into spheres of influence. Its director was A. N. Bakh, who in the 1930s was only tangentially involved in scholarship. Born in 1857, he was expelled from Kiev University for participation in the activities of revolutionary Populists. In 1883 he emigrated to Geneva, where he organized a private laboratory to conduct biochemical research. In 1917 he returned to the Soviet Union and became an enthusiastic supporter of the Bolshevik regime. In 1928–1929 he was actively engaged in the Communist attack on the Academy of Sciences, and in 1929 he was elected a member of the Academy. In 1935, at the age of seventy-eight, he founded the Institute of Biochemistry and became its director. Although he then spent most of his time writing popular articles in favor of a technological orientation in the Academy and assisting various Communist front organizations in the Academy—he was also the head of VARNITSO—he kept the Institute of Biochemistry chained to his own narrow interests in ferments and their role as catalysts in the chemical processes taking place in living cells and tissues.[191] While the institute's research interests were welded into organic unity, they were gravely limited in scope of operation, depth of vision, and effectiveness of leadership.

While the institutes gave the Academy inner stability and a firmly established mechanism for institutional adaptation to new developments in science and technology, commissions and committees were transitory bodies in search of solutions to specific problems, be they of a theoretical or a practical nature. They also provided an efficient base for interdisciplinary cooperation and exploratory work in new research areas or in matters of administrative significance. The Astronomical Council, founded in 1937, represented yet another institutional component of the Academy. It consisted of the leading astronomers in the country, who gathered annually to coordinate research in all the branches of their science, to plan future developments, and to maintain contact with the international community of astronomers. Its numerous commissions handled the problems of individual disciplines and organized scientific sessions at various astronomic centers. By avoiding institutional rigidity, the Academy sought to develop an organizational structure that was adaptable to the dynamics of modern science and to the ever-increasing ties between science and society.

On the eve of World War II, the Academy was a center of rich, exciting, and eminently productive research in many leading areas of modern science. In many disciplines—typified by physics, mathematics, crystallography, and evolutionary biology—young scholars, trained in and inspired by the relatively unrestricted atmosphere of the 1920s, were making tangible contributions to the mainstream of modern scientific thought. At this time, Soviet science provided conclusive proof that it had entered the atomic age. The physicists formulated the theory of the combined scattering of light (L. I. Mandel'shtam and G. S. Landsberg), established the theory and mechanisms of chain reactions in combustion (N. N. Semenov), anticipated the discovery of the neutron (D. D. Ivanenko), built a diminutive cyclotron in 1937, presented the "drop model" of atomic structure (Ia. I. Frenkel'), and penetrated into the secrets of the spontaneous fission of uranium nuclei (G. N. Flerov and A. G. Petrzhak).[192] At this time also, Petr Kapitsa published the remarkable findings of his pioneering experimental work in the superfluidity of helium-2, which brought him a Nobel Prize in 1980. The biophysicists, led by N. K. Kol'tsov, produced pioneering studies of the macromolecular basis of heredity.[193]

In an effort to reach beyond the theoretical and methodological limits of Pavlov's legacy, P. K. Anokhin constructed the first "cybernetic" model of the brain as a complex dynamic system of automatic action based on the feedback of self-generated information—or "reverse afferention," as he called it. Soviet mathematicians received international acclaim for their work on set-theoretical probability theory, topology, theory of transcendental numbers, mathematical logic, and the foundations of mathematics. It was during the 1930s that A. N. Kolmogorov attracted the attention of the international community of mathematicians by presenting the first comprehensive and logically integrated axiomatics of probability theory. And in 1937 I. M. Vinogradov enriched the theory of numbers by solving the famous Goldbach problem, postulated in 1742, and by indicating a series of applications for his contribution. L. V. Kantorovich's pioneering work on the mathematical apparatus of linear programming was awarded a Nobel Prize. A group of Soviet scholars made bold efforts to extend the theory of numbers to crystallographic research. Resounding international recognition came also to A. I. Oparin, whose theory of the abiogenetic origin of living forms laid the foundations for pioneering work on the ever-evasive but immensely attractive topic of the origin of life. Notable efforts were also made to lay the theoretical and methodological groundwork for the study of the biosphere, the foundation of modern ecology.

During the 1920s political authorities stayed out of scientific controversies and left it to the scientists and philosophers to search for a consensus in the interpretation of modern theories. During the 1930s they encouraged the press to give wide publicity to preferred orientations, but they did not as a rule take direct actions against unfavored

views. The unlimited scope of Communist propaganda in favor of Pavlov's conditioned-reflex theory, Lysenko's Michurinism, and Oparin's
biochemical theory of the origin of life did not produce an outright
suppression of heterodox views in neurophysiology, biology, and chemistry, although it did place such views in a disadvantageous competitive
position. In general, however, the expression of ideological preference
affected only a limited number of theoretical orientations in science.

In physics, for example, the authorities did not become involved in
the bitter feud between the philosophers and scientists over the "correct"
dialectical-materialistic interpretation of the theoretical body of quantum
mechanics and the theory of relativity. There were three major orientations in physics. One orientation was articulated by A. K. Timiriazev
and V. F. Mitkevich, who advocated a return to the legacy of Faraday
and Maxwell as the soundest way of resolving the compounded difficulties of modern physics.[194] Its adherents were inclined to scrap Einstein's theories altogether. Another orientation was advanced by
physicists close to the dominant breed of Marxist philosophers, typified
by A. A. Maksimov, E. Ia. Kol'man, and M. B. Mitin, who favored a
major overhauling of the theoretical foundations of modern physics to
make them harmonious with dialectical materialism. They considered
modern microphysics and the theories of special and general relativity
a blatant attack on four pillars of Marxist scientific philosophy: the law
of causality, materialistic ontology, objective epistemology, and the principle of concreteness (according to which all scientific abstractions have
an empirical base). Unlike Mitkevich and Timiriazev, typical representatives of this orientation did not seek salvation in a return to classical
physics.[195]

The third orientation was articulated by a solid contingent of leading
physicists led by V. A. Fok, Ia. I. Frenkel', and A. F. Ioffe, who argued
vehemently that all the leading ideas of quantum mechanics and relativity theory were fully compatible with dialectical materialism—and
that there was no need for Marxist philosophers and their followers
among physicists to interfere with the work of scientists. In 1938 Fok,
internationally known for his major contribution to the mathematical
apparatus of Schrödinger's wave function, turned with relentless vigor
against the philosophers and their satellites in the scientific community.
He emphasized the failure of philosophers to detect the fundamental
agreement of the general principles of quantum mechanics with dialectical materialism. He argued that (1) the principle of complementarity

was an unchallengeable component of quantum mechanics and "a firmly established law of nature"; (2) the thesis of the incompatibility of quantum mechanics, in general, and the principle of complementarity, in particular, with materialism was an idealistic thesis; (3) the difference between the philosophical views and physical theories of the pioneers of quantum mechanics must be clearly stated and accepted; and (4) the interference of philosophers with the work of theoretical physicists was injurious to the best interests of the nation.[196]

In biology five major theoretical views of organic evolution existed side by side: the morphological and comparative-anatomical theories of A. N. Severtsov, centered in the Academy of Sciences; the genetic theory, strongly represented in the Institute of Experimental Biology and the Institute of Plant Breeding; the synthetic theory of I. I. Shmal'gauzen, an elaborate plan to unify the morphological-anatomical and the genetic theories, centered in Moscow University;[197] Lamarckian theory—particularly in the form of Lysenko's Michurinism—with a strong representation in the Lenin Academy of the Agricultural Sciences; and the theoretical orientation of E. S. Bauer, a unique effort to extend the methods of modern microphysics to the study of living matter.[198] These orientations were not variations on a theme but fundamentally different systems of theoretical construction.

Although the party press clearly favored Lysenko, most of the scientific community continued to be skeptical about his extravagant promises and claims.[199] Unlike physics, where diversity of opinion continued unabated until the end of the decade, biology showed abundant signs of the rapid growth of Lysenkoism and mounting pressure on the representatives of theoretical orientations criticized by Lysenko to withdraw from the avenues of scientific communication. A similar but more limited situation existed in neurophysiology. During the early 1930s certain Marxist theorists were willing to come out publicly against the efforts of the Pavlovian school to place sociology on biological foundations.[200] At the end of the 1930s, however, no serious attack on Pavlov's legacy was aired either in the party press or in professional publications.

Political strategists looked at science through two different but equally important prisms: one showed science as the surest path to prosperous, enlightened, and virtuous life; the second evoked Lenin's warning about the new physics and its overelaborated mathematical formalism as a particularly luring invitation to all kinds of idealistic philosophies.

Since 1918 the idea of a "Soviet science" had occupied the minds of many philosophers and commentators on current developments of social import. A. A. Bogdanov, the leader of the Proletkul't movement, went so far as to claim that in the future Soviet science—as "socialist science"—would differ qualitatively from "bourgeois science" in substance, theory, and method. During the 1930s no writer held such an extremist view. Discussions about "Soviet science" were quite common at this time, but there was no consensus on what it stood for. The new extremists in the Communist Academy—led by A. A. Maksimov and E. Ia. Kol'man—viewed "Soviet science" as a modern effort to disencumber major theoretical postulates of all residues of "idealistic" epistemology. Far from rejecting quantum mechanics, they treated it as a core of modern science. Nevertheless, they wanted a quantum mechanics unencumbered by the principles of uncertainty and complementarity, which they considered philosophical presuppositions unsupported by experimental data. Soviet science, in their view, was in a position to effect a dialectical synthesis of many contradictions in modern physics and mathematics— such as the recognition of both continuous and discrete natures of motion—and thereby to give the modern scientific revolution a broader perspective and richer social ties.[201] In criticizing the leaders of Western physics, the new extremists operated on the assumption that a "wrong philosophy" could produce only a "wrong science." This view found a feeble but growing echo in the scientific community.

According to another view, the emphasis on "Soviet science" meant an emphasis on continued building of the strong national traditions in science that had made the Soviet Union a unique and important contributor to world science, and that did not demand isolation from scientific developments in capitalist countries. These traditions included such dominant interests as structural orientation in organic chemistry and crystallography, Pavlovian neurophysiology, probability, number and set theories in mathematics, interdisciplinary study of the biosphere, biological orientation in soil science, and evolutionary morphology.

According to yet another view, Soviet science derived its uniqueness not from its theory and method but from its style of work. It placed a particularly strong emphasis on coordination of scientific research with the national economic plans, on group research, and on extensive use of scientific knowledge for atheistic propaganda.[202]

The persisting ambivalence in the interpretation of "Soviet science" as a unique social and cultural reality helped to reduce the effectiveness—and the immediate damages—of ideological interference with the daily work of scientists. But it did not do away with portentous challenges to the traditional right of the scientific community to be the supreme master in charting the course of scientific development, in upholding the moral code of science, in forging the tools of scientific inquiry, and in defining the social obligations of the scientific community. Despite annoying perturbations on the ideological front, the Academy grew into a scholarly empire covering vast domains of modern science. After the 1939 elections of new members, the empire grew even larger, for now it included also representatives of the agricultural and medical sciences, as well as practitioners of belles-lettres.

Although deeply scarred by the reign of terror, the Academy entered the 1940s in a state of rapid rebuilding and with a generally optimistic outlook. Communist propaganda in favor of practical research did not divert the Academy from fundamental research; despite endless equivocations in favor of direct ties of science to socialist construction, the government left it much to the leading scientists to detect and spell out the elements of applied value in their research. The inauguration of practically oriented annual and five-year plans of research wrought only minor changes in the traditional pattern of the Academy's commitment to science; this ambitious undertaking was seriously hampered by the pronounced lack of experience in dealing with the theory and methodology of planning scientific work on a large scale and with indices of measuring scholarly output. Although the unceasing campaign to transform the Academy into a bastion of Stalinist ideology—as articulated by philosophers—had hampered the general work in some of the key sciences, it left wide areas of scientific theory uncontrolled and open to debate.

As viewed by Stalinist ideologues, science performed two basic functions: it played an active part in the development of modern industrial technology, and it served as an intellectual pillar of Soviet ideology. As a source of technological advances, it required a strong involvement in applied research and a strict coordination of scientific activity with national five-year plans. As a pillar of ideology, it required theoretical unity—a job entrusted to the philosophers. During the 1930s neither technological nor ideological goals of science were achieved. The quest for technological orientation did not seriously challenge the Academy's allegiance to the idea of basic theory as the surest path to advances in practical knowledge. The government, ever enamored with practicalism in science, found it advantageous not to raise serious objections to the primacy of theoretical exploration. The aspiration of ideological unity—as the backbone of the theoretical unity of science—was brought into bold relief: if it was not realized, the main reason must be sought in the inability of philosophers to cope with the complexities of modern science and in the resistance of scientists to all efforts to subordinate the internal diversity of scientific thought to the external unity of ideological dictates. Nevertheless, the general trend was clear and seemingly irreversible: the future development of the Academy was in expanding efforts to tie scientific research to national economic plans and in a relentless quest for unanimity in the interpretation of the relation of scientific theory to Marxist philosophy.

IV

THE TRIUMPH OF IDEOLOGY (1941–1953)

WORLD WAR II ADAPTATIONS

World War II inflicted staggering losses on the research facilities and human resources of the Academy of Sciences.[1] The enemy destroyed countless scientific workshops, including such major and clearly visible objects as the astronomical observatories in Pulkovo and Simeiz (Crimea). Many young scientists joined the armed forces, and the number of persons working toward advanced degrees dropped precipitously, particularly in the early years of the war. Most leading scientists were relocated to safer areas; removed from major laboratories and libraries, they worked in teams engaged in thousands of research assignments related directly to the war effort. Academy personnel helped to transfer hundreds of industrial plants to safe areas in the east.[2] "Many of our activities," reported A. F. Ioffe in 1942, "take place not in laboratories but in factories, where we help in constructing prototypes or in developing new manufacturing methods."[3] Physicists concentrated on improving existing weapons and on designing new ones; they also worked on reducing the destructive effects of the many new weapons used by the enemy, such as acoustic and magnetic mines.

At the end of the summer of 1941, the Academy organized the Commission for the Mobilization of the Resources of the Urals for National Defense, which was quickly recognized by GOSPLAN as a key link in the organization of the wartime economy. Led by the leading members of the Technical Department of the Academy, the commission fielded many research teams—embracing a grand total of 800 experts—united by the common goal of making the Urals the key military-industrial base of the country. The German occupation of the southern base of heavy industry resulted in a 68 percent decrease in the production of cast iron and 58 percent in the production of steel; one of the basic tasks of the commission was to find substitutes for these losses—to expand the production of the existing foundries and to build new ones.[4] The findings of the commission made it possible for the Kuznets metallurgical complex "to acquire local sources of iron and manganese ores," to make use of zinciferous iron ore and to increase the production of heat-resistant materials.[5] Special units of the commission conducted extensive geological surveys that led to discoveries of new deposits of bauxite, wolfram, cobalt, nickel, zinc, lead, mercury, antimony, and molybdenum. Other teams played a vital role in establishing the "Second Baku"—the area between the Volga and the Urals—as a center of petroleum production to make up for the loss of Baku to the enemy. In 1943 the Urals and western Siberia yielded more aluminum and magnesium than the entire country had in 1940.[6] Teams of chemists worked on new types of steel alloys, partly to substitute for reduced supplies of certain strengthening materials (because of the German occupation) and partly to meet the need for new materials required by the machinery of modern warfare.

The Scientific and Technical Council, a subsidiary of the State Committee on Defense, guided and coordinated the participation of research organizations in the war effort. Its membership included such distinguished academicians as A. F. Ioffe, P. L. Kapitsa, and N. N. Semenov.[7] The research coordinated by the council dealt primarily with the weapons of war and the mobilization of natural resources for the needs of national defense. The Academy research staff was split into adaptive and mobile commissions, detached from the more rigid and unwieldy institutes.[8] The institutes continued to exist more as administrative frames of reference and budgetary units than as functional research bodies. While the war pressure encouraged practical inventions, the dislocation and instability of the institutional base of science worked against theoretical work. By bringing scholars closer to the technical processes of production, the experience of World War II helped the scientific community to acquire valuable skills in producing complex instruments of scientific research.

During the war Soviet scientists became deeply involved in the work on nuclear accelerators and rocket technology. Soon after the start of the war, N. G. Flerov, a discoverer of the spontaneous fission of uranium-235 nuclei, wrote to the State Committee on Defense pleading for immediate work on atomic weapons. The leading physicists, however, were generally unimpressed with Flerov's plea. Some physicists thought that such a project was generally unreasonable and unfeasible; others thought that its realization would require many years of uninterrupted and intensive research. Nevertheless, the committee decided to go along with the proposal and appointed I. V. Kurchatov to organize research in an isolated building in the vicinity of Moscow. Although under-equipped and undermanned, the laboratory started a chain of developments that enabled the Soviet Union to explode an atomic bomb in 1949. Work on atomic fission did not become intensive until 1944 when V. I. Veksler suggested the original principle of charged-particle acceleration by an oscillating field, which served as a basis for new types of accelerators—including the synchrotron, synchrocyclotron, and microtron. The same principle was advanced independently by the American physicist E. M. MacMillan. In 1946 Kurchatov's institute produced a chain reaction in what was the first European reactor: in 1948 the country acquired the first industrial uranium-graphite reactor.[9]

During the war E. K. Zavoiskii produced the first measurement of electron paramagnetic resonance, or electron spin resonance (ESR), which laid the foundations for modern magnetic radiospectroscopy, a basis for quantum paramagnetic amplifiers.[10] This discovery belonged to the category of dramatic achievements in science that opened significant new vistas in both scientific theory and industrial technology. At the same time, the physicist Ia. I. Frenkel' completed *The Kinetic Theory of Fluids*, a major effort at physical synthesis, subsequently published in the prestigious series *The Classics of Science*. According to Ioffe, the wartime need for a productive coupling of practical perspectives and theoretical insights was manifested in the ongoing work in "the theory of numbers, topology, theory of elasticity and aerodynamics, as well as in . . . the superfluidity of Helium II, semiconductors, fluorescence, scattering and propagation of waves, theory of polymers, strength of solids and new classification of comets."[11] It was at this time that L. D. Landau published his famous papers on the superfluidity of helium-2 that brought him a Nobel Prize in 1962.

World War II brought about a major detour in the evolution of Soviet philosophy of science. The ideologically inspired crusades against "idealistic" aberrations in modern science lost much of their belligerency and fervor immediately after the beginning of the war. The philosophers, led by men like M. B. Mitin, now concentrated more on the positive scientific contributions of the leaders of Western physics than on the "pathology" of "physical idealism." Scientists displayed disciplined consistency in avoiding philosophy altogether, and philosophers pondered the fundamental compatibility of modern science with dialectical materialism. S. I. Vavilov represented the majority of his colleagues when he fully accepted three ideas advanced by the Western leaders of quantum mechanics, termed "idealistic" by Stalinist philosophers: the principle of uncertainty as a major deviation from the Laplacian notion of causality; the role of macroscopic research instruments in constructing microphysical reality; and the limitless power of mathematics in the development of modern science.[12] Vavilov did not overlook "physical idealism," the archenemy of dialectical materialism. But he came to see "physical idealism" in the philosophical excursions of Arthur Eddington and J. H. Jeans, not in the scientific work of Bohr, Heisenberg, and Einstein.

World War II was the period of truce on the philosophical front, achieved primarily by the withdrawal of philosophers from positions of maximum ideological sensitivity, pronounced inactivity in the epistemological domain, and a shift of emphasis from the elaboration of the Marxist world view to the evocation of Russian nationalism. While philosophical withdrawal was the norm, a small group of Stalinist stalwarts, led by A. A. Maksimov and E. Ia. Kol'man, showed little inclination to retreat from their war on "physical idealism." In 1943 Kol'man was still busy reminding his readers that the philosophical views of such modern scientists as Einstein, Bohr, Heisenberg, and Fok could not be separated from their theoretical views in physics. Philosophical positions of individual physicists, he said, determined their scientific approaches and shaped their theoretical orientations. Both Kol'man and Maksimov continued to remind their readers that Soviet physics had a long way to go toward freeing itself from deeply ingrained admixtures of idealistic philosophy. Nevertheless, their writing was now less biting and less bellicose.[13]

The World War II experience provided strong support for the idea of building stronger and more versatile scientific centers in the areas east of the Urals. The new centers were not mere auxiliaries of the Moscow and Leningrad scientific institutions or transitory adaptations to the contingencies of war; they were vital steps in the national effort to achieve a geographical deconcentration of basic and applied research. The West Siberian Branch of the Academy of Sciences, one such center, was founded in 1944 and served as a model for regional research organizations concerned with applying the most advanced methods of modern science to the study and exploitation of natural resources. "The first large and complex Siberian institution of the Academy of Sciences," the West Siberian Branch undertook a wide variety of scientific tasks, ranging from the expansion of the technological base of regional ferrous and nonferrous metallurgy and a modernization of various branches of the chemical industry to a systematic exploration of western Siberian flora and a comprehensive economic study of the industrial and agricultural potential of the region.[14] The East Siberian Branch in Irkutsk, founded in 1949, concentrated its research activities on geology, chemistry, biology, and economic geography.[15] The Far Eastern and Komi scientific bases also were established at this time.

The creation of new scientific centers was not limited to the areas east of the Urals. New research units were set up to meet the needs of the economy and defense, to elevate the cultural standards of individual national groups, and to tap new sources of scientific manpower. Gradually, most national groups acquired modern institutions that, together with selected science complexes, placed strong emphasis on the study of regional archeology, language, geology, literature, folklore, and history.

In 1941, on the eve of the country's entrance into World War II, the Georgian Academy was founded, a significant event that marked the beginning of what proved to be an unwavering effort to introduce basic research in locations far removed from traditional scientific centers.[16] In quick succession similar academies were established in the Uzbek (1943), Armenian (1943), Azerbaidzhanian (1945), and Kazakh (1945) republics. In addition to studying local cultures and regional economic resources, these organizations became elaborate systems of research units dedicated to fostering and creating strong national traditions in individual sciences. Georgian mathematicians and Armenian astrophysicists, for example, quickly became vital and widely recognized contributors to mainstream Soviet scientific thought. Among other research centers that grew rapidly during the war, the Kazakh Branch occupied a particularly important position; in 1945, shortly before it was elevated to the rank of academy, it combined sixteen institutes and employed 1,400 scholars.[17] The Kirghiz Branch was at first strongly influenced by the evacuated members of the biological institutes of the Soviet Academy of Sciences; soon, however, it spread its activities to many fields ranging from geological exploration and physical chemistry to history, linguistics, and economic geography.[18]

Nationalist fervor, brought to a high pitch by World War II, stimulated Academy scholars to undertake an intensive study of the historical roots of Russian culture. The history of Russian science provided a particularly inviting universe of inquiry. V. L. Komarov, president of the Academy, wrote in 1944:

> The history of science is one of the most important branches of scientific scholarship concerned with the scientific world view in its most general orientation. The war has shown the value of preserving the cultural and scientific values of the past in the national memory, of helping the general public to understand and esteem the creative work of the leaders of science and of teaching young people to view science in its historical aspect.[19]

The role of I. M. Sechenov and I. P. Pavlov in the evolution of "the Russian physiological school," the broader scope of P. L. Chebyshev's contributions to the founding of "the St. Petersburg School in Mathematics," D. I. Mendeleev's part in setting the stage for the emergence of modern microphysics, and A. M. Butlerov's work on structural theory in organic chemistry received particularly extensive treatment. The historical role of Mikhail Vasil'evich Lomonosov as the father of Russian science in general and as a pioneering contributor to a wide array of individual sciences from physical chemistry to botany became a topic of the most extensive and exuberant research. The work and life of N. I. Lobachevskii, the founder of the first non-Euclidean geometry, received the first book-length treatment by a Russian—or Soviet—scholar.[20] A group of leading mathematicians prepared a collection of essays covering the role of Russian scholars in the evolution of probability theory, theory of numbers, topology, and theories of metric and descriptive functions.[21]

These and similar studies resulted in the first comprehensive effort to systematize the historical study of Russian science, to integrate the history of Russian science into the mainstream of the national intellectual culture, and to make a full record of the contributions of Russian scholars to "world science." The new enthusiasm for the history of Russian science led to a series of studies devoted to the life and work of the most eminent scholars who served as a bridge between tsarist and Soviet science. The aerodynamics pioneers N. E. Zhukovskii and S. A. Chaplygin, the plant physiologist K. A. Timiriazev, and the mathematician and expert in naval mechanics A. N. Krylov were among the most idolized scientists who were active in both pre-Soviet and Soviet eras and who testified to the unbroken growth of scientific thought in Russia.

During the war the Academy was deeply involved in commemorating great dates in the history of scientific thought. In 1942 it observed the three-hundredth anniversary of the death of Galileo and used this opportunity to enrich Galilean studies in the Russian language. In 1943 several festive convocations marked the three-hundredth anniversary of the birth of Newton.[22] Newton's life and work received sympathetic treatment in two biographies, by S. I. Vavilov and by A. N. Krylov, a volume of essays, and numerous articles in professional and popular journals. Several papers dealt exclusively with the role of Newtonian ideas in the early development of the Russian scientific community.[23] In the same year, the Academy celebrated the four-hundredth anniversary of the publication of Copernicus's *De revolutionibus*. The purpose of these and similar celebrations was both to honor the masters of classical science and to point out Russia's intellectual debt to—and extensive affiliation with—world science.

Clearly, the wartime climate favored a general view of science as an amalgam of national and international effort. L. S. Berg, a member of the Academy and president of the All-Union Geographical Society, cited P. P. Semenov-Tian-Shanskii, a nineteenth-century naturalist, in support of this view. In 1859 Semenov stated that science was both national and cosmopolitan. It was cosmopolitan because it was the product of a cumulative growth of ideas originated in many parts of the world—it was an indivisible property of all mankind. It was national because it was a pivotal reality of the life of every nation. Without science no modern nation could create the conditions indispensable for cultural progress and improvements in social well-being. A scientist, "if he did not want to be a cold cosmopolite," must work on both sides of the scientific equation: on advancing the frontiers of universal science and on weaving the facts of science into the fabric of national life.[24] Like most of his colleagues, Berg viewed science as the most potent vehicle for bringing the Soviet Union closer to Western culture.

THE SEARCH FOR ACADEMIC AUTONOMY

Soon after the defeat of Germany, the Academy of Sciences celebrated the two-hundred-twentieth anniversary of its birth—a momentous occasion that brought a solid contingent of foreign scholars to Moscow, including, among others, the American astronomer Harlow Shapley, the German physicist Max Born (who spent the war years in England), and the French mathematician Jacques Hadamard. In the words of a visiting Swedish scholar, the Moscow celebration marked the first postwar international congress of scientists.[25] Speaking on behalf of the government, V. M. Molotov reaffirmed the dedication of Soviet authorities to creating "the most favorable conditions" for a rapid development of science and technology, for "a broader organization of the training of young scientific cadres," and for "closer ties of Soviet science with world science."[26] Indeed, the recurrent emphasis on the need for a firmer and broader integration of Soviet science into the mainstream of "world science" precipitated the most enthusiastic applause from the scholarly audience. The festivities were dominated by tedious recountings of the achievements of Russian—and Soviet—science and by ebullient praises of the unity of science and socialism. In individual departments of the Academy, Soviet and foreign scholars presented papers on the current state of individual sciences, from atomic physics to soil science. Leading members of the Academy came out of World War II convinced that the future growth of Soviet science required a consistent recognition of the preeminence of basic research, the expansion and intensification of intellectual ties with the Western scientific communities, a systematic deconcentration of the institutional base of science, and a clearer and more functional separation of science from ideology. Such leading scholars as V. I. Vernadskii and P. L. Kapitsa also advocated a wider scope of academic autonomy as an essential requirement for a more vigorous and diversified engagement in scientific work.

Digging out from the ravages of the recently ended war, the Academy moved with astonishing speed in returning to normal life. With substantial increases in government budgetary allocations, the Academy was in a position to expand its institutional base and to modernize its laboratories. The founding of the Kazakh, Latvian, and Estonian Academies of Sciences in 1945–1946 marked a key step in the process of giving each republic the same kind of centralized body devoted to scientific research. The last in the series of republic academies was established in Moldavia in 1963. The vast expansion of the network of academic institutions necessitated the creation of a central coordinating body at the Soviet Academy, which began to function in 1945. At the beginning, the new council did not have a clearly defined function and met at irregular

intervals. At this time, the attention of the central Academy was concentrated on getting its own house in order.

The staff of the Academy grew at a fast pace: in 1947 it included over
5,000 persons occupying various positions in the hierarchy of scholarly
ranks.[27] Most notable among the new research units of the central Academy was the Institute of Atomic Energy, the product of a modern coordination of theoretical inquiry and technical work on complex research
tools. V. I. Veksler's construction of the synchrocyclotron, the first in a
series of powerful particle accelerators, provided the most convincing
proof of the growing ability of the Soviet scientific community to cope
with the rapidly rising need for huge and labyrinthine research
machinery. The new institute was only the first step in the rapid growth
of atomic research centers anchored in and outside of the Soviet Academy. In 1949 the Academy completed the construction of a synchrocyclotron with a capability of accelerating protons to energies of up to 680
million electron volts.[28] In addition to coordinating the work of republic
academies of sciences, the Soviet Academy now emerged as the chief
coordinator of atomic research conducted in mushrooming laboratories
subordinated directly to individual ministries or organized in selected
universities. While the Academy dominated the overall Soviet scientific
endeavor, physics ruled supreme in the Academy.

Moved by the new optimism, the members of the Academy worked
on many fronts to strengthen the position of Soviet scholarship in the
international community of scientists, to emancipate science from rigorous ideological controls, and to restore the legitimacy and dignity of
those branches of science that fell on hard times as a result of discrim-
inating influences exerted by forces external to the pursuit of science.
Three developments gave graphic expression to the prevailing sentiment
in the Academy.

In the speech opening the festive convocation in honor of the two-
hundred twentieth anniversary of the founding of the Academy, V. L.
Komarov, president of the Academy, placed special emphasis on the
close ties of Russian science with the development of scientific thought
in the West. He noted "the strong scientific and social echo of Darwin's
ideas in Russia"; he observed that Lomonosov deserved much credit for
pursuing the rich scientific vistas opened by Benjamin Franklin; and he
underscored the close ties of such leading Russian scientists as P. L.
Chebyshev, I. I. Mechnikov, and V. I. Vernadskii with the French sci-

entific community. Komarov voiced the opinion of his eminent colleagues that the most promising future of Soviet science lay in a broad and unalloyed cooperation with the scientific communities of the wartime allies and all other nations opposed to fascism.[29] Soon after Komarov's pronouncement, G. F. Aleksandrov published a volume on the history of Western philosophy that, in addition to standard Marxist criticism, contained an abundance of references to the historical debt of Russian thought to the great thinkers of the West, even though they were not customarily placed into the categories of "utopian" or "scientific" socialism.

The second major development was of a different order: the Academy made a bold effort to restore the right of genetics to exist as a full-fledged, legitimate discipline. Since the 1939 conference on biology sponsored by the journal *Under the Banner of Marxism*, a gathering that went a long way toward making Lysenkoism the official Marxist biology, genetic research in the Academy fell on hard times, even though it was not yet officially abolished. In 1946 the Academy—in an obvious effort to restore genetics to a position of full academic respectability—elected N. P. Dubinin a corresponding member, a position that traditionally marked a major step on the way to full membership. Dubinin was one of the most prominent surviving members of the Soviet school of genetics, which had received wide international acclaim during the 1920s and 1930s. He was also one of the main targets of Lysenko's persistent attacks on the scientific foundations and ideological bearings of genetics. In urging the General Assembly of the Academy to vote against Dubinin, Lysenko could state only that "Dubinin is the leader of the anti-Michurinist group of geneticists, who represent a conservative ideology in our genetic science and are closely allied with ideologically reactionary foreign biologists."[30]

The election of Dubinin was so much more remarkable because it occurred at the same time that the party press was portraying Lysenko as the greatest of all living Soviet scientists—a paragon of pure dedication to science, boundless intellectual resourcefulness, exemplary patriotism, and unmatched loyalty to the cause of socialism. Although supported by a contingent of vociferous academicians, led by B. A. Keller and N. A. Maksimov, Lysenko stood on shaky ground. The election of Dubinin showed clearly that in his crusade against genetics, Lysenko could succeed only by a total dependence on political authorities. These authorities were most generous in showering him with the privileges of academic power (for example, he was a ranking member of the Supreme Certification Commission, which made the final decisions in the granting of higher degrees, and of the Commission on State Prizes, which selected scientists for special financial and honorific awards). The Academy was not the only source of resistance to Lysenko's insatiable thirst for power. In 1946 the *Scientific Journal* of Moscow University published an article on the development of Russian experimental biology that praised S. S. Chetverikov, N. K. Kol'tsov, and A. S. Serebrovskii, the three towering figures in Soviet genetics.[31]

The third development in the Academy's search for autonomy was related to physics. Encouraged by successes in generating and harnessing atomic energy and in setting the stage for constructing much-improved particle accelerators, the physicists were now ready to mount a counteroffensive on the philosophers of science who for years had waged a degrading and debilitating war on the complementarity and uncertainty principles of modern physics. In 1947 the theoretical physicist M. A. Markov published an article in the newly founded journal *Questions of Philosophy (Voprosy filosofii)*, in which he complained about the rigidity and sterility of the objective view in epistemology, the mainstay of the reigning Soviet philosophy of science and a position that disregarded the role of the "subjective" element in constructing "physical reality" as treated by quantum mechanics. Quantum mechanics, he contended, relied on "macroequipment" (fully controlled by the subject) to acquire knowledge about microphysical objects. "Quantum theory," wrote Markov in a Bohrian vein, "did not add anything new to 'the subject-and-object division' in its general philosophical sense, but it added something essentially new to the question of the interrelation of the subject and the object: it gave broader meaning to physical reality and to the macroscopic form of our knowledge of the microworld." Physical reality, as studied by physics, is not a mere reflection of an external object in the mind of the subject, as Leninist epistemology had implied; it is a product of the interaction of the object and the subject. The subject both reflects and creates physical reality. Markov's carefully worded essay had two goals: to remove the unnecessary philosophical barriers that stood in the way of a closer cooperation between Soviet physicists and their Western peers; and to restrict the efforts of Marxist philosophers attempting to use physics as the main funnel for pouring ideological content into the body of modern science.[32] In a short introduction to Markov's paper, S. I. Vavilov, the new president of the Academy of Sciences, emphasized the need for dialectical materialism to keep in step with the advances in science.

All these developments showed that the Academy was ready to broaden the scope of its autonomy and ideological detachment and to protect science from the ominous effects of the successive waves of philosophical interference. Moreover, they gave new vibrancy and substance to the rapidly growing optimism in the promising future of science in the Soviet Union—an optimism that pervaded the scientific community from top to bottom. The climate of great hopes for new measures of independence did not go unchallenged. In 1942 the government engineered the promotion of N. G. Bruevich, an Old Bolshevik and a minor scientist, from the rank of corresponding member to that of full member and immediately made him academic secretary of the Presidium of the Academy, a resurrected but modified position of permanent secretary. V. I. Vernadskii noted that the elevation of Bruevich to such an exalted position was done in "an irregular manner."[33] The academicians took no part either in promoting Bruevich or in making him academic secretary. Thus, they suffered a major loss on one of the key fronts in the struggle for institutional independence.

Before the effort for broader scientific autonomy and for liberal science policy could take deeper roots and cover a wider range of professional activities, another setback of serious proportions took place. In 1946 Petr Kapitsa, the most determined and eloquent defender of academic autonomy and of international cooperation in science, was dismissed as director of the Institute of Physical Problems, his own creation. Because of his great reputation, no public attack was made on him. Many years later, an unofficial source stated that "the method for the production of oxygen proposed by Kapitsa was unjustly condemned" and, as punishment, he was not only dismissed as director of his institute but was also forbidden to work in its laboratories.[34] He remained under house arrest until 1955.

THE WAR ON IDEOLOGICAL IMPURITIES AND COSMOPOLITANISM

Immediately after Kapitsa's dismissal, the government moved to squelch the liberal stirrings in the Academy. In doing this, it transformed the Academy into the focal point of a new dedication to the world of rigid conservatism and mounting opposition to all digressions from the sacred tenets of ideological orthodoxy. Unfolding swiftly and with blinding fury, the government action helped to usher in a period of stark oppression, intellectual isolation, and devastating ideological vendetta.

In 1947 A. A. Zhdanov, the authority on culture in the Politbureau of the Central Committee of the Communist Party, gave the new ideological crusade a sense of direction and urgency. At a specially convened session of selected party loyalists representing the major areas of intellectual and artistic endeavor, he took the philosopher G. F. Aleksandrov, a full member of the Academy, to task for two flagrant errors contained in his newly published *History of Western Philosophy*. The book was described as a distorted interpretation of the basic philosophical postulates of Marxist theory and an unpardonable exaggeration of the role of Western philosophy in the evolution of Russian secular thought.[35] Zhdanov's speech received wide publicity and triggered the politically manipulated hysteria that culminated in drastic actions engaging every branch of intellectual endeavor in a war against ideological impurities and cosmopolitanism. The demand for ideological purity was a demand for Marxist orthodoxy as a basis for a united front of Marxist philosophers and scientists. The demand for a frontal war on cosmopolitanism was a demand for greater emphasis on the historical uniqueness of Russian culture—including science—and for a crusade against "servility" to Western culture. Mass meetings of specialists in the main branches of scientific knowledge served as the chief vehicle for airing the guiding ideas and expressing the fervor of the new ideology, for exposing culpable scholars to public censure, and for announcing institutional changes aimed at giving the new science policy a firmer footing.

Zhdanov's declaration of war on idealism and cosmopolitanism did not immediately shut the door on the discussion invited by M. A. Markov's 1947 article in *Voprosy filosofii*. The discussants were clearly divided into two irreconcilable groups. One group, led by the philosopher A. A. Maksimov, was unwilling to make any concession to the "physical idealism" of Bohr and Heisenberg and rejected every argument advanced by Markov in favor of building a bridge between dialectical materialism and the high theory of the Copenhagen school. Maksimov argued that an acceptance of Markov's views would be the same as a rejection of dialectical materialism.[36] The second group, with no philosophers included, favored a careful and sympathetic examination of Markov's suggestions and found many of his arguments acceptable in principle. The mounting mass hysteria, however, acted quickly to silence this group.

In March 1948 an editorial in *Voprosy filosofii* stated that "M. A. Markov's inclination toward idealism, under the influence of Bohr's views, is an expression of cosmopolitan vacillation among a segment of Soviet physicists," and that Markov, "not armed with the great ideas of Marxist-Leninist theory and not guided by the materialistic tradition of Russian natural science and Russian philosophy," was the victim of "a servile acceptance of the reactionary idealistic philosophy."[37] The editors of *Voprosy filosofii* went on to apologize for publishing Markov's paper in the first place. They stated that the paper did not analyze or explain the achievements of modern physics and that it contributed only to "the weakening of materialism." The editorial brought the discussion to an abrupt end. Unidentified authorities granted victory to the philosophers, whose views became the basis of the unity of high theory in physics as well as of the unity of science and philosophy in general.

T. D. Lysenko lost little time in tying his campaign against genetics to the crusade inspired by Zhdanov.[38] The fight against genetics, Lysenko contended, was a fight against unpardonable digressions from Soviet ideology, against violations of the basic postulates of dialectical materialism, and against servility to metaphysical ideas in science that were alien to the Russian intellectual tradition. Inspired by Zhdanov's speech and encouraged by the new atmosphere, he concluded that the time had come for the final and crushing assault on the real and imaginary enemies of Michurinist biology.

To carry out his design, Lysenko organized the Session of the All-Union Lenin Academy of Agricultural Sciences in the summer of 1948. Intended as an exposure of flagrant abuses of the scientific community's traditional right to stand up for academic autonomy and to serve as the unchallengeable guardian of the scientific legacy, the session achieved Lysenko's goal: it ordered the discontinuation of all inquiry into localized agents of heredity, all work toward a synthesis of Mendelian and Darwinian traditions in the theory of evolution, and all studies that required a combination of biological and physicochemical methods. It made Lysenko's antigenetic orientation the official biological theory of the Soviet Union. It proclaimed modern genetics an illegal science, antithetical to the cherished values of Soviet culture. A month later, the Academy of Sciences held a three-day session to reassure the government that it was ready to eradicate the last vestiges of genetics from its institutes and laboratories. In a letter to Stalin, the Presidium of the Academy of Sciences humbly apologized for the sins of the past:

The Presidium of the Academy of Sciences and the Bureau of the Department of Biological Sciences committed a most serious error when they gave support to the Mendel-Morgan orientation at the expense of the progressive Michurinist theory. In guiding the institutes of the Academy, the Presidium did not perform its duties in a satisfactory manner. It allowed the enemies of the Michurinist theory to occupy positions of leadership in the biological institutes of the Academy, limiting their usefulness in socialist construction.[39]

To show its readiness to go along with the new course in biology, the Presidium discharged the leaders of the anti-Lysenko orientation from all administrative positions. L. A. Orbeli, one of I. P. Pavlov's most eminent disciples, was the first to go: he was dismissed from the coveted position of academic secretary of the Department of Biological Sciences and was divested of "almost all" administrative functions in the Academy. He was found guilty of conspiring to make genetics a basic concern and orientation of his department. Among other indiscretions, he had found a place in the Institute of Evolutionary Physiology in Koltushi, which was under his direct control, for R. A. Mazing, who conducted experiments on *Drosophila melanogaster* aimed at throwing new light on "the genetics of behavior," and for M. E. Lobashev, after he was dismissed from Leningrad University on charges of close affiliation with the genetics of Weismann and Morgan.[40] With all the research facilities in the Academy made inaccessible to him, Orbeli and a dozen of his coworkers moved to the physiological laboratory of the Lesgaft Institute of Natural Science, a component of the Academy of Pedagogical Sciences. His new research base accommodated only a handful of scholars and was in full disrepair.

I. I. Shmal'gauzen was the second major victim of Lysenkoist purges in the Department of Biological Sciences; he was dismissed as director of the Institute of Evolutionary Morphology. Shmal'gauzen was an internationally recognized authority on synthetic evolution, a comprehensive branch of investigation that attempted to bring together Darwin's evolutionary views and the theory of heredity advanced by modern genetics. In his recently published books—*Problems of Darwinism* (1946) and *Factors of Evolution* (1947)—he had foreseen little future for Lysenko's reworking of Darwin's theoretical legacy.

The third leading scholar to suffer heavy punishment was the geneticist N. P. Dubinin. He lost both his administrative position and a place to do research when the Lysenkoist steamroller erased the Laboratory of Cytology, Histology and Embryology, which he had headed and which was the last stronghold of experimental research in genetics.[41] The last link with the legacy of S. S. Chetverikov, a pioneer in population genetics during the 1920s, Dubinin was noted particularly for experimental work in the genetic foundations of selection and in the contributions of genetics to a modern theory of evolution.

The hurried reorganization of the biological research centers in the Academy—and in the nation—caused a heavy dislocation of professional talent as well as many personal demotions. Immediately after the two sessions, *Bol'shevik* found it opportune to clarify the new course in biology and to reassure all concerned scholars that the new line was official:

> The ranks of workers in biology now face the task of bringing all general and specialized sciences related to agriculture into harmony with the Michurinist evolutionary theory, based on dialectical materialism. The reactionary and idealistic theory of Weismann and Morgan—which has no basis in science and is engaged in spreading the disease of agnosticism and in crippling the work of scholars engaged in experimental research—has been exposed, rejected and removed from the paths chosen by Soviet science. This step is a turning point in the history of our science; it signifies the beginning of unprecedented acceleration of advances in the biological sciences, united in the grand stream of Michurinist materialistic biology.[42]

V. M. Molotov, who occupied top positions in both the party and the government and who was totally alien to the craft and spirit of science, used the celebration of the thirty-first anniversary of the October Revolution to sum up the Lysenkoist crusade:

> The discussion of heredity has raised questions of principal theoretical significance in the struggle of true science, based on the principles of materialism, against such reactionary and idealistic survivals in scientific work as Weismann's theory. It has underscored the importance of materialistic principles for all areas of science and has helped to accelerate the progress of scientific theory in our country.[43]

A good theory could not but improve social well-being. As presented by the swelling chorus of professional propagandists, Lysenko's theories—Michurinist biology—had taken pseudoscience, alien ideology, and mysticism out of science; but they had also produced richer harvests in Russian fields. Relying on information contrived and disseminated by Lysenko and his aides, *Bol'shevik* wrote:

Many facts have attested to the great help that the Michurinists have given to our agriculture. The 1948 plan has called for planting cereals vernalized by Lysenko's methods on seven million hectares of agricultural land. The "Odesskaia 3," a new breed of frost- and drought-resistant wheat, gives three to four more centners of wheat per hectare than usual varieties. Lysenko's measures have increased the output of millet by fifteen centners per hectare. Lysenko's method of scorching seed as a way of accelerating the germination process and his method of introducing winter wheat in the Siberian steppe have contributed to higher agricultural yields in the eastern regions of the USSR. Summer planting of potatoes has opened a reliable path to achieving high yields of this staple in our southern regions.[44]

With the Department of Biology under their full control, Lysenko's forces could now issue authoritative statements supporting the scientific legitimacy of the new theory. Hundreds of sessions presented Lysenko as the destroyer of both "formal genetics" (the genetics of Gregor Mendel and William Bateson) and "neo-Darwinism" (a modern amalgam of the theories of mutation and natural selection). He was proclaimed the creator of an original synthesis of Lamarck's theory of the hereditary transmission of acquired characteristics and Darwin's theory of natural selection. The "synthesis"—presented as "creative Darwinism"—had no room for Lamarck's belief in predetermined progress. Nor did it go along with Darwin's Malthusian bias, his emphasis on random change in living forms, or his disregard of qualitative "leaps" in biological processes. It "succeeded" where Darwin's theory had failed: it offered a full and consistent "explanation" of the origin of variation as the main agent of adaptation. It also "advanced" techniques for the human control of evolution.[45]

The search for legitimation concentrated also on showing the congruence of Lysenko's theories with the basic postulates of dialectical materialism. Lysenko's theory stood for absolute causality as the basic tool of scientific explanation; it recognized the existence of both qualitative (revolutionary) and quantitative (gradual) changes in living nature; it treated "internal contradictions" as the basic source of vital impulses; and it made "practice" the exclusive judge of scientific theory. Critics who had claimed that Lysenko's theories were actually contrary to dialectical materialism were now completely silenced. These critics had contended, for example, that Lysenko's negation of intraspecific struggle for survival was an outright denial of the role of contradictions in vital processes. His denial of a corporeal base of heredity, they argued, went against the line of reasoning advanced by Engels and brought his theory closer to vitalism than to the materialist tradition in natural science.

Lysenkoism spearheaded a politically articulated movement to achieve the ultimate goal of Stalinism: to make the philosophical unity of science the basic precondition for a broader unity of science and ideology, as advanced by Marxist philosophers. To achieve all this, it was necessary to extend the Lysenkoist-type attacks to "the fields of physics and chemistry, geography and geology, and even medicine." Under the pressure of political forces outside the scientific community, the Academy vowed to play the leading role in the current war against ideological impurities and "alien influences."[46] Ideological attacks, administrative purges, and institutional recastings became daily occurrences. To add to the gravity of the situation, many opportunists—in search of self-aggrandizement— became loyal soldiers in Lysenko's army. Among them the best known was academician A. I. Oparin, the internationally reputed expert in chemical studies of the origin of life. Having been made the new academic secretary of the Department of Biological Sciences, he expressed his gratitude by making Lysenkoism a theoretical pillar of his own biological thought and by joining the merciless campaign against ideological indiscretions in the scientific community.

Lysenko's war on the leading Western biological theories contributed to the rise of wild doctrines that combined full submergence in ideological pathos with total disregard for laboratory experiments. One such theory was advanced by O. B. Lepeshinskaia, an old Communist stalwart and a strong and vociferous supporter of the war on cosmopolitanism. During the waning years of the Stalinist era, the Lysenkoist circles made her theory of the noncellular origin of some cells—which contradicted Rudolf Virchow's claim that all cells came from cells—an organic part of Michurinist biology.[47] She based her claim on her own "experimental" research, on the Marxist theory of "qualitative leaps" in the development of living nature, and on extensive citations from Engels's *Dialectics of Nature*. According to a contemporary statement:

> The entire Virchowian cytology stands in absolute opposition to Michurinist theory. By contrast, Lepeshinskaia's theory reinforces Michurinist explanation of the causes and mechanisms of qualitative leaps in the transformation of living organisms. Indeed, one of the basic propositions of the Michurinist theory is that living bodies originate in inorganic substances.[48]

Virchow's "theoretical generalizations," according to a Lysenkoist source, "were thoroughly metaphysical and idealistic"; they brought much harm to biology by "hindering its growth."[49] Changes in environment, according to the new theory, were direct causes of changes in cells. T. D. Lysenko was quick to recognize the accumulation of extra-specific noncellular substances by organisms as the primary source of new species. B. G. Kuznetsov summed up the ideological argument when he stated in 1951 that "the picture of the origin of cells from noncellular living matter has contributed to a further development and enrichment of our world view and to a defeat for reactionary Weismannism and Virchowianism."[50] In addition, he said, Lepeshinskaia's ideas raised biological theory to unprecedented heights. They "enriched" both science and dialectical materialism, and they "vindicated" a strong national tradition in biological thought: that is, D. I. Pisarev's defense of the theory of "spontaneous generation" against the "reactionary arguments" advanced by Pasteur in the 1860s.[51]

Impressed with her sudden rise to eminence, Lepeshinskaia wrote a long essay presenting her ideas as an organic component of the materialist tradition in science created by Pavlov and Michurin/Lysenko.[52] She was particularly interested in showing a fundamental agreement of this tradition with dialectical materialism. Numerous biological conferences were held in 1950–1953 for the purpose of unveiling the full scope of Lepeshinskaia's "revolutionary" discoveries and showing their roots in dialectical materialism. Oparin thought that the new "theory" offered key leads to decoding the mysteries of the origin of life.[53] In 1952 a conference sponsored jointly by the Biological Department of the Academy of Sciences and the Academy of Medical Sciences celebrated the second anniversary of the "triumph" of Lepeshinskaia's theory. All papers presented at the conference had one theme: the "incorrectness" of Virchow's theory and the "correctness" of Lepeshinskaia's materialistic position.

Michurinism—as elaborated by Lysenko—was only one theoretical pillar of Soviet biology as seen through the prism of Stalinist ideology: the Pavlovian neurophysiological theory was the second pillar. Pavlov died in 1936. L. A. Orbeli, after his appointment as the new head of the Institute of Physiology, announced that in the future the institute would not be dominated by a single person and that scholars would be encouraged both to advance Pavlovian ideas along previously uncharted paths and to search for new insights and ideas unrelated to Pavlov's scholarly legacy. To the traditional Pavlovian concerns he personally added a strong interest in the evolution of the nervous system, which attracted much attention in scholarly circles, as did the physiology of sense organs.[54]

Orbeli's job, however, was made difficult by the presence of a strong group of institute members who were unhappy about the choice that made him director of the institute.[55] At the end of the 1940s it became clear that the Pavlovian school had evolved along two different lines: the scholars who stuck closely to Pavlovian ideas and experimental procedures and the scholars who, while paying homage and expressing loyalty to the great man, tried to expand, enrich, and recast both Pavlovian theory and Pavlovian methodology by moving physiology toward new areas of research.[56] The latter group encountered strong opposition from two quarters: from the defenders of Pavlovian orthodoxy, who felt that the new theoretical notions and methodological designs went against the spirit of Pavlov's legacy; and from the ideologues, who saw in the new heterodoxy a threat to the unity of Pavlovian neurophysiology and dialectical materialism, a precariously balanced combination.

All this preceded the joint session of the Academy of Sciences and the Academy of Medical Sciences held in Moscow on June 28 to July 4, 1950. The session was greeted by *Bol'shevik* as a continuation of the "Michurinian conference" held in August 1948, which made Lysenko the indisputable master of Soviet biology.[57] The relentless fury and vast scope of the attacks directed at the scientific community during this joint session were surpassed only by the assault that led to the consolidation of Lysenkoist power two years earlier. A long procession of speakers, leaning heavily on nationalist pathos, exaggerated statements and out-of-context citations, attacked selected representatives of the "deviationist" school, led by L. A. Orbeli, P. K. Anokhin, and I. S. Beritashvili, who were found guilty on many counts: trying to substitute functional and holistic interpretation for causal analysis of neurophysiological processes; disregarding the dangers of—or flirting with—the theory of psychophysical parallelism; making an effort to bring together neurophysiology and genetics; and generally tampering with Pavlov's "materialistic straightforwardness."[58] The victims were not forewarned that the session was coming so that they could have ample time to prepare defense statements. Their voices were muffled by boisterous and insulting remarks from carefully selected gallery crowds. The victims were quickly demoted, their books withdrawn from circulation or proclaimed undesirable reading.

As the campaign against alleged deviationists grew in size and intensity, the criticism became more sweeping and acrimonious. Beritashvili was found guilty of having claimed that the "principle of conditioned reflexes" could not explain the total picture of psychic activity. P. K. Anokhin was attacked for his alleged claim that Pavlov's work suffered from "analytical limitations," responsible, in his opinion, for the incompleteness of the theory of conditioned reflexes.[59]

These and other unfortunate neurophysiologists were found guilty not only of maintaining fallacious views and of having misguided substantive interests but also of harboring unpatriotic sentiments. A resolution passed by the session ordered the holding of annual meetings of physiologists for the sole purpose of protecting the sanctity of the Pavlovian legacy and the unity of neurophysiology and dialectical materialism. Psychologists were instructed to abandon all theories and research activities that did not adhere to the basic premises of Pavlov's theory of conditioned reflexes. Orbeli was told directly that neither his "high authority in science" nor his "real scientific contributions" were sufficient to elevate his "ideological errors" above "severe criticism."[60] The triumph of Lysenkoism made "ecological physiology"—concerned with the influence of planned changes in the environment on physiological processes—a special concern of the Institute of Physiology. Pavlov's theory became a closed system, ironclad and ideologically pure. Its authority grew larger and more unbending; its rigidity quickly closed many openings inviting methodological innovation and philosophical challenge.

The triumph of Michurinism was also a triumph for V. R. Vil'iams's theory, which Lysenko proclaimed the "socialist method of soil conservation." Unlike Pavlov's theories, however, Vil'iams's doctrine needed revisions to become an inextricable component of Lysenkoism; in particular, it needed a stronger emphasis on climatic conditions and a more detailed concern with idiosyncratic features of individual regions. Once the need for readjustment was spelled out, Vil'iams's school was reassured of a continued position at the heights of socialist science, even though the scientific community continued to be unimpressed with the scientific merit of Vil'iams's theory.

Vil'iams, dead since 1939, continued to serve as a model scientist of the new society. The *Great Soviet Encyclopedia* (second edition) stated:

"Vil'iams's agronomic theory is an integral part of Michurinist agrobiological science, successfully advanced by Soviet scholars headed by
T. D. Lysenko."[61] The government decision of October 24, 1948, which
called for the forestation of the Russian steppe, wide application of
grassland rotation, and construction of reservoirs and ponds as parts of
a soil conservation move, relied heavily on Vil'iams's legacy for suggestive ideas and technical guidance. The corrections to which his general
theory was subjected did not come from a consensus of scholarly opinion; instead, they were specific responses to a ruling of the Central
Committee of the Communist party.[62]

Lysenkoism gave the empire of biology inner unity, single purpose,
national spirit, and religious fervor. It made biology part of an exaggerated anthropocentric orientation by tying interest in basic theory to
the immediate needs of socialist agriculture as formulated in five-year
plans. It made agrobiology a synonym for the fundamental theory of
the sciences of life. And it achieved all this by discouraging disciplined
skepticism, critical digression, challenge to authority, and exploratory
zeal.

It did not take long for the crusade for ideological unity to assert itself
outside the huge dominion of biology. In December 1948 the Leningrad
section of the All-Union Astronomical and Geodetic Society sponsored
a conference, attended by over 500 delegates, for the purpose of criticizing cosmological ideas alien to dialectical materialism. A long procession of speakers subjected relativistic cosmology and mathematical
cosmogony to acrimonious criticism.[63] The conference greeted the emergence of Soviet cosmogony as a unique discipline dominated by three
principles: historicism, practicalism, and materialism. Historicism
reached a high point in the work of the Armenian astronomer V. A.
Ambartsumian, who argued in favor of making the evolution of galaxies
a central problem of astronomy and who concentrated on protostellar
and other nonstationary phenomena. In full consonance with Ambartsumian's historicist bent, the philosopher B. M. Kedrov disputed Jeans's
theory, which perceived the universe in a static state interrupted occasionally by "sudden jolts and explosions"; the Soviet cosmogonic theory,
according to him, emphasized the constancy of cosmic change.[64] Practicalism found strong expression in the stern criticism of the astronomical
"school" of D. N. Moiseev, accused of heavy reliance on a kind of mathematical formalism far removed from the practical needs of the five-year

plans. Moiseev was also reminded that "formalism"—mathematical or otherwise—was a specific expression of "astronomical idealism."

Cosmogonic materialism received strong support from the newly formulated hypothesis of the origin of the planet Earth, advanced by O. Iu. Shmidt.[65] In an elaborate argument, Shmidt saw the origin of the planets in rotating protoplanetary "swarms" of gas and dust particles that at one time surrounded the Sun. After having "captured" this mass, the Sun gave it the unique planetary distribution of matter and angular momentum—and shaped it into a planet. According to this cosmogonic hypothesis, the Earth was never in a molten state; the high temperature inside the Earth came from the disintegration of radioactive substances. Since Shmidt was a national hero and a "revolutionary scientist," all his ideas were predestined to receive wide publicity. He presented his theory as the result of a collective effort of a specially selected research team, and he went out of his way to point out the deep roots of his cosmogonic ideas in the mainstream of Russian and Soviet astronomical tradition.

Shmidt's theory was presented as a dialectical synthesis of modern knowledge and the Kant-Laplace hypothesis. It made broad references to the work of gravitational force in the evolution of planets; it was based on a thermodynamic model, which helped it to avoid the pitfalls of the Kant-Laplace mechanistic explanations; and it made much use of analogies borrowed from nuclear physics. Most of these excursions, however, were used primarily as decor; the theory had no experimental backing, no support from the open vistas of astronomical knowledge, and no consistency in moving from one level of reasoning to another. Shmidt did not hesitate to ally his theory with the Lysenkoist emphasis on the role of environment in the history of natural phenomena. The temper of the time, rather than the burden of scientific evidence, led him to take a particularly critical view of Western cosmology, in which he saw a paradoxical contrast between the richness of observatories and instruments and the poverty of fundamental theory.[66] He admitted that his theory required much additional work, but that its dialectical moorings needed no additional buttressing.

With the Markov controversy brought to an abrupt end, the time had come for a public demonstration of the philosophical and ideological unity of the rapidly growing community of physicists. In 1949 the journal *Advances in Physics* revealed that "the unmasking of idealism in the biological sciences has exercised a favorable influence on scholars working in theoretical physics," and that "the Leninist materialistic theory of knowledge can resolve all the philosophical problems of modern physics and can serve as the fulcrum of the war effort against the idealistic strivings of foreign scholars."[67] A series of articles in the same journal, featuring the main lines of attack on "physical idealism," prodded scholars to work for a full victory of "mass" over "energy," "causality" over "indeterminacy" and "complementarity," the Leninist notion of the "inexhaustibility of electrons" over epistemological "agnosticism," and "materialistic relativity" over "idealistic relativism."

In 1948 V. L. L'vov, a science reporter, took it upon himself to transmit the new criticism of "physical idealism," articulated by the leaders of the campaign for the ideological purity of science, to the readers of *Literary Gazette*. He called for a new quantum mechanics—a Soviet quantum mechanics—free of the principles of indeterminacy and complementarity, the notion of measuring instruments as key factors in determining physical reality, "exaggerated" reliance on mathematical formalism ("mathematical mystification"), and similar "negations" of the objective nature of physical processes.[68] The ominous pressure of the new crusade was so overpowering that even V. A. Fok, the most consistent Soviet follower of Bohr's views, was now compelled to state that the principle of complementarity had given birth to so many "erroneous notions" that it would be advisable to drop it.[69]

Marxist critics also struggled against what they viewed as lingering elements of "neopositivism" in Soviet physical theory. S. E. Khaikin, author of a popular college textbook in mechanics, was bitterly attacked as a follower of operationalism; the critics noted that he preoccupied himself with techniques for measuring experimental data rather than with scientific abstractions. Furthermore, Khaikin was accused of defending an ahistorical position in physics and of manifesting a clear tendency to isolate physical theory from philosophical—or dialectical-materialistic—considerations.[70] Dialectical materialism, reaffirmed as the only reliable philosophical tool of modern physical theory, was seen as performing two vital functions: meeting the urgent need for a way out of the "crisis" in physical theory and offering a theoretical framework unencumbered by alien idealistic and positivistic elements.

A national conference of physicists, planned for 1949, was intended to give public airing to the newly forged unity of physics and Marxist philosophy—to the rejection of scientific ideas inconsistent with dialectical materialism. Although the congress was never held, for reasons not made public, the papers prepared for it made up the chapters of a much-heralded volume published in 1952.[71] The volume presented the most vitriolic Soviet attack yet on the general theoretical principles and philosophical views of the Copenhagen school and on Einstein's scientific contributions. The contributors to this symposium volume did not even spare such leading Soviet physicists as L. D. Landau, Ia. I. Frenkel', L. I. Mandel'shtam (dead since 1944), and A. F. Ioffe, popularly known as the dean of Soviet physicists. These scholars, and many others, were accused of a lenient attitude toward the "idealistic" biases of Western physicists.[72] The philosopher M. B. Mitin took time off from his ebullient praises of Lysenkoism to draw the attention of recalcitrant physicists to some choice "incriminating" citations from Niels Bohr, J. H. Jeans, Werner Heisenberg, and Bertrand Russell, passages that illuminated the "agnosticism," "idealism," and "mysticism" of Western physical theories. He mentioned, for example, the idealistic foundations of Bohr's principle of complementarity, the "mysticism" of Jeans's attack on causality as a key component of scientific explanation, and the "agnosticism" of Russell's claim that electrons and protons are complex logical constructs rather than the building blocks of the physical world.[73]

Spearheading the attack on Einstein, M. E. Omel'ianovskii, a fast-rising Stalinist philosopher, noted that Einstein's interpretation and exposition of the theory of relativity suffered from "idealistic falsifications of a number of the most important problems."[74] He lamented the failure of Soviet scholars to produce a systematic study of the theory of relativity, unhampered by the role of "observer" in the construction of physical reality and free of "thought experiments" as research devices. A. A. Maksimov no longer hesitated to subject the theory of relativity to public ridicule and to proclaim that "many physicists" were now convinced that Einstein's theory of relativity was "the blind alley of modern science."[75] I. V. Kuznetsov, one of the angriest and most bellicose standard-bearers in the war on ideological impurities in science, asserted categorically that Einstein's scientific theories, built on idealistic foundations, could not by any stretch of the imagination be considered correct. A faulty philosophy, in his opinion, could only result in a faulty scientific theory. "The example of Einstein shows how pernicious are the effects of a reactionary idealistic philosophy on the creative work of a scholar and on the development of natural science."[76] This criticism went against the prevalent opinion of Soviet physicists, who favored

keeping separate accounts of Einstein's physics and Einstein's philosophy. But this was not a time when the physicists could actively oppose
views of the kind expressed by Kuznetsov.[77]

To prove the correctness of Kuznetsov's dictum that a "faulty philosophy" could produce only a false scientific theory, the philosophers faced
the difficult task of proving that Einstein's physical theories were based
on false premises. No philosopher, however, was prepared to undertake
such a task, for there was too much in Einstein's thought that had become
part of standard knowledge in modern physics. To resolve the dilemma,
E. Ia. Kol'man offered a modified version of Kuznetsov's interpretation.
Noting that the objective core of the laws of science was much stronger
than the subjective whims of individual scientists, he advanced the argument that an idealistic bias should not prevent a scholar from constructing "objective" theories. He hastened to add, however, that the
combination of erroneous philosophy and correct scientific generalizations was a transitory phenomenon; sooner or later, an idealistic bias in
philosophy led inevitably in the direction of false "scientific" theories.[78]

In 1949 *Voprosy filosofii* carried two articles on the relation of quantum
chemistry to the structural theory formulated by A. M. Butlerov in 1860.
While the Soviet critics did not agree on many points of quantum chemistry, they were united in arguing that Butlerov's theory was solidly
grounded in materialistic philosophy and that Linus Pauling's resonance
theory was a product of thinly masked philosophical idealism.[79] A year
later, an editorial in the same journal stated that Pauling's theory must
be rejected because it was based on E. Mach's principles of "economy
of thought," it reduced chemical phenomena to "physical and mechanical laws," and it created confusion in science by allowing for the use
of many formulas to describe the same substances.[80] At the same time,
Moscow University, the Academy Institute of Organic Chemistry, the
Academy Institute of Inorganic Chemistry, and the Karpov Institute of
Physico-Chemical Sciences held special conferences condemning the
"uncritical attitude" of individual chemists toward the "idealistic" trend
in Western chemistry. *Pravda*, in the issue of August 10, 1950, singled
out "Anglo-American" idealistic aberrations in chemistry for special
attack.

In 1951 the campaign against the idealism of "Anglo-American quantum chemistry"—and particularly the resonance theory—reached a
peak. The Department of Chemical Sciences sponsored the All-Union
Conference on the Theories of Chemical Structure to look into the grow-

ing popularity of structural orientations in organic chemistry. Attended by 400 persons, the conference unleashed a furious attack on individual scholars who accepted and worked on advancing Pauling's "idealistic" resonance theory, an intricate application of quantum-mechanics principles to the study of chemical structures.[81] The conference made it clear that ideological aberrations in chemical theory were inseparable from "alien theories" in biology and physiology, that all were parts of a united front against materialism.[82] In early 1953 a noted chemist stated that the resonance theory, which "erroneously" identified "valence schemes" with "molecular structures," was merely a distortion of A. M. Butlerov's theory of chemical structure formulated in the 1860s.[83] This scientist was eager to show not only that Butlerov's was the only scientifically established theory of chemical structure but also that the Russian scholars deserved the main credit for making it a vital part of modern science.

The social sciences did not escape the calamities of ideological expurgation. The Institute of Ethnography held a conference in 1949 for the purpose of pointing out and condemning ideological digressions in the published work of its members. The session found P. G. Bogatyrev, an eminent folklorist, guilty of expressing himself favorably about the "functional-structural method" in ethnography, a creation of Western scholarship. He was reminded that this was an idealistic method because it placed more emphasis on cultural forms than on social content, that every kind of formalism led inevitably to a servile attitude toward alien ideas and that it was impossible to be a formalist and a patriot at the same time.[84] In the meantime, the institute sponsored a symposium on the major theoretical orientations in Western cultural anthropology, showing their close affiliation with colonial policies and Freudian psychopathological views. Similar sessions were held in many other institutes. All sessions followed the same pattern: there was only one victorious group—the group selected and supported by political authorities external to the scientific community. In all of them the victors were cheered and the villains taunted by galleries carefully selected by organized groups with no interest or competence in the pursuit of science.

The Institute of Economics received its share of publicly aired attacks on ideological nonconformity, and these attacks led to distracting institutional readjustments and extensive personnel dislocations. The institute was also the locus of one of the most tragic Stalinist attacks on the Academy: in early 1950 N. A. Voznesenskii, an academician since 1943, was imprisoned by the secret police and was promptly executed. There were no conferences debating his "crimes," no open trials, and no related events in the public campaign for Stalinist orthodoxy. A special case, this tragedy went beyond the issues surrounding the Academy: Voznesenskii was not only an academician but also a member of the Politbureau, the most powerful political body in the Soviet Union. The tragedy of Voznesenskii began to unfold in 1949 when he was suddenly dropped from all government and party positions. The trouble probably began with the appearance of his book on the Soviet economy during World War II, published in 1947, which earned him the reputation of a leading economist of Soviet socialism—an honor that Stalin most likely did not want to share with any other person.[85] According to some sources, he was implicated in the so-called Leningrad affair, an alleged anti-Stalin conspiracy.[86]

In addition to conferences that centered on the ideological impurities in individual sciences, there were other conferences that dealt primarily with the evils of cosmopolitanism and concentrated on a ceremonial recounting of original Russian contributions to science and technology. Cosmopolitanism, according to *Bol'shevik*, "is a negation of patriotism, its very opposite. . . . It preaches full indifference toward the future of the motherland, [and it] recognizes no civic or moral debt to the nation."[87] Worst of all, "modern bourgeois cosmopolitanism invites a rejection of national sovereignty." As viewed by Stalinist ideologues, cosmopolitanism undervalued the historical uniqueness of individual nations and, in the Soviet case, bred a servile attitude toward the cultural achievements of the West.

In March 1948 the Leningrad department of the All-Union Astronomical-Geodetic Society held a huge conference whose agenda was taken up by detailed reviews of the original contributions of Soviet scientists in their efforts to surpass the achievements of astronomers outside the Soviet Union.[88] Two months later, the Academy sponsored the First Conference on the History of Russian Chemistry in Moscow, which attracted 500 scholars representing the leading scientific centers of the nation, all inspired by the patriotic duty to uncover or reestablish the hidden priorities of Russian chemistry. The purpose of the conference was set forth by the introduction to the published proceedings:

> The conference participants have learned to appreciate the enormous and honorable task we face in the study of the national history of chemistry. We must examine the scientific legacy of our leading scholars—the founders of chemistry and its particular orientations—on a

much broader basis than we have done in the past. It is necessary to reestablish the historical truth, distorted by numerous falsifiers of history—to bare the names and contributions of the leading Russian chemists, buried in various archives by reactionaries and obscurantists, servile idolizers of the foreign world, who occupied high positions in scientific and educational institutions of tsarist Russia. We must constantly defend the scientific priorities of Russian scholars in important discoveries and inventions which are so often attributed to scholars or pseudoscholars in capitalist countries.[89]

The war on cosmopolitanism was a war against real and imaginary efforts to deny Russian priorities in science and technology. According to a leader of the anticosmopolitan crusade:

The struggle to establish national priorities in science is part of the war on cosmopolitanism. Representatives of militant cosmopolitanism preach a false "theory" of the non-national character of world science. Under the guise of this "theory," they ascribe the unearned glory of discovery to their own nations and deny other peoples a role in the development of world science. As our press has shown clearly, this aggressiveness has been particularly evident in the interpretation of Soviet contributions to science. The cosmopolitan conception of world science is theoretically unsound and politically reactionary. World science is not non-national; it does not grow outside concrete historical forms. Every science, like every culture in general, is national in form and class-oriented in content. Every scientific discovery belongs to a specific nation; every new theory is created by scholars belonging to a particular people. "World science" is only a summation and a synthesis of contributions made by scholars of different countries. The struggle for priorities in science is of great political and theoretical significance. Without it, we cannot meet the task, placed before Soviet science by Comrade Stalin, of "not only emulating the scientific achievements beyond the boundaries of our country, but also surpassing them in the near future."[90]

In the search for new Russian priorities, Lenin received the laurels for the most eminent achievement. According to B. M. Kedrov:

A hundred years ago, Le Verrier discovered Neptune, . . . giving Newtonian mechanics its greatest triumph. A quarter of a century later, Mendeleev predicted the existence of gallium, scandium and germanium, giving the periodic law of elements its greatest triumph. At the beginning of the twentieth century, Lenin, endowed with the power of a genius, pointed to new physics as a confirmation of dialectical materialism. By predicting the inevitable coming of a deeper and more comprehensive understanding of the development and transformation of matter, he anticipated the birth of a science of elementary particles as the foundation of modern physics. Lenin's achievement marked the greatest triumph for Marxist-Leninist dialectics and was superior to the first two discoveries.[91]

The work toward lengthening the list of Russian priorities in science went hand in hand with the mounting effort to show the superiority of Soviet science. New priorities were discovered across the entire field of scientific knowledge, but particularly in the social sciences. The "inferiority" of Western ethnography—or cultural anthropology—attracted most attention. Not a single school in Western ethnography escaped the wrath of the aroused critics. Franz Boas was chastised for making the "agnosticism" of neo-Kantian epistemology the foundation of an idiographic orientation in ethnography. Ruth Benedict's search for distinct patterns in the integration of individual cultures was condemned as psychological reductionism; according to a typical criticism, she treated cultural leitmotifs as ontological entities untouched by the vicissitudes of history.[92] Gregory Bateson evoked criticism for having made Benedict's ethnopsychology part of the current ideological campaign in favor of a world-state dominated by "American monopolists."[93] Bronislaw Malinowski, the founder of functionalism in cultural anthropology, was seen as a charlatan moved by stylish thought in science and total subservience to colonial authorities. He too was chastised for his contempt for history as a method of inquiry and as a scientific perspective.[94]

As might have been expected, the Institute of Philosophy was particularly active in the seemingly unending upsurge of anticosmopolitan sentiment. It sponsored a succession of internal meetings dominated by acrimonious attacks on associates whose thinking was marred by cosmopolitan digressions. B. M. Kedrov was found guilty of insufficient appreciation of the "uniqueness of Russian science and philosophy" and their progressive orientation. He denied "the importance of priorities in scientific discoveries and fought for 'one world science.'" Z. A. Kamenskii, another associate of the institute, was accused of denying the existence of a materialistic tradition in Russian philosophy and of exaggerating the role of Schelling's idealism in the emergence and development of Russian philosophical thought. M. Z. Selektor advanced the "antipatriotic idea" of a full absence of ties between the revolutionary-democratic tradition in Russian thought and Leninism.[95]

Nor were other institutes in the humanities spared this externally imposed soul-searching and self-denunciation. Individual associates of the Institute of History were ordered to answer the charges that they had undervalued the international role of the October Revolution, had overlooked the leading part of the Russian nation in building Soviet socialism, and had given little attention to the originality of Russian culture and revolutionary tradition. *The History of the U.S.S.R.*, a college textbook prepared by the institute, claimed that A. N. Radishchev, "the pride of Russian literature, a great materialist, and a fighter against serfdom," borrowed his ideas from the eighteenth-century French materialists and Leibniz, and his literary style from the English writer Sterne. Similar charges were leveled against individual associates of the Institute of the History of the Arts and the Institute of the History of World Literature. V. Ia. Kirpotin, a member of the latter institute, was attacked for identifying the literary work of Lomonosov, Radishchev, and Krylov as "pseudoclassical."[96] In a special reference to the Islamic and Oriental studies, W. S. Vucinich has noted:

> From 1949 until 1951 leading Soviet newspapers and journals often published warnings to historians and literati, as well as to the institutes sponsoring them, and offered acceptable interpretations of controversial issues in the history of the Soviet Muslim and certain other Asian peoples. At various professional conferences Pan-Islamism and other forms of "cosmopolitanism" and "bourgeois nationalism" were condemned. A number of Orientalists were reprimanded or purged and their works condemned. New studies with ideologically acceptable interpretations were produced. In their writings Asian authors were obliged to refrain from expressing any ideas or interpretations that were anti-Russian, and were told to honor and extol the many virtues of the "Great Russian people," under whose leadership the Soviet peoples would attain a common supranational culture for the entire "Soviet family" of nations.[97]

The new ideological line made it imperative for all scholars to adhere to a monolithic system of ideas that viewed the leading Russian intellects of the past as builders of materialistic philosophy and a unique national culture and as fighters against tsarist autocracy and institutions based on serfdom. One of the basic tasks of scholars in the humanities was to present the Russian people as the cultural and political vanguard of the family of Soviet nations. Responding to this dictate, I. V. Kuznetsov supplied a suitable explanation:

> Independence and originality, the spirit of innovation, have been the basic characteristics of Russian natural science. Russian science is not a mere repetition of Western European thought; it speaks a new and weighty language, not limited to secondary details and fragmented

knowledge. The annals of science show that the leaders of Russian natural science express broad views and tackle general problems. Among Russian scientists we find the creators of new disciplines, new orientations, new methods of scientific inquiry and new technological systems.[98]

With the intent of unifying the diffused interest in the national origins of Russian science, the Academy of Sciences sponsored a general session on January 5–11, 1949, dedicated to "the history of national science." A long procession of speakers dealt with the national history of all the major sciences and their leading branches.[99] To set the tone, S. I. Vavilov, the new president of the Academy, stated that the time had come to give the national effort in science the credit it deserved and to eradicate the last residues of the intolerable tradition, with roots traceable to the eighteenth century, that encouraged a "contemptuous attitude" toward national achievement in science and a servile admiration of Western scientific thought.[100] In the past, he said, the Russians had committed many unpardonable sins in appraising the work of the national leaders in science. M. V. Lomonosov's pioneering work in several sciences waited a century and a half to receive national recognition. N. I. Lobach-evskii, the creator of the first non-Euclidean geometry, waited fifty years to be recognized; when he was finally recognized, he shared the laurels with K. F. Gauss. Guided by "pseudoobjectivism," some native historians of science looked at D. I. Mendeleev as only one of several discoverers of the periodic law of elements.[101] The name of M. S. Tsvet, the discoverer of the chromatographic method, "which can be favorably compared with spectral analysis," remained unnoticed in Russia for a long time. In his eagerness to make an impression on the gathered scholars, Vavilov relied heavily on dramatic overstatements and a free interpretation of the historical record.

The resolution passed by the conference was both sweeping and categorical. In requesting the Academy to undertake the publication of studies "illumining the basic questions of the history of science and technology from Marxist-Leninist positions," it stated:

> By relying on the theories of Marx, Engels, Lenin and Stalin, these studies must present a correct appraisal of the motive forces, historical causes and social importance of scientific and technical discoveries. They must produce a correct periodization of the history of science and technology, show the struggle of the materialism of our national science

against the idealism and metaphysics of the bourgeois history of science and technology and protect national priorities in scientific discoveries and inventions.[102]

The resolution did not state who had the authority to choose the "correct" approach to the history of science. The question of authority was answered in August 1948, when T. D. Lysenko announced publicly that his antigenetics orientation was a "correct theory" because it was approved by the Central Committee of the Communist party.[103] The session attended to yet another matter: it decided to erase the name of the American geneticist H. J. Muller from the roster of corresponding members.[104] Muller was among the most astute and uncompromising Western critics of Lysenkoism as a pseudoscience and as an attack on the moral code of science.

Marxist scholars found the prospect of working in many fields of the history of science particularly attractive. They hoped that this discipline would give them a more responsible—and a more respected—place in the scientific community. In their view, science was propelled by an inner impulse, or an internal logic, as well as by external (sociocultural) conditions. While specialists in individual sciences were called upon to study the inner logic of the growth of science, Marxist scholars made the interaction of "internal" and "external" factors their legitimate field of study. External factors, as they saw them, consisted of two sets of determining conditions: the needs of technology, at the structural base, and the dictates of ideology, at the superstructural apex. Technology influenced the experimental base of science; ideology determined the theoretical orientation of science.

A. A. Maksimov defined the natural sciences as "a totality of objective truths discovered on the basis of production practices and experiments, and united, directed, and interpreted on the basis of a definite world view."[105] The Marxist historians of science contended that they alone could be relied upon to make the correct use of the Marxist dictum that science was a unique expression of ideology and social consciousness. They alone could be trusted to study the history of science as a progressive confirmation and enrichment of the basic postulates of dialectical materialism. They assumed positions of leadership in the rapidly spreading effort to make the history of science a weapon in the war on cosmopolitanism. In this enterprise they were guided by the idea that "to ignore the national element in the study of the growth of science is to separate science from the social soil on which it feeds."[106] While world science was an "abstraction," national science was a historical "reality." To ignore this reality was to fall into the trap of cosmopolitanism.

The work on the national and materialistic character of Russian science proceeded in many directions. The Institute of the History of Natural Science was the center of this activity. It adopted an ambitous plan consisting of three types of activity: editorial management of *Proceedings*, containing papers on Russian contributions to science and technology; publication of the series *Scientific Legacy*, presenting previously unpublished correspondence of eminent Russian scholars; and republication of the major works—or complete works—of leading Russian scientists.[107] The publication of the works of such giants of Russian science as M. V. Lomonosov, N. I. Lobachevskii, A. M. Butlerov, S. M. Sechenov, D. I. Mendeleev, and I. P. Pavlov was carried out with exemplary efficiency. Despite the questionable motivation behind it, this publication spree was a cultural event of memorable proportions. The book market was inundated with studies of the life and work of the masters of Russian scientific tradition.

Not so beneficial or commendable were the extensive studies that concentrated on refuting selected Western interpretations of Russian scientific thought. Criticism—often misdirected and distorted—dealt with Western writings that were interpreted as deliberate efforts to belittle Russian contributions to science or to minimize the materialistic leanings of Russian naturalists. The anticosmopolitan historians were particularly resentful of an apparent tendency in the West to underestimate the role of A. M. Butlerov in the formation of structural chemistry, to place Lobachevskii's non-Euclidean geometry into the framework of an idealistic philosophy, and to scoff at Pavlov's neurophysiological analysis of behavior.

The spirit of anticosmopolitanism called for a critical examination of the current work on the history of Russian science. Most criticism concentrated on digressions from the anticosmopolitan path. Even A. A. Maksimov, the most ardent and pugnacious philosopher of Stalinist persuasion, did not escape criticism: B. M. Kedrov discovered in his *Essays on the History of the Struggle for Materialism in Russian Natural Science*, published before the actual start of the anti-Western campaign, an inexcusable number of cosmopolitan indiscretions.[108] Ironically, Kedrov, too, became a target of similar criticism. The physicist D. D. Ivanenko found his *Engels and Natural Science*, published in 1946, much in need of a broader treatment of Russian contributions. Kedrov, for example, had nothing to say about Soviet physics, and his statement that the search for priorities in science was alien to the spirit of the scientific community was "a completely erroneous judgment." [109] B. E. Raikov, an eminent historian of science, was the target of a specific line of anticosmopolitan criticism. In 1947—on the eve of the war on cosmopolitanism—he published a noted volume on the history of heliocentric ideas in Russia; in 1952, when the war on cosmopolitanism had reached maximum intensity, he was criticized for emphasizing the Russian resistance to, rather than acceptance of, heliocentric ideas.[110] Furthermore, he made the error of placing more emphasis on the conservative strain in the Russian intellectual tradition than on the progressive and the materialistic traditions.

From 1943 to 1952 the work on the history of Russian science had passed through three phases. In 1943–1947 the historians were particularly concerned with identifying and analyzing Russian contributions to the world pool of scientific knowledge. Russian science was viewed as an important tributary of the mainstream of international science. The historian's job was to establish the record of Russia's rapidly growing participation in the evolution of modern science. But equally important was the task of studying the progress of science in general. On the occasion of the establishment of the Institute of the History of Natural Science in 1945, V. L. Komarov noted:

> The real historical role of Russian science can be shown only by efforts to place it within the framework of the universal history of natural science. . . . Therefore, the Institute of the History of Natural Science, in its monographs, periodical publications and general surveys, must reconstruct the concrete course of the historical development of science in its totality and study the influence of Russian natural science on world science and the reflection of world science in Russian scholarship.[111]

During this period the historian of science was in a position to resist the mounting ideological pressure to attack or fully ignore the Soviet biologists who were bitterly attacked by Lysenko and his followers. True, the name of N. I. Vavilov was a name that most historians and scientists tried to avoid; after all, they did not know the charges leveled against him by the secret police. M. M. Zavadovskii chose not to ignore Vavilov, but he criticized his identification of genetics as a general science of "individual development, variation, heredity, and evolution."[112] While recognizing the immense value of genetics as a specialized branch of biology, Zavadovskii rejected the "English model" advanced by W. Bateson and accepted by Vavilov, which viewed genetics as the most general and most comprehensive biological discipline. Another writer, however, showed no inhibition in referring to Vavilov's work on the main centers of the origin of cultivated plants as a valuable contribution to modern science.[113] Nor did L. S. Berg hesitate in 1947 to write several pages on Vavilov's work as president of the Russian Geographical Society.[114] Zavadovskii, likewise, had no reservation in giving credit to N. K. Kol'tsov, A. S. Serebrovskii, or S. S. Chetverikov, all treated by Lysenko and his followers as archenemies of Michurinist biology, for making major contributions to the modernization of biological research in Russia.

From Zhdanov's speech in 1947 to the session on the history of national science sponsored by the Academy of Sciences in 1949, the historians shifted their emphasis from the unity of world science to the distinctive attributes of Russian science. Although the historians accented the essentially materialistic orientation of Russian naturalists, they were encouraged not to overlook inconsistencies in their materialistic views and, in specific cases, their antimaterialistic digressions. The historian was not allowed, for example, to overlook A. M. Butlerov's flirtation with mediumism, despite his generally materialistic outlook.[115] B. G. Anan'ev, the author of the first historial survey of Russian psychology, was praised for his broad concern with materialistic tradition but was criticized for his lenient attitude toward "idealistic" psychologists.[116] During this period, N. I. Vavilov and hundreds of other victims of the successive purges of the 1930s were not referred to in print. Other enemies, typified by the biologists N. K. Kol'tsov, A. S. Serebrovskii, M. M. Zavadovskii, B. M. Zavadovskii, N. P. Dubinin, and I. M. Poliakov, continued to be referred to but always with utmost contempt.[117]

During the third phase, 1949–1952, the historians were clearly discouraged from indicating inconsistencies and digressions in the professional work and general behavior of the pre-Soviet leaders of Russian science. Now they were encouraged to emphasize the pristine purity of natural science materialism in pre-Soviet Russia. The famed neurophysiologist I. P. Pavlov was depicted as a consistent dialectical materialist; the historians of science who had grown accustomed to recognizing strong elements of "mechanical materialism" in his general theoretical views had been silenced.[118] This was the period of the most reckless and bizarre war on cosmopolitanism. The taboo of "objectivism" was now deeply ingrained in historical studies of science. "Objectivism," an excessive concern with the objectivity of scientific knowledge, was viewed as an effort that recognized the superiority of "impartial facts" over "party loyalty" *(partiinost')*. According to B. M. Kedrov:

> Objectivism ignores class struggle in the domain of ideology; it treats Party tendentiousness as an antithesis to genuine scientific inquiry. Bourgeois objectivism strives to recruit supporters from among scientists inexperienced in ideological and political struggle. In reality, every move of bourgeois objectivism expresses bourgeois *partiinost'*, idealism and fideism. Thoroughly decadent, objectivism is fully subservient to reactionary ideology, which has infiltrated the camp of materialism with surreptitious activities. The special task of bourgeois objectivism is to advance the perverted idea of the possibility of effecting a reconciliation of materialism and idealism—of the progressive Soviet ideology and the decadent and reactionary bourgeois ideology. The aim of bourgeois objectivism is to disarm the camp of materialism—to divert the adherents of materialism from a sustained struggle against bourgeois reaction. The basic task of bourgeois objectivism is to blunt the weapons of revolutionary Marxist-Leninist criticism and to weaken the principle of *partiinost'*, which demands a struggle against the champions of old and reactionary ideas in science and ideology and for the triumph of communism. Bourgeois objectivism is so much more dangerous because it masks its loyal service to a universally reactionary ideology by feigned impartiality and deceptive efforts to throw objective light on available facts.[119]

The scholar could no longer be satisfied merely to note the polarization of modern thought between scientific materialism and pseudoscientific idealism.[120] He was expected to be a visible and relentless fighter on the front of ideology—to use every opportunity to indicate and condemn the intellectual sterility and social decadence of idealism in all its manifestations. The current studies of the scientific legacy and philosophical views of N. I. Lobachevskii provided the best illustration of anticosmopolitanism at work. The *Progressive Ideas of N. I. Lobachevskii,* by S. A. Ianovskaia, generally known as a champion of mathematical logic, contained particularly bitter attacks on various Western "distortions" of Lobachevskii's philosophical views on non-Euclidean geometry—particularly Henri Poincaré's reliance on Lobachevskii in constructing a theory of the "arbitrary" or "conventional" nature of scientific laws and axioms.[121] In Ianovskaia's view, Lobachevskii represented an ideal scientist measured by Stalinist standards. He opposed "subjectivist" and "apriorist" epistemology, he believed in the dialectical growth of knowledge, he admired Francis Bacon's utilitarian view of science, and, by having built geometry on an empirical basis, he showed that even the most abstract branches of scientific knowledge are firmly rooted in the reality of the world existing outside and independently of the human mind. Another philosopher advocated a complete reconstitution of Einstein's general theory of relativity by fastening it to Lobachevskii's— rather than to Riemann's—non-Euclidean geometry. He said that only an alliance with Lobachevskii's orientation could bring the general theory of relativity into the fold of philosophical materialism.[122] The writer took solace in the fact that some of the most respectable Soviet scholars, including V. A. Fok and G. I Naan, claimed a much closer link between Lobachevskii's ideas and Einstein's physical theory than was generally assumed by others. Another commentator on the strengths of national science noted that the beginnings of Bohr's correspondence principle were traceable to Lobachevskii's theory.[123]

I. V. Kuznetsov, an unrelenting warrior for the cause of purity and originality in national science, took on the job of reinterpreting the work of M. V. Lomonosov, the versatile scientist rightfully called the father of Russian science, in the light of anticosmopolitanism. In 1950 he wrote that Lomonosov's ideas were superior to those built into Newtonianism and Cartesianism, and he scolded P. S. Kudriavtsev, a historian of physics unmoved by the spirit of anticosmopolitanism, for having written that Benjamin Franklin's studies in atmospheric electricity antedated and inspired Lomonosov's work in the same field. Nor was he happy with Kudriavtsev's claim that Lavoisier—rather than Lomonosov—gave original formulation to the law of the conservation of mass, that John Dalton—rather than Lomonosov—was the true founder of the "atomic-molecular theory" of the structure of matter, and that the French chemist C. F. Gerhardt—rather than A. M. Butlerov—was the founder of the theory of chemical structure.[124]

The war on cosmopolitanism, according to a *Herald of the Academy of Sciences* editorial, showed that in most cases unwelcome digressions were made by scholars who did not have sufficient grounding in the theory of Marxism-Leninism. This insufficiency, it was alleged, led to a mechanical repetition of bourgeois cliches that, in turn, produced "a nihilist rejection of every notion of the cultural originality of individual peoples." The *Herald* also pointed out that individual scholars were not mindful of the Bolshevik party ruling on fundamental differences between Soviet socialist science and bourgeois science and on the imperative need for applying the principle of *partiinost'* to every phase of scientific work.[125]

The campaign to remove ideological impurities from science and to fan nationalist fervor in science created a climate favoring intellectual isolation of the Soviet Union from the West. It forced Soviet scientists to disregard, or to condemn, the challenging breakthroughs made by Western scientists. Norbert Wiener's cybernetics, which emphasized information feedback as the key to understanding complex, dynamic, and self-governing social, biological, and technical systems, could find no support in the Soviet Union until the mid-1950s. The guardians of ideology contended that cybernetics violated the principles of "scientific psychology" and Pavlov's "progressive science of nervous activity," and that it erred in trying to replace man by computer and in preferring "functionalism" to causal explanation.[126] Furthermore, it was found guilty of denying the objective validity of the laws of social development.[127] When the philosophers rejected cybernetics, the scientists had no choice but to follow suit. This was a time when philosophers played the determining role in the war on ideological digressions in science.

In 1944 the physicist Erwin Schrödinger published *What Is Life?*, a book that directed the attention of physicists to the study of life. It stood in the background of a revolutionary application of the tools and insights of microphysics to the world of biology and led to the emergence and triumph of molecular biology. Thanks to the great reputation of the author, the book was quickly published in a Russian translation. As could have been expected, it caught the immediate attention of Stalinist philosophers, who saw in it an effort to combine the idealistic view of physical reality with the basic theory of "formal genetics," the main target of Lysenko's wrath. Schrödinger, they contended, tried to extend the philosophy of indeterminism to the study of life and to combine a mechanistic interpretation of life with agnostic psychology. E. Ia. Kol'man saw in Schrödinger's ideas a return to "the philosophy of Brahmanism, medieval mysticism, and Schopenhauer."[128] Schrödinger's recognition of the corporeal basis of heredity went against the main tenets of Lysenkoist thought. A Lysenkoist spokesman noted that Schrödinger's work marked a high point in modern efforts to achieve a unity of Machian thought in physics and the Weismannist-Morganist orientation in genetics.[129] As soon as the philosophers rejected Schrödinger's suggestive ideas, they became taboo for the scientific community, which, in turn, seriously reduced the role of the Soviet Union in the developments that laid the foundations of molecular biology.

At this time, Soviet scientists were too preoccupied with protein as a primary source of life to pay sufficient attention to the ongoing research in nucleic acids, research that opened the way to the discovery of DNA. The biochemist A. I. Oparin, who tied his lifelong interest in the origin of life to the study of protein, became a Lysenkoist and showed only scorn for the burgeoning studies that saw the basis of life in "the structure and spatial arrangement of atoms in a molecule."[130] In an article published in 1951, Oparin, true to the spirit of anticosmopolitanism, tried to show that the most reliable leads for a biochemical study of the origin of life came from Russian scientists—from D. I. Mendeleev and A. M. Butlerov in chemistry, V. I. Vernadskii in geochemistry, A. N. Bakh in biochemistry, and O. Iu. Shmidt in cosmology.[131] Lepeshinskaia's noncellular theory gave him an opportunity to introduce qualitative "leaps" at critical points of the evolution of "proteinlike" substances—and to add new support to the close ties of his theory with dialectical materialism.[132]

As soon as Lysenko's incredible power began to show signs of weakening, Oparin lost no time in disassociating himself from the Michurinist tradition.[133] He continued, however, to attack the theory that associated the origin of life with the emergence of nucleic acid as the first living molecule; he contended that life was not bound up by an elementary substance but by the "whole system" of the living body in its adaptation to the environment.[134] His alliance with Lysenkoism represented a brief detour from an otherwise careful and notable study of the origin of life, an attractive and legitimate inquiry in the area where physics, chemistry, and biology met. The high positions that Oparin occupied in the academic hierarchy gave him a preeminent position in Soviet studies of the origin of life. For more than three decades, he was the only person who contributed critical surveys of origin-of-life studies to the leading symposia on the history of Soviet science. When the first international congress on the origin of life met in Moscow in 1956, one of its functions was to honor Oparin's long, dedicated, and original work in this challenging field.

The rejection of Schrödinger's ideas did not signify a total loss of interest in the physical processes of life. At this time much discussion was centered on the place of biophysics within the general framework of biological sciences: in 1952 the Academy went so far as to establish the Institute of Biological Physics for the purpose of placing the new effort in this area on solid footing. But true to the spirit of Lysenkoism and Michurin's theory, the biophysicists concentrated almost exclusively on the physical and physicochemical influence of the environment on living forms. The influence of changing air pressure and gravitational pull on organisms, biotic effects of ultraviolet light, and radiation capacity of living organisms were typical topics attracting the attention of biological physicists. Biophysics was also assigned the task of employing "the new physical methods" in planned efforts to change the existing species according to predetermined plans.[135] Lysenkoist "biophysics" left little room for the use of the methods of modern microphysics and quantum chemistry in the study of the structure of living matter at the subcellular level. By avoiding the "structural" approach to heredity, Lysenko and his school avoided a structural approach to variation. Their theory allowed no room for a corporeal and localized base of heredity; it was built on a mystic communication between the external environment and individual members of a species.[136]

The struggle for the ideological purity of science and the effort to overcome the cosmopolitan interpretation of the history of Russian science went hand in hand with a patriotic campaign to involve the Academy's staff in the practical application of scientific knowledge. This drive was part of the officially instigated mass "movement" to establish "active contact between the workers of science and the workers of production." In 1949 the Academy prepared the first annual "implementation" plan— a plan of completed research projects to be carried out cooperatively. The plan was one of many new strategies for bringing science closer to production. The Academy became involved in dispatching special "brigades of scholars" to various construction projects to supply scientific advice. For example, in the second half of 1951 numerous teams of scholars went to construction sites of the Volga-Don, South Ukrainian, and North Crimean canals and hydroelectric power stations in Kuibyshev and Stalingrad. "The task of these brigades was to provide on-the-spot scientific assistance to the designers and builders of hydroelectric power stations, canals and irrigation systems, to trace and elucidate problems whose scientific study is indispensable for blueprinting and building these projects and to maintain close contact with construction."[137]

Huge construction projects were heralded as catalysts of coordinated research on a massive scale and examples of new forms of interdisciplinary approach. The "Great Stalinist Plan for the Transformation of Nature," which included projects for the forestation of the steppe and for improving the quality of agricultural land, gave closely related research topics to countless institutes and laboratories. It also gave the Academy a broad and well-defined basis for joint research with scientific collectivities throughout the country. The new campaign for practicalism in science reached a high point in 1952 when A. N. Nesmeianov, president of the Academy since 1951, recommended that brigades of scholars involved in construction projects be made permanent components of the internal organization of the Academy. Nesmeianov also suggested that the Academy undertake an immediate organization of commissions to funnel scientific information to specific branches of the national economy.[138] At the same time, the Academy expanded its sponsorship of "production conferences" at which scientists met with engineers to discuss specific problems of industrial technology. In the early 1950s, 600 scientists employed in various branches of industry were temporarily transferred to the Academy for advanced training. The new practicalist orientation took many forms and deeply affected every phase of the Academy's activities.

The new practicalism was not merely a strategy for organizing scientific work. It was also a vehicle, a measure, and an expression of patriotism. For this reason it was engineered and built not only by administrative bodies of the Academy but also by numerous mass organizations, led by the party and trade-union locals. The function of these organizations was "to conduct educational work among scientific workers, to equip them with Marxist-Leninist theory and to direct their commitment to the fulfillment of the key projects of economic and cultural construction." Mass organizations assumed the task of "helping scientific institutions to strengthen their ties with practical endeavors, to raise the quality of scientific work, and to imbue scientific workers with ethical values of scientific work, advanced by criticism and self-criticism."[139]

This activity opened the professional work of scientists to daily interference by an alarmingly large number of zealots, self-seekers, and opportunists. It also added new political complexities to the Academy, particularly by strengthening the position of individuals, like Lysenko, who built private empires by exploiting close personal ties with high government officials. Constant pressure for practically oriented research caused much consternation among scholars who worked in such abstruse sciences as mathematics and astronomy and who were often hard-pressed to relate their work to the immediate needs of the national economic plans. Indeed, mathematicians working in the theory of numbers, group theory, set theory, and topology were asked in 1949 to do something about overcoming the impracticality of their main lines of interest.[140] Mathematical logic was tolerated only inasmuch as it did not occupy the full attention of individual scholars.

The growing emphasis on direct participation in practical projects caused grave concern among leading men of science who favored basic research as the most vital function of the Academy. These men thought that, in its plans for future activities, the Academy should be guided primarily by long-range perspectives opened by the internal impulse of science rather than by external dictates of politically directed technology. The Academy could best help the national economic plans, said S. I. Vavilov in 1947, just before the new push for applied science was under way, by moving ahead of current needs and "storing" knowledge that had no immediate applicability. He said that the activities of research centers should be focused not only on the present application of knowledge accumulated by the scientific workers of the past but also on creating fresh reserves of knowledge for future application. "Science has its own specific logic of development—a significant fact which must be taken into account."[141]

Stalin stuck closely to the Marxist notion of science as a response to the technical needs of society and as a generalizing of practical experience. A typical academician, in contrast, showed no enthusiasm for the moves to broaden the "engineering orientation" of the Academy.[142] He favored the view of science as a force moving forward by its own impulse. He reasoned that the Academy could best meet its social obligations by adhering to the principle that "perspectives opened by uninterrupted growth of science were often considerably broader than perspectives opened by economic plans."[143] Even in the thick of the anticosmopolitan furor, A. N. Nesmeianov warned against attacks on "astronomy, astrophysics, and cosmogony" as sciences far removed from the practical needs of the day.[144] "Impractical" sciences of today, he said, could easily become "practical" sciences of tomorrow. In this view Nesmeianov was guided by the principle that the higher the level of technological advancement, the larger the number of sciences responsible for it.

Most Academy scholars did not resent the emphasis on applied science, particularly in the face of enormous war damages and the critical
need for accelerated rebuilding of the wheels of industry. They never
deviated from Bacon's view of science as the mainspring of social well-
being. Nor were they ever actually forced to abandon fields of inquiry
unrelated to the needs of current economic plans, even though they
were occasionally reminded of the remoteness of their specific research
projects from real life. What they resented was the proliferation of the
"brigades of scholars," who interfered with their research by taking them
to mushrooming construction projects to give technical advice and
express patriotic fervor. Although Stalinist crusaders neither denied nor
affirmed the vital importance of pure science, they were sure to measure
the social consciousness—a key notion of historical materialism—of each
scholar by his participation in selected engineering ventures. Nesmeianov's carefully worded commendations of pure science were aimed at
preserving a sense of realism without turning against the ardor of
patriotism.

The campaign for practical science undercut the traditional right of
the scientific community to serve as a key judge of the social utility of
scientific theory. This right was instead usurped by political authorities
who were external to science and alien to the scientific community. The
government held supreme authority in determining the social worth of
individual research activities and of pure science vis-à-vis applied science. It did not rely on the consensus of scholarly opinion when it opted
for V. R. Vil'iams's "biological" method of soil conservation, based on
the grassland system (travopol'e) of crop rotation, reducing D. M. Prianishnikov's "chemical" method (the use of chemical fertilizers) to a minimum of support.

To compound the plight of the scholarly community, Stalin decided
to make a direct entrance into the arena of science. In 1951 he wrote a
scornful critique of the linguistic theory of N. Ia. Marr, which emphasized the social-class nature of language. This ended a long period of
intellectual hegemony for Marr's theories, usually identified as Marxist
linguistics. Only a short time before Stalin's attack, the journal Voprosy
filosofii wrote: "Based on the theory of Marx, Engels, Lenin and Stalin,
the new theory of N. Ia. Marr and his followers has been a powerful
weapon in the struggle for Marxism-Leninism and dialectical and historical materialism, and against idealistic pseudoscience in the field of
linguistics."[145] With one stroke Stalin was made the country's most

authoritative linguist, whereas Marr's theories were proclaimed a major digression from the true postulates of Marxist social theory.

In 1952 Stalin made another move to straighten out the logic and the method of social science. In a lengthy article on the economic foundations of Soviet socialism, he underscored the objective nature of the laws of economics and, as a result, condemned so-called voluntarism in the interpretation of the inner workings of social societies. He asserted that the basic economic laws of socialism were the laws of the structure and dynamics of Soviet society, and he reiterated the old Marxist rule that the laws of society were no less objective—that is, no less independent of human consciousness—than the laws of nature.[146] Each society is governed by two kinds of laws: the universal laws of society and the specific laws of individual socioeconomic formations (such as feudalism, capitalism, and socialism). Stalin also emphasized the need for recognizing the distinction between social "forms" and social "content."[147] In stable societies (typified by Soviet society, according to Stalin) most social change must be understood as a gradual placement of new content into old forms. The social scientist must not be led by the persistence of forms to overlook the need for a deeper study of the changing content.[148]

There is no need here to present a detailed account of Stalin's theory of the economic foundations of Soviet socialism. Its real importance was not in the ideas it presented but in the authority vested in it: it became an official blueprint for Soviet society, and it made true research in general economic and sociological theory—dying since the early 1930s—totally unnecessary. Stalin gave a clear, complete, and conclusive "scientific" explanation of the structural principles of Soviet society. His ideas were not presented as topics for critical discussion but as state edicts not subject to challenge by the scholarly community. Stalin's notion of Soviet society as an entity unburdened by conflict between the "forces of production" and the "relations in production" implied that Soviet social scientists could not learn anything from bourgeois sociologists and economists, who live in a socioeconomic formation dominated by "irreconcilable conflict" at its very base. "The Economic Problems of Socialism" eliminated an arterial bridge between Western and Soviet thought.

Stalin became the supreme master of scientific truth, the authority protected from any kind of challenging criticism. He became the chief architect of the total unity of science and ideology, and he achieved this, according to Maksimov, by waging a relentless war against bourgeois—

that is, cosmopolitan, metaphysical, and idealistic—"vestiges" in scientific thought and by making the method of dialectical materialism the basic tool of scientific analysis and explanation.[149] Endless conferences dealing with the methods for applying and amplifying Stalin's "scientific" pronouncements followed a well-established pattern: all scrutinized the evils of nonconforming thought, all subjected errant scholars to scorching criticism, and all demanded that "guilty" scholars make public admissions of, and public apologies for, their unpardonable sins. The linguist I. I. Meshchaninov and many others were paraded from conference to conference to hear their crimes recounted time and again and to make public apologies for their past sins. And at every conference they were told that their apologies had not begun to cover all their evildoing.

It did not matter what Stalin said about linguistics and economics; what mattered was that his intrusions precipitated an endless and profitless succession of academic conferences in search of a theoretical reorientation of every social science.[150] What mattered even more was the unwanted and dangerous interference of an authority external to science with the principle of academic autonomy in the interpretation of the procedures and the logic of scientific work and with the moral obligation of scientists to maintain an open mind toward all ideas making up the theoretical inventory of individual disciplines. Stalin's ideas were forced on the confused scholarly community as political edicts—as legally promulgated and enforced norms—rather than as topics for free discussion. The scientific community lost its right to serve as the sole authority in the certification of scientific knowledge.

The most symptomatic—and the most portentous—product of this period was the full crystallization of what subsequently became known as the cult of personality, a cluster of myths idolizing Stalin and ascribing to him an unmatched wisdom and moral rectitude comparable to the holiest of all saints. The Academy played no small part in building the cult of personality to its staggering proportions. All academic conferences that dealt with the relations of science to Soviet ideology found it appropriate to send telegrams to Stalin thanking him for his leading role in advancing Soviet science and promising to abide by his instructions. The antigenetics conference, held on August 24–26, 1948, which acknowledged and "ratified" the victory of Lysenkoism, thanked Stalin for his help in fighting reactionary and alien forces and offered an apology for the past errors of the Department of Biology in defending the genetic theories of Gregor Mendel and T. H. Morgan. At the end of 1949, the Academy held a lavish celebration of Stalin's seventieth birthday. A congratulatory telegram sent to Stalin on this occasion stated that the gathered scientists were fully convinced that every science reflected his "guiding ideas" and "creative genius."[151] Mitin noted that Stalin's writings marked "a new phase in the development of Marxist-Leninist philosophy." He argued that they advanced the guiding principles of the Marxist-Leninist interpretation of "the *Weltanschauung* problems of physics, biology, and other modern sciences" and supplied the most effective ammunition for the war against bourgeois idealism.[152] The cult of personality gave Stalin unchallengeable authority in science, which he exercised either by direct pronouncements or by giving support to crusaders of Lysenko's kind.

Government and party leaders claimed full credit for forcing the academic community to hold soul-searching conferences for the purpose of ridding science of ideological impurities. They presented these conferences as special vehicles of free and spontaneous "criticism" and "self-criticism," aimed at eradicating dogmatism, breaking up special-interest groups in the scientific community, bringing scholars closer to the practical needs of the country, and assuring a smoother and richer influx of younger scholars to positions of authority. These conferences, however, were neither free nor spontaneous. Prepared in secret, they confronted unsuspecting victims who were given no chance to study the charges against them or to prepare a defense. The galleries were packed with tested Communists ready to malign the victims or to greet patriotic pronouncements of Stalinist heroes with approving outcries that edged on hysteria.

While the Academy and the scientific community in general suffered under the attacks of frenzied mobs, the government announced proudly that its policy had gone a long way toward forging the unity of Soviet ideology, Marxist philosophy, and science. G. M. Malenkov, a ranking Communist, made this statement to the Nineteenth Congress of the Communist party, held in October 1952:

> The intervention of the Central Committee of the [Communist] Party in many fields of science has helped to unveil practices and traditions alien to the Soviet people, to expose elements of caste exclusiveness and intolerant attitudes toward criticism, and to unmask and uproot various manifestations of bourgeois ideology and all kinds of vulgar distortions. The well-known discussions in philosophy, biology, phys-

iology, linguistics and political economy have unveiled serious ideological digressions in various fields of science, have stimulated and developed criticism and controversies and have played an important role in the advancement of science.[153]

A. N. Nesmeianov gave a typical summary of the Stalinist position on the harmony of Soviet society and Soviet science when he stated laconically: "The birth of agrobiology as a new science is closely linked with socialist practice. Without a doubt, the emergence and development of Michurnist [Lysenkoist] agrobiology would have been impossible without the *kolkhoz* system."[154] Without the collectivization of agriculture, he added, there would have been no unity of science and production.

The government led the attack on the Academy, on the traditional ethos of science, and on the exclusive role of the scientific community in forging the paradigmatic unity of science. To achieve ideological unity, it fought the residual efforts—crystallized during and immediately after World War II—to strengthen the associative autonomy of scientific institutions and to increase ties with the world of science outside the Soviet Union. Various government moves provided particularly clear and startling indications of official efforts to bring the Academy in line with Stalin's vision of an airtight society. From 1947 to 1953 the government did not allow the Academy to elect new and full corresponding members. No doubt it tried to create the impression that the continued existence of the Academy was under serious deliberation—and that only blind loyalty to the dictates of Stalinist ideology and political strategy could save it. The feeling of apprehension and uncertainty was compounded in 1949 when the government forced the Academy to "elect" A. V. Topchiev—a member of the Communist party since 1932—to full membership and to make him the chief scientific secretary of the Presidium of the Academy. A chemist who could show only mediocre achievement in his field, Topchiev acquired the job of protecting, coordinating, and implementing Stalinist designs for an ideological and nationalistic bastion of Soviet science. His annual progress reports contained sharp criticism of academic behavior unacceptable to government authorities as well as summaries of achievement on the ideological front.

The campaign to make science ideologically pure, nationally pristine, and practically oriented produced many violations of the moral code of science and threatened to undermine the very foundations of the scientific community—its esprit de corps, sense of autonomy, intellectual independence, and professional identity. The Stalinist war on the scientific community was in essence a war on the ideals and realities of academic autonomy. Scientists were denied the right to challenge the authority of the Stalinist leaders of Marxist thought, to interpret dialectical materialism in the spirit of new scientific developments, or to proclaim philosophical neutrality. No statement coming from Stalin was open to challenge, even when it caused untold destruction of research units and inflicted heavy penalties on scientists with heterodox theoretical views. Opportunists, typified by Lysenko, could uproot entire branches of knowledge simply by securing Stalin's sponsorship. The Stalinist war on the scientific community sought to impose drastic limitations on the rights and duties of scholars as guardians of scientific legacy and masters of the methods of scientific inquiry. Stalinism made science a special expression of ideology and made ideology the prime force in shaping the high theory of every science. By substituting the fiat of political authority for the freedom of scientific discussion, it narrowed the scope of scientific exploration, at least in such disciplines as quantum chemistry, neurophysiology, psychiatry, sociology, linguistics, evolutionary biology, soil science, and economics. It threatened to make scientists politically responsible for their theoretical views in individual disciplines. And it empowered the government to bestow full academic rights on favored pseudosciences, as well as to subordinate all sciences to a philosophy that expressed the will of the government rather than a consensus of the scholarly community. In all this, the scientific community lost its ancient right to defend the international spirit of science. The pursuit of "national priorities" came increasingly to violate both the ethic and the logic of science.

ACADEMIC RESISTANCE

The crisis that left deep scars on the institutional network of Soviet science was a transitional phenomenon: before it had released all its fury, it began to ebb under the pressure of developments more favorable to the national commitment to science. Soon after Stalin's death in 1953, the personality cult became an object of critical scrutiny. But even before the death of Stalin there were forces at work that helped the scientific community in the difficult task of reducing the intensity of the Stalinist onslaught.

It came as no surprise to the scientific community that the first traces of resistance surfaced among the leading physicists. Particularly notable in the efforts of physicists to preserve the intellectual integrity of their discipline was V. A. Fok's counteroffensive against the philosophers, represented by I. V. Kuznetsov and A. A. Maksimov, who in criticizing Einstein—as part of a general campaign for ideological purification—were more successful in showing their unfamiliarity with the subtleties of modern science than in detecting serious flaws in Einstein's thought. Fok was firmly convinced that Einstein's contributions to science—particularly his formulation of the general theory of relativity—needed serious modifications;[155] however, he was just as convinced that Einstein's work was the work of a true genius and a revolutionary step in the development of modern science. His own suggestions for a reinterpretation of the general theory of relativity were, he said, elaborations of suggestions made by Einstein himself in his writings after 1927. While the philosophers argued that a "bad" philosophical stance inevitably produced an "incorrect" scientific view, the physicists took a radically different position: in their view, the scientific ideas of a scholar were not necessarily connected with his philosophical work. They considered it imprudent to assume that Einstein's scientific elaborations of relativity were distorted by a "wrong" stance in philosophy. Philosophical relativism and scientific relativity were two qualitatively different categories of thought. In the words of G. I. Naan: "It should be emphasized at the outset that the physical problem of relativity can in no way be identified with the problem of relativity in philosophy—even in reference to motion, space and time."[156]

Fok's and Naan's war on orthodox philosophers was made easier by the absence of unity among the chief spokesmen for the Marxist philosophy of science. Kuznetsov's statement that Einstein's "idealistic" views in philosophy were responsible for his "errors" in science was not shared by all the philosophers. In 1952 the philosopher I. P. Bazarov attacked Kuznetsov's rash criticism of Einstein and stated directly that dialectical materialism must recognize the differences between "the results of physical investigation" and "philosophical conclusions" drawn from these results.[157] Bazarov accused Kuznetsov not only of misreading Einstein but also of misinterpreting dialectical materialism. He found much in Einstein's "idealism" open to criticism, but he also gave a clear impression that the dilemma of a dialectical-materialistic interpretation of the theory of relativity had not been resolved. He resented the tendency of a group of Soviet philosophers to deal exclusively with idealistic "distortions" of Einstein's contributions to science. While Fok tried to stir up scientific controversies in the interpretation of Einstein's legacy, Bazarov contributed by helping to give new life to old philosophical controversies. The proclivity of Stalinist philosophers to place a tag of idealism on theoretical elaborations of the theory of relativity continued to be obfuscated by conflicting and equivocal pronouncements.

The mathematicians in the Academy worked slowly but continuously, and with great determination, to protect their discipline from ominous attacks by the apostles of Stalinist practicalism. The philosophical critics of excessive formalism in mathematics never gained much strength. Speaking for mathematics, A. D. Aleksandrov, a mathematician of note and a Marxist, acknowledged the sins of "formalism" in his discipline; however, his primary interest was to show that even the most abstract branches of modern mathematics were clearly identified with practical problems. He extended the notion of practical utility far beyond the dictates of the day: the "unreal" mathematical theories of today and the "reality" of tomorrow are connected by long or short series of transitional phases, each bringing the apparent "aloofness" of mathematics closer to easily detectable realities. With this in mind, Aleksandrov was in a position to point out that topology, one of the most abstruse branches of modern mathematics, had "practical" applicability in four distinct fields: exploration of the concept of continuity and its multiple practical uses; transposition of major notions and procedures of topology to other branches of mathematics (typified by L. A. Liusternik's work on the calculus of variations); direct solution of practical problems, as shown by L. S. Pontriagin's work on topological manifolds; and axiomatic topology, whose practical potentials were just beginning to be discerned.[158] In a later article, Aleksandrov determined that the work of some leading mathematicians—including N. N. Luzin and A. N. Kolmogorov—showed signs of tolerating residual idealism of dominant Western schools in the foundations of mathematics.[159] But this time, too, his criticism was mild and oblique and did not presage either personal punishment or institutional revampings. Nor did the mathematicians hold a conference devoted to "criticism" and "self-criticism" as a means of ideological purification.

Ideological interference did not prevent Soviet mathematics from making rapid and universally noted advances. When the authors of a general survey of Soviet science asserted in 1952 that no branch of modern mathematics was without serious representatives in their country, they were generally correct.[160] The champions of Marxist orthodoxy in philosophy who defended Lenin's statement that too much mathematics opened the gates of science to various currents of idealistic thought were much more vociferous and hostile during the 1930s than they were during the 1940s and the 1950s. All this, however, did not mean that ideological interference did not pose a constant—and sometimes humiliating—threat to mathematicians who worked in the most abstruse branches of their science.

The attack on the resonance theory in chemistry was a combination of open hostility to quantum chemistry on ideological grounds and a belligerent expression of nationalism in science. In Pauling's structural theory Soviet critics saw both an exaggerated emphasis on the quantum-mechanical principle of indeterminacy and a negation of Butlerov's contributions. In their criticism of the resonance theory, however, most chemists were much less categorical and sweeping than the philosophers. Indeed, these chemists made two major concessions: they noted that a quantum theory of chemical structure was both possible and necessary, and they expressed no qualms about relying on approximative mathematical methods in describing the chemical structure of organic molecules, a mode of operation treated by orthodox philosophers as an unpardonable attack on absolute causality as the sine qua non of scientific explanation. There were chemists ready to admit cautiously that Butlerov's theory was only the beginning of a structural orientation in organic chemistry, a beginning that still awaited the solution of many key problems.[161]

In the thick of anticosmopolitan furor, which imposed ominous limitations on intellectual contact with the world abroad, the Soviet scientific community, led by the Academy of Sciences, conducted a quiet campaign to ensure a constant, though limited, flow of Western scientific ideas to the Soviet Union. At a time when mathematical "formalism" was the main target of ideological attack, the Academy published a Russian translation of the *Foundations of Geometry* by David Hilbert, the founder and leading articulator of formalism, an influential orientation in the foundations of mathematics. At a time when the principle of indeterminacy was roundly attacked as an expression of "physical idealism," the Academy published Russian translations of three works by Heisenberg—the formulator of the indeterminacy principle—including his *Philosophical Problems of Nuclear Science*.[162] And at a time of the rapid growth of Lysenkoism, the Academy published Erwin Schrödinger's *What Is Life?*, a work that envisioned the future of biology in a synthesis of microphysics and genetics.

During the Stalinist ideological crusade, the Academy was involved in preparing and publishing the second edition of the *Great Soviet Encyclopedia*. The editorial board received careful instructions on how to treat the long list of scholars who had fallen into disfavor. For example, it was instructed to omit the names of N. I. Vavilov, N. K. Kol'tsov, A. S. Serebrovskii, and most other pioneers of genetics from the new publication. N. P. Dubinin, who was elected a corresponding member of the Academy in 1946, was also ignored. Quietly and unobtrusively—and in obvious defiance of the authorities—the editors of the encyclopedia followed their own standards for the treatment of the more notorious campaigners for ideological and national purity in science. I. I. Prezent, Lysenko's closest follower and collaborator, merited only seven lines in the new edition and no mention of his scholarly contributions. A. A. Maksimov and E. Ia. Kol'man, the most ubiquitous and ardent Stalinist philosophers during the 1930s, were not accorded special articles. I. V. Kuznetsov, author of a book on the principle of correspondence in modern physics and the most inflexible and aggressive critic of Einstein's scientific theories, also was ignored.

Academic resistance to the Stalinist war on ideological digressions came from so many sources and in so many forms that it is impossible to place them into clearly hewn logical categories. The best that can be done is to take note of some of the more obvious and more common modes of resistance. A number of leading and influential scholars resisted the Stalinist onslaught by refusing to apologize for committed "sins." L. D. Landau, a leading physicist, answered his philosophical critics by total silence, even after he was warned that it was his moral obligation to make a public announcement of his withdrawal from the positions that made his work unacceptable to Marxists. The biologist I. I. Shmal'gauzen, the physiologist I. S. Beritashvili, and the geographer A. A. Grigor'ev—all academicians and all vilified by the purifiers of ideology—followed the same line of quiet resistance. There were also scholars, typified by N. N. Semenov, the internationally recognized creator of the kinetic theory of combustion, who made themselves very conspicuous by their refusal to participate in the proceedings of ideologically motivated conferences.[163] Still others, represented by L. A. Orbeli, lashed out against the common practice of not giving attacked individuals a listing of charges in advance and sufficient time to rebut these charges. There were also those who—like V. A. Fok—advocated shifting the center of attack from the philosophical flaws of the pioneers of modern physics to the appalling unfamiliarity of the ideological crusaders with the intricacies of modern physical theory. This was one of the many reasons for Loren Graham's justified assertion that Fok was "a brave leader in the effort to prevent Soviet physics from being subjected to the kind of perversion that occurred in Soviet genetics with the victory of Lysenkoism."[164] Small wonder, then, that *Bol'shevik* warned at the end of 1952 that there were still "groups of physicists" who stayed away from all efforts to subject "idealistic currents in modern physics" to critical examination.[165]

There were also scholars who risked their careers by making direct attacks on Lysenko's empire. The statistician V. S. Nemchinov, an eminent member of the Academy of Sciences, fought with courage and determination all pressures to make him renounce his emphatic statement, made at the August 1948 session of the Lenin Academy of Agricultural Sciences, that the chromosome theory belonged to "the golden fund of scientific knowledge."[166] N. A. Maksimov represented the group of scholars involved in an effort to prevent the Lysenkoist forces from fully eradicating research based on condemned scientific theories. Lysenko's aides had attacked Maksimov in the mid-1930s for serious deviations from the Michurinist tradition in biology. Immediately after the attack, he apologized for indiscretions he had committed and became a vociferous defender of Lysenko's creed. He reiterated his allegiance to Lysenkoism in an article published in the *Herald of the Academy of Sciences* in 1943, and at the August 1948 session of the Academy of Sciences he first apologized for harboring the defenders of the chromosome theory at the Institute of Plant Physiology, which he had headed, and then promised to carry out a thorough housecleaning.[167] Maksimov died in 1952, before the post-Stalin thaw. Yet, as David Joravsky has pointed out, his apparent allegiance to Lysenkoism was a ploy: he accommodated himself to the Lysenkoist forces "at points of great pressure" for the purpose of preventing a total destruction of science. For example, by devoting a few laudatory passages to Lysenkoist thought, his textbook on plant physiology, a scholarly work of international acclaim, continued to keep in print valuable scientific knowledge that otherwise would have been totally uprooted by the new authorities in biology.[168]

In 1952 the *Botanical Journal* dropped the first bomb by publishing a paper pinpointing serious errors in Lysenko's theory of the origin of species. N. D. Ivanov adduced extensive evidence showing that Lysenko not only misinterpreted the basic premises of Darwin's theory and of modern genetics but also misread Michurin. Unlike Michurin, for example, he did not recognize natural and artificial evolution as distinct processes irreducible to the same dynamics.[169] Furthermore, he placed much more faith in artificially induced "leaps" in the evolutionary process than Michurin had done. In general, Lysenko's idea that directed evolution described the true nature of the inner dynamics in the creation of new species was alien to Michurin's thinking. Ivanov made a clear impression that, in his opinion, Lysenko did not represent a new step in the development of modern evolutionary theory but a fruitless digression. Views similar to Ivanov's were expressed by N. V. Turbin, an erstwhile defender of Lysenkoist theory, who claimed forthrightly that Lysenko's biological constructions did not qualify to be labeled a Michurinist theory.[170] Unlike Michurin, Lysenko rejected Darwin's theory of the origin of species without adducing sufficient proof for his position. By rejecting Darwinism, according to Turbin, Lysenko rejected dialectical materialism.[171]

A fairly large group of maligned scholars worked hard to regain the confidence of the new masters in the Academy. Most notable in this group was G. F. Aleksandrov, the main target of Zhdanov's attack on cosmopolitanism, who wrote profusely on the newly discovered unity of science and philosophy and who became an articulator of the dialectical-materialistic foundations of Pavlov's neurophysiological theories.[172] Immediately after Lysenko announced at the August 1948 session of the Lenin Academy of the Agricultural Sciences that his attack on genetics was approved in advance by the Central Committee of the Communist party, several attacked scientists recanted and asked for forgiveness. At the 1951 conference on structural theories in organic chemistry, M. E. Diatkina, accused of having been one of the chief supporters of Pauling's chemical theory in the Soviet Union, admitted her "errors" in undervaluing the contributions of A. M. Butlerov and Russian chemistry in general and in overvaluing the scientific merits of resonance theory.[173]

There were also scientists, typified by the academician V. N. Sukachev, head of the Forest Institute, who first joined the Lysenko bandwagon and then turned against the new regime in the Academy. At the session of the Academy of Sciences held in August 1948, Sukachev made a public apology for his limited participation in the war against "reactionary, metaphysical, and idealistic tendencies" in biology and promised to intensify his efforts on behalf of Michurinist biology. But as early as

1950—with Lysenko still busy consolidating his empire—he began to publish papers that clearly diverged from Lysenkoist thought.[174] In January 1953 Sukachev published an essay in which he defended the Darwinian notion of intraspecific struggle and its role in the origin of species and in organic evolution in general.[175] The discussion opened by Sukachev's paper was not limited to theoretical considerations; it also led to the reexamination of a broad compass of agrotechnical problems. But above everything else, it started "the general criticism of a number of biological concepts which led to the liberation of Soviet biology from ideas forced on it by administrative pressures."[176]

Not only individuals speaking on behalf of their conscience but also organized groups, such as departments, institutes, and laboratories, resisted the campaign to transform the Academy into a bastion of Stalinism. These organizations became obliged to carry out the new designs for institutional adaptations of the Academy to new ideological dictates—a task that also gave them a widely exploited opportunity to modulate these designs in such a way as to make them less damaging to the interests of science and to the scientific community. They worked toward delaying the implementation of these designs by sponsoring endless discussions on procedural matters. In the summer of 1950, for example, an interdisciplinary academic conference ordered the creation of stronger institutional backing for Pavlovian neurophysiology and psychiatry, treated by Stalinist philosophers not only as a complex of incontrovertible scientific truths but also as a full triumph of dialectical-materialistic methodology in science. In reality, however, the implementation of this order made it possible for the Academy to engineer wider institutional support for the Pavlovian orientation, as well as to remove, under multiple disguises, broad areas of research in physiology from close proximity to the theory of conditioned reflexes. By ordering the creation of a separate institute for the study of higher nervous activity, the Academy planned to make it possible for the I. P. Pavlov Institute of Physiology to branch off into many types of research not associated with the scientific legacy of the founder of the theory of conditioned reflexes. The Academy satisfied the order of government authorities to expand Pavlovian research horizontally and vertically, but it also sought to satisfy the demand of the scientific community to make wide areas of physiological research completely independent of the Pavlovian tradition.

The intense campaign to remove ideological impurities from science encountered particularly serious obstacles in chemistry. The chemists were ordered to voice open enmity toward Pauling's resonance theory, but it was up to them to work out the strategy of attack. Isolated chemists "agreed" that Pauling must be attacked, but many of them engineered and carried out their own attacks. The result was a profusion of views on quantum chemistry. At one extreme were scholars eager to "save" quantum chemistry from Pauling's idealistic aberrations; at the other extreme were scholars eager to dismiss quantum chemistry in its totality, including Pauling's contribution.[177] At times it appeared that the attack on Pauling was lost in the tangled web of bitter internal feuds.

Cosmopolitanism, idealism, and objectivism were the main targets of the Stalinist crusade for the ideological purity of science. Politicians and their ideological aides defined these labels in general and simple terms; they left it to the philosophers of science to establish how they applied to particular developments in scientific theory. It was this tedious assignment that evoked controversy and dulled the weapons of ideological warfare. Often the attacks on true or imaginary ideological adversaries were overshadowed by feuds among the contending champions of ideological purity—that is, among the builders of Stalinism. In 1946 A. A. Maksimov, the most ardent Stalinist philosopher, published a book on the materialistic roots of Russian science. He was immediately attacked for an inexcusable disregard of the nationalist perspective.[178] A year later M. E. Omel'ianovskii published a philosophical study of the relation of V. I. Lenin to modern physics, a direct effort to rewrite the history of science in the spirit of currently emphasized ideological dictates. It did not take long for critics to accuse Omel'ianovskii of undue worship of Western science and of the denigration of Russian physics. The claim was made that his book was written from the position of objectivism.[179]

While the philosopher B. M. Kedrov criticized A. A. Maksimov for ideological indiscretions in his study of Russian science, he was the target of similar criticism by the physicist D. D. Ivanenko.[180] In 1948 I. V. Kuznetsov published a monograph on the principle of correspondence in quantum mechanics and its relevance for dialectical materialism. The direct purpose of the book was to make a contribution to the current ideological crusade by exposing the fallacies of physical idealism. In March 1950 a special session of the Institute of Philosophy found Kuznetsov's book inadequate in presenting the superiority of physics grounded in dialectical materialism over Western physics, lost in a morass of various idealistic currents.[181] Despite these criticisms, Maksimov, Omel'ianovskii, and Kuznetsov continued to be the stalwarts of Stalinist orthodoxy. They heeded the criticism by making their attacks more sweeping and unbending. Unceasing squabbles lessened the destruction wrought by the ideological war and added a bizarre quality to the entire enterprise. Their effects, however, were not strong enough to force important readjustments in Stalin's onslaught on the scientific community.

The campaign for philosophical orthodoxy, ideological purity, and national priorities in science dislodged many scholars from their academic posts: some were demoted from higher positions, others were simply let go outright. But the vastness of the Academy and the fuzzy edges of its labyrinthine organization made it possible for academic administrators to give many dismissed scientists new employment. N. P. Dubinin, the last link between the Academy and genetics, accepted the offer of the Forest Institute to join a forestation commission by filling the position of ornithologist.[182] I. I. Shmal'gauzen, who was dismissed from the position of chairman of the Institute of Evolutionary Morphology for his internationally noted contributions to the synthetic theory of evolution, found a position in the Institute of Zoology and helped the Academy to maintain a low-key but continuous interest in the work of Western laboratories that were ushering in the age of molecular biology. Some research collectivities fought the forces of Stalinist reaction in yet another way: the Institute of Mathematics, for example, provided screens for a number of scholars to undertake intensive research in the mathematical apparatus of cybernetics by discussing entropy as a statistical notion related to the theory of information.[183] At the moment when cybernetics became a lawful discipline in the mid-1950s, Soviet scientists had already come a long way toward meeting the mathematical needs of the new orientation and adding to the strength of mathematical logic as the science of the age of electronic computers.

Although the many strategies of resistance helped to soften the blows of Stalinist oppression, the Academy was nonetheless left with a sense of misplaced identity, ambivalent hopes, and uncertain ideals. It was not until the death of Stalin—and the war on the Stalinist cult of personality—that Soviet science and the Academy could undertake extensive actions to heal their wounds and to regain the lost qualities of intellectual integrity. The scientific community eventually moved into full action with vigor and determination, for the Stalin-Lysenko coalition had failed to break its spirit or to weaken its adherence to the moral law of science as a professional activity and a social force of enormous magnitude.

V

THE THAW AND THE SCIENTIFIC AND TECHNOLOGICAL REVOLUTION (1954–1970)

The death of Stalin in 1953 marked the beginning of sweeping changes in the Academy of Sciences that affected every component of its intricate structure and every link between science and Soviet ideology. The main lines of this transformation were related to two developments of vast proportions: the post-Stalinist thaw and the mid-century revolution in science and technology. The coalescence of the two during the second half of the 1950s and during the 1960s provided the main force behind growing efforts to free the Academy from the burden of Stalinism, which undermined the traditional right of the scientific community to protect the moral code of science, to certify scientific knowledge, to safeguard the scientific legacy, and to measure the social utility of science. It also set the Academy on a course of rapid expansion and institutional modernization. It made the Academy the pivotal force in rapid changes in the national administration of Soviet science, guided by the principles of geographical deconcentration and administrative centralization.

THE THAW

Six months after Stalin's death, the Academy held a special conference in celebration of the fiftieth anniversary of the Communist party of the Soviet Union. It was on this occasion that F. V. Konstantinov, a noted member of the Institute of Philosophy, stated that "the cult of personality"—which had made deep inroads into Soviet thought during recent years—was an unwelcome relic of Populist and Social Revolutionary thought. It was a phenomenon "totally alien" to the Communist party's position on the historical role of the "masses of people" and on collective leadership in politics.[1] The attack on the cult of personality was quickly followed by a pronounced loss of militancy on the anticosmopolitan front. Soon the boisterous war on cosmopolitanism became a quiet war on anticosmopolitanism, as it had been defined by the ideologues of Stalinism. The idea of Russian thought as a phenomenon sui generis gave way to the idea of close interrelations between Russian and Western thought.

The *Herald of the History of World Culture,* a new journal, undertook the task of answering the dictates of the new orientation. It carried essays on such assorted topics as the early influence of Kant and Schelling on Russian philosophical thought, the echoes of Benjamin Franklin's scientific ideas in Russia, and the "progressive" aspects of Auguste Comte's positivist theory of the evolution of science. In a full rejection of anticosmopolitanism, B. M. Kedrov stated in 1957:

> The question of the mutual influence and interrelations of national cultures is of great importance to historians. In the natural sciences, this implies the interaction of scholars from different countries in acquiring and applying knowledge. There is nothing more absurd than the idea of scientific discovery or technical invention as a result of fully independent creative activity of an eminent scholar or inventor. This view attributes individual inventions to single persons, usually from a selected national group, at the expense of contributions from other countries. As a result, a distorted picture of the history of a discovery or an invention is presented. The spirit of nationalism and chauvinism distorts the real historical process by exaggerating the contributions of some nations and minimizing those of others.[2]

Proceeding in many directions, the de-Stalinization of science concentrated on two activities: rehabilitating theories and disciplines banned by Stalinists on ideological grounds; and opening the doors of the Soviet Union to current revolutionary developments in Western science, even when orthodox philosophers treated them as contaminated by ideological impurities.

The process of rehabilitating the sciences and scientific theories banned by the political authorities took a slow but steady course. When in 1954 *Kommunist,* the party's theoretical journal, made a surprise attack on Lysenko for not answering the *Botanical Journal*'s criticism of his views on the origin of species and for the use of "administrative pressure" in silencing nonconformists, it triggered a sequence of developments that culminated in the total collapse of an empire built on the shaky foundations of Michurin's Lamarckism and on gross violations of the moral code of science. The decline of the Lysenko empire was a slow and uneven process. The ultimate fall of the bizarre doctrine came after a long succession of dialectical whirlwinds that, at first, moved from one extreme to another.

The government did not hide its displeasure with the scanty and unreliable output of the Institute of Genetics, Lysenko's academic fortress; but, at the same time, N. S. Khrushchev was ready to defend those of Lysenko's theories that continued to promise quick solutions to critical agricultural problems. To complicate the matter even more, Khrushchev was also ready to listen to the leaders of the scientific community who—particularly after the triumph of molecular biology in the West—did not hesitate to express their firm conviction that the future of biology was in its alliance with physics and chemistry, a union that had no place in Lysenko's calculations. While protecting Lysenko, Khrushchev did not prevent the Academy from building, at first cautiously, a series of new institutes in biophysics and biochemistry. Nor did he stand in the way of the rehabilitation of N. I. Vavilov—an unmistakable sign of the beginning of a new era in Soviet biology. While Lysenko's fortunes wavered, many of his followers searched for new loyalties. His once unchallengeable hold on the Academy and his invincible power in the Department of Biological Sciences were visibly eroding. On one occasion he could not muster enough support to be reelected president of the Lenin Academy of Agricultural Sciences; at other times, however, he had enough power to force the dismissal of N. P. Dubinin as director of the newly founded Institute of Cytology and Genetics in the Siberian Department of the Academy, and to vanquish the entire editorial board of the *Botanical Journal,* which published a series of articles illustrating grave flaws in his theories.

The scientific community, encouraged by the thaw and bolstered by the triumphs of Soviet science in atomic physics and space exploration, became the bastion of anti-Lysenkoism. Its struggle was carried on by eminent scholars acting as individuals or in small groups. Public seminars at Petr Kapitsa's Institute of Physical Problems heard speakers conversant with—and ready to discuss—the revolutionary developments in Western genetics and molecular biology. It was at one of these seminars, in 1956, that N. V. Timofeev-Resovskii and the physicist I. E. Tamm recounted the developments that led to the triumph of molecular biology, a science for which there was no place in Lysenko's plans and expectations.[3] Tamm adduced strong arguments supporting the thesis that biology, strengthened by the methods of modern physics, was destined to play a pivotal role in the fast-approaching new revolution in science.[4] The message of these and similar reports was clear and forceful: the time had come to rebuff Lysenko, who had removed Soviet science from research activities leading to revolutionary developments in modern biology. N. N. Semenov, a Nobel laureate, stated forthrightly in 1959 that Lysenko's claims were discredited by new "physico-chemical and biochemical experiments."[5] In 1961 M. V. Keldysh, the new president of the Academy, issued a stern warning that the future of biology was in its alliance with physics and chemistry.

Lysenko responded to the mounting criticism by doing what he knew best: he took Khrushchev and a select group of party and government officials for a tour of "Gorki Leninskie," his experimental farm. He received from this tour what he wanted: a featured article in *Pravda*— and in the *Herald of the Academy of Sciences*—underscoring the government's approval of his work. Despite this favorable publicity, the government showed clear signs of moving in the opposite direction: responding to academic pressure, it decided to finance accelerated expansion of biophysical and biochemical research in the Academy and to build new institutes in this field.[6] Although Lysenko continued his monopolistic hold on the Institute of Genetics, which he transformed into a center of Lamarckian biology, all branches of modern genetics and molecular biology found homes in other, mainly new, institutes. The Institute of Molecular Biology, the M. M. Shemiakin Institute of Bioorganic Chemistry (established in 1959), the Institute of Biological Physics (established in 1952), and the Institute of Protein (founded in 1967) grew rapidly into huge centers of modern biological research. In 1962 the Academy announced plans to build a modern biological center in Pushchino, near Moscow. Built without delay, the new academic city consisted of several institutes concentrating on the physical and chemical analysis of life.

When in 1964 Khrushchev was toppled from power, Lysenko lost his chief defender and main spokesman in the government. The Academy wasted no time in appointing a special commission to make an inquiry into the operations of "Gorki Leninskie." In the meantime, V. N. Sukachev, a distinguished member of the Academy, made the attack on Lysenko more direct and devastating. The earlier critics concentrated on Lysenko's ruthless use of "administrative pressure" in dismissing actual or potential enemies of his "theories" and in reshaping the biological institutes into centers of Lysenkoism. Resorting to a new mode of attack, Sukachev charged Lysenko with fraudulent claims on the scientific front. He concentrated on Lysenko's method of reforesting the Russian steppe, which emphasized the planting of trees in clusters and was applied on a gigantic scale in the late 1940s and early 1950s. By 1954, according to Sukachev, who was the director of the Academy's Forest Institute, more than half of the clustered trees were dead; by the fall of 1956, only 15.6 percent of these trees were still alive, with only 4.3 percent in average health.[7]

Two months after Sukachev's attack on the fraudulent practices of Lysenko's institutes, the popular journal *Science and Life*—with a circulation of over 3 million—carried an article by N. N. Semenov that centered on Lysenko's intolerable and far-flung violations of the rules of the scientific method and the ethos of science. Lysenko, according to Semenov, substituted "subjective fabrications for unambiguous and well thought-out experiments"; he relied on "observations which in most cases were erroneous, and on the arbitrary handling of material which led to absurd conclusions, often allowing for conflicting interpretations." In Semenov's view, Lysenko's methodology belonged to the distant past of science, when it was still heavily weighted by metaphysics. In Lysenko's endorsement of and heavy reliance on Lepeshinskaia's unfounded theory of the noncellular origins of cells in explaining the origin of species and in denying intraspecific struggle for existence, Semenov saw "subjective devices" that produced negative results: they misdirected the search for agrotechnical improvements, and they directed biological theory into channels irrelevant to those research developments that led to the triumphs of molecular biology. Semenov summarized Lysenkoism in this way:

The history of a typical Lysenko innovation followed a set course. He usually began by making a widely publicized announcement that he had come close to making, at minimum cost, a major breakthrough in agrobiology. Then, after some time, he announced that the prediction had been generally fulfilled and that the time had come for a large-scale application of the new method in agriculture. As a rule, all this was accomplished in the midst of much fanfare about the new achievement. However, the application of the recommended method soon began to lose ground and to receive little notice from the press, a sure indication of its economic ineffectiveness. But Lysenko concealed this failure by sensational press releases about a new prediction, the history of which differed from previous predictions only in minor details. This practice persisted from 1932 until the present time. This happened to vernalization, to the creation of new varieties in short periods of 2 to 2.5 years, to the improvement of varieties by intravarietal and intervarietal crossing, to the introduction of spring crops in the southern regions and winter crops in Siberia and Kazakhstan, to the utilization of stubble for soil improvement, to the introduction of intervarietal hybrids of corn and highly productive branching-stalk wheat, to the cluster method of forestation, to mixtures of organic and mineral fertilizers and to compost consisting of mixed manure and earth.[8]

In September 1965 the commission investigating the state of affairs in "Gorki Leninskie" presented its report to the Academy. The report confirmed what was already well established: the Lysenko empire was built on a combination of planned deceit, unfounded hopes, and transparent flaws in scientific premises. It stated that the farm had produced no evidence showing that the use of "advanced techniques" was the source of its much-publicized prosperity. Indeed, it showed that only special marketing and tax privileges had saved "Gorki Leninskie" from recording a financial loss in 1964.[9] Lysenko's claim to have developed a new breed of cattle that produced fat-rich milk and that transmitted this acquired characteristic to the offspring had no basis in fact. The records showed that the fat content of milk actually decreased from generation to generation. Experiments were conducted in a most unprofessional fashion. Records were kept in a slipshod manner. The control group of animals did not meet the announced specifications. There were no technical accounts of consumed fodder. There was no record of planned selection of crossbreeding animals. No biometric analysis of any kind was done. Nor was there a record of the use of genealogical analysis.[10]

The evidence was now in. It showed that Lysenkoism was not a usual fraud but a fraud of immense magnitude. The evidence did not show—because no one sought to point out—the conditions and maneuvers that made a hoax of such great proportions possible in the first place. As soon as the verdict was in, the Academy proceeded expeditiously to dismantle the institutional base of Lysenkoism. The Institute of Genetics was abolished, giving way to the newly established Institute of General Genetics, which made modern research in genetics its primary duty. N. P. Dubinin, an eminent geneticist suppressed by Lysenko, was appointed director of the new institute. *Agrobiology*, Lysenko's key "theoretical" journal, was abolished and was immediately replaced by *Genet-*

ics. The word agrobiology, which warranted a 5,000-word article in the second edition of the *Great Soviet Encyclopedia,* was now dropped from the Russian languge.

Before the drama of Lysenkoism had run its course, a few last fragments were added to the general scenario. The philosopher B. M. Kedrov, relying on fact and logic, showed that Lysenkoism was not only a pseudoscience and a flagrant attack on the moral code of science but that it was also a crude distortion of dialectical materialism. According to Kedrov, Lysenko relied too heavily on philosophers steeped in excessive dogmatism. He divided all scientific theories into materialistic and idealistic and advocated an uncritical acceptance of the former and an equally uncritical rejection of the latter. He disregarded the rule that the validity of individual scientific theories should not be judged by the philosophical views superimposed upon them. In every statement made by Michurin, Lysenko had seen a confirmation of materialism and dialectics, and in every statement made by Michurin's enemies he had seen an exercise in idealism. Believing that there was only one scientific path to truth, he rejected the possibility of various theories offering partial— but complementary—explanations of a natural phenomenon.[11] Lysenkoism, in Kedrov's opinion, was the highest point in the efforts of Stalinist philosophers to turn dialectical materialism against some of the most sublime achievements of modern science.

Before the drama of Lysenkoism became history, a last fragment was added to it: V. V. Skripchinskii published an article showing how Lysenko and his associates took unlimited liberty in twisting Michurin's teachings to make them fit the designs of Lysenkoist Michurinism. For example, Michurin, particularly during the last phase of his career, did not rule out the "theory of genes" or "Mendel's method" as useful— though probably limited—tools in the study of heredity; Lysenko made it a point to cite only Michurin's unfavorable references to the pioneers of genetics, whose work he rejected *in toto.*[12]

Lysenkoism brought staggering losses to Soviet biology. It forced the community of biologists to steer clear of research paths that brought biology in close contact with modern physics and chemistry and led, in the West, to the triumph of molecular biology. It stood in the way of bringing mathematics to biology. Lysenko condemned Mendel not only as the creator of modern genetics but as the first serious user of mathematical methods in biological research. The outlawing of Mendel's genetic theory in the Soviet Union meant also an abandonment of a serious concern with the advancement of mathematical methods in biology in general. During the reign of Lysenkoism, most universities made no effort to initiate young biologists into mathematical methods. Under the sway of Lysenkoism, the country produced "an entire generation of mathematically illiterate biologists."[13]

An important by-product of the decline and fall of Lysenkoism was the full rehabilitation of the synthetic theory of evolution, which combined the Darwinian conception of natural selection and the modern theory of mutations—particularly as elaborated by I. I. Shmal'gauzen's work in evolutionary morphology and by N. V. Timofeev-Resovskii in his theory of microevolution. The synthetic theory became the rallying point for biologists concerned with combining the historical bent of traditional theories with the structuralist orientation of modern biology, particularly after the emergence of molecular biology.

Plant ecology, particularly as advanced by V. N. Sukachev, and the general studies of the biosphere conducted by the students and followers of V. I. Vernadskii, now made the ideas of population genetics part of their basic theories. The I. M. Sechenov Institute of Evolutionary Physiology, established in 1956 and headed by L. A. Orbeli, made genetics a partner of Pavlovian studies of the nervous basis of animal behavior. The Institute of Plant Physiology, which previously was almost completely overrun by Lysenkoist brigades, concentrated on the structure of cells, which, in turn, brought it into the domain of the chromosome theory of heredity and variation. In 1958 Khrushchev called for a broad application of the methods of physics and chemistry in biological research. This move prompted A. V. Topchiev, chief secretary of the Academy, to announce that "in the coming years attention will be focused on meeting the broad possibilities of advancing biology by relying on such research methods as X-ray analysis, molecular spectroscopy, electronic paramagnetic resonance, radioactive indicators, electronic microscopy, and physico-mathematical analysis."[14]

Recently scorned as the very negation of science, dialectical materialism, and Russian intellectual history, genetics now became the queen of the biological sciences. The new situation was best described by N. P. Dubinin:

> In recent years, science has undergone radical changes. Genetics has unveiled the material base of heredity in the form of the molecular structure of DNA. It has established the chemical nature of genes. The

genetic code, regulating the synthesis of protein, has been deciphered and the nature of the molecular mechanisms of mutations has been fully explained. Genetics has elevated Darwinism to new heights and has enriched it by explaining the mechanisms of microevolutionary processes. It has produced elaborate methods for a study of the inter-action of environment and organism and it has shed new light on the role of natural selection in evolutionary processes and the role of arti-ficial selection in improving domestic plants and animals. . . . Genetics has become a central discipline in biology and a most advanced branch of modern natural science. It has become the cornerstone of the study of the basis of life and of the control of heredity. In all this, it has relied on the latest achievements in physics, chemistry and mathematics. We are entering the period of a strong alliance of genetics and selection, genetics and medicine, genetics and education, genetics and space biol-ogy and genetics and the biological problems related to the use of atomic energy.[15]

The effects of the post-Stalinist thaw reached far afield. Genetics was only one of several research activities that benefited from the relaxed situation in the academic world. The rebirth of sociology—a discipline that met apparent death in the 1930s and was buried in 1951–1952, when Stalin published his thoughts on N. Ia. Marr's linguistics and on the fundamental principles of socialism—was another major product of the thaw. Stalin's writings—as interpreted during the years of anticosmo-politanism—obviated the need for special inquiries into the sociological laws of the structure and dynamics of Soviet society. Sociology became known as a tool of capitalism, built on an ideology of colonialism, racism, and social inequality.

In 1955 V. S. Nemchinov—a defender of genetics at the August 1948 session of the Lenin Academy of Agricultural Sciences—published an article in the journal *Questions of Philosophy*, pointing out the rapidly growing need for sociology as a recognized discipline closely related to, but clearly separated from, historial materialism.[16] Historical material-ism, he argued, deals both with the unique features of individual socio-economic formations—such as capitalism and socialism—and with gen-eral laws, applicable to all formations. As a study of universal laws, historical materialism *is* "Marxist sociology." In addition to these laws, sociology, taken in a more general sense, deals with laws and regularities of specific domains of social life—with class relations, formations of nations, cultural dynamics, and interactions of various branches of social consciousness. Sociology encompasses a research domain that extends far beyond historical materialism. Nemchinov pleaded not only for a full rehabilitation of sociology but also for the creation of a new sociology dependent on the most advanced tools of statistical analysis. This conceptualization brought Nemchinov close to modern behaviorism—to the recognition of the vital importance of quantitative methods in the study of individual behavior as a significant expression of regularities inherent in mass behavior.

Although Nemchinov's plea encountered serious opposition from the old guard who held a monopoly on preparing widely circulated textbooks on historical materialism, it was justifiably considered the first step in a succession of fast-moving developments that made the Soviet Union a prominent home of modern sociology, ranging from simple empirical studies to intricate efforts to reconcile the structuralism of Western sociology with the historicism of Marxist theory and to advance a mathematical methodology for the study of the infinite complexity of modern social existence. The Academy's newly founded sociological research centers pursued three lines of activity: the Institute of Concrete Social Studies in Moscow, founded in 1968, worked on the methodology of empirical studies; a special research unit in Leningrad devoted primary attention to the sociology of work, institutions, and social stratification; and the newly founded Siberian Department in Novosibirsk concentrated on the problems subsumed under the general category of mathematical sociology.

Under Stalin, the deviations from social norms—crimes of all kinds, bureaucratic aberrations, and low work motivation, for example—were treated as survivals of capitalist mentality and were not subjected to scientific scrutiny. Now the time had come to recognize that digressions from social norms must be treated not as mere relics of the past but as concrete indicators of structural anomalies generated and harbored by the existing society. Stalin's ideologues, who dominated the social sciences, "knew" the causes of delinquent behavior: they were to be found in the survivals of capitalist attitudes. Post-Stalinist social scientists admitted no a priori explanations; they recognized that these causes could be found only in the realities of socialism and could be discovered only by "concrete sociological studies."[17]

These efforts were bolstered in 1963 when the procurator of the USSR established the All-Union Institute for the Study of Causes of Crime and Preparation of Recommendations for Its Prevention.[18] To fight crime it was necessary to identify its causes, and to identify its causes it was necessary to undertake empirical studies of the functioning of society in its many subsystems and to take a closer look into the dynamics of both anticipated and unanticipated results of social planning. The new sociological studies of religious practices in the countryside became important sources of illuminating information on social stratification in collective farms, occupational aspirations of rural youth, persistent elements of patriarchal social arrangements, the symbiosis of old and new customs and family rites, concrete effects of cultural isolation, and distinctive social attributes of rural families.

No science can be fitted into the Marxist scheme of useful knowledge with more ease than social psychology, for, after all, it is based on the principle that human behavior is molded by social forces and cultural values, a fact stated with equanimity in Lenin's criticism of the social theory of Russian populism. Yet for over two decades social psychology—which reached an apogee in the work of L. S. Vygotskii and A. P. Luriia—did not exist in the Soviet Union as a respectable and lawful discipline.[19] In the first place, by its very nature it was essentially an anti-Pavlovian discipline, for it shifted the focus of research from the neurophysiological moorings of behavior to the role of sociocultural and historical factors in molding human personality. In the Pavlovian view, every effort to study behavior as a special reflection of social life was considered unscientific, for there could be only one scientific psychology: the physiology of "higher nervous activity." The source of the awesome power of the Pavlovian school was not only in its huge size (and international standing) but in its newly acquired position as an integral part of Marxist theory. In the second place, Marxist scholars had appropriated "social consciousness"—the crucial problem of social psychology—as a key topic of historical materialism.

The emancipation of psychology from the restrictive influence of Pavlovian supremacy and the reemergence of social psychology as a legitimate area of scientific inquiry were products of the rising waves of opposition to the resolution of the joint session of the Academy of Sciences and the Academy of the Medical Sciences, passed in the summer of 1950, which allied Pavlov's science with Lysenko's theories and made psychology and psychiatry subsidiaries of physiological studies of conditioned reflexes. The psychologists were paralyzed since they were given the impossible task of subordinating the historical problem of the social dynamics of human psychology to the ahistorical analysis of the physiology of "higher nervous activity." In an effort to find a way out of the anomalous situation, a writer suggested—in an article published in *Voprosy filosofii* in 1954—that psychology, as an "independent" discipline, be made up of two distinct components: one based on Pavlov's neurophysiological theory and the other on the basic principles of historical materialism.[20]

The first step in the official retreat from the policy of a full subordination of psychology to Pavlovian neurophysiology was made in August 1955, in Kiev, at the Eighth All-Union Congress of Physiologists. The resolution passed by this body censored the Scientific Council on the Problems of the Physiological Theories of I. P. Pavlov, established by the 1950 session, for its failures in fighting the "negative forces" in the development of Pavlov's scientific legacy. It also criticized the Central Council of the Physiological Society for its role in suppressing "every healthy discussion of the fundamental problems of physiology." The crux of the new criticism was that the abnormal situation in physiology had created strong disharmony in the major research concentrations: while some areas received undeserved attention, others were given a minimum of attention. The best and most unmistakable harbinger of the changing time was the election of L. A. Orbeli—the main victim of Stalinist attacks on impurities in the interpretation of Pavlov's theories in 1950—to membership in the Central Council of the Physiological Society.[21]

In 1956 the Soviet Academy of Sciences published two classic studies of Vygotskii—"Thought and Word" and "The Problems of Mental Development of Children"—which received plaudits as major contributions to the study of "higher psychic activity," irreducible to Pavlovian "higher nervous activity."[22] The rehabilitation of Vygotskii, who died in 1934, heralded the beginning of a new era in the professional study of social behavior. Immediately, such capital works as *Being and Consciousness* (1957) by S. L. Rubinshtein and *The Problems of the Development of Psychic Life* (1958) by A. N. Leont'ev gave the "new discipline" a momentous start and a broad perspective.[23] The 1950 resolution of the joint session of the Academy of Sciences and the Academy of Medical Sciences stipulated that all psychological research must be anchored in Pavlovian neurophysiology; the resolution of the All-Union Conference on the Philosophical Questions of the Physiology of Higher Nervous Activity and Psychology, held in Moscow in 1962, recognized psychology as a discipline with large areas of inquiry irreducible to Pavlovian postulates.[24]

A. R. Luriia, in an effort to codify the changes in the attitude toward the Pavlovian tradition, developed a clear distinction between neurophysiology and neuropsychology. Both disciplines, he maintained, recognize reflexes as building blocks of human behavior. But neurophysiology is concerned exclusively with reflexes as natural material and as natural processes of behavior; it traces reflexes to their physiological moorings. Neuropsychology, in contrast, considers society a creator of "artificial stimuli" that place reflexes into new structures irreducible to the laws of nature. The new psychology, according to Luriia, is a discipline dealing with "qualitatively new formations" in the area where the natural and social sciences meet.[25] Unlike Pavlov's neurophysiology, the new neuropsychology took into account the cybernetic properties of behavior, particularly purposefulness and self-regulation of psychic processes.[26] The new developments have led Loren Graham to conclude that at this time Soviet psychologists and physiologists came closer to the views and approaches of their Western colleagues than at any other time. Graham also noted that, at the same time, American and British scholars came closer to the positions held by Soviet psychologists.[27] Western interest in Soviet psychology found clearest expression in the wide-ranging scope of English translations of Vygotskii's and Luriia's "post-Pavlovian," or neuropsychological, studies.

Several disciplines, or theoretical orientations, that were suppressed or severely crippled by the punitive measures called for by the joint session of the Soviet Academy of Sciences and the Academy of Medical Sciences in 1950 found their way back to respectability and intensive study. Evolutionary physiology, non-Pavlovian psychiatry, and several orientations that placed primary emphasis on physicochemical approaches to neurophysiological processes quickly entered a phase of accelerated growth, often providing strong challenges to the fundamental ideas of the Pavlovian tradition. The Pavlovian orientation was essentially biologistic: it was dominated by the principle that neurophysiological processes can be explained only with the help of the tools of biology, a science dealing with a reality sui generis, irreducible to the laws of physics and chemistry. Pavlov relied on the metaphors of Newtonian mechanics, but he used them as formal analogies rather than as explanatory principles. N. A. Bernshtein acknowledged the coming of a new age when he stated at the All-Union Conference on the Philosophical Questions of Higher Nervous Activity, held in Moscow in 1962, that in adhering to the principle of "the unity of the world and its laws," physiology, no less than the other sciences, must rely on the revolutionary ideas of modern science rather than on Newtonian analogies.[28]

The rehabilitation of resonance theory in structural chemistry was a slow process. In 1961 Linus Pauling visited Moscow and delivered a public lecture on the theory of resonance.[29] On that occasion he was informed that as early as 1954 a special commission erased most of the previous objections to his theory, a development, if it really took place, that did not receive public airing. Despite all this, A. N. Nesmeianov, president of the Academy of Sciences, complained in 1957 that the state of Soviet studies in structural and quantum chemistry was at a low ebb. He said that the discussions of 1951, which proclaimed Pauling's resonance theory an idealistic aberration, "turned scientists away from this field."[30] It was not unusual in the early 1960s to encounter outright dismissals of Pauling's contributions to modern chemistry, and some writers chose not to make any reference to Pauling's scientific work.

The post-Stalinist thaw reached far beyond the rehabilitation of condemned scientific disciplines. It also helped to remove barriers that stood in the way of a smooth and uninterrupted flow of the most recent theoretical advances from Western centers of scientific scholarship to the Soviet Union. Among the new Western developments, cybernetics occupied a place of particular importance and attractiveness. A system of intertwined principles describing the behavior of complex, self-organizing technical, social, and biological systems, cybernetics—formulated by Norbert Wiener in the 1940s—did not reach the Soviet Union during Stalin's reign. Indeed, it was not until the mid-1950s that the first serious steps were taken to make Wiener's ideas a component of scientific thought. A skeptical attitude toward the new theory prevailed, even though Wiener acknowledged the contributions of the Soviet mathematicians A. N. Kolmogorov and N. M. Krylov and the neurophysiologist I. P. Pavlov to the foundations of the new science. It was clear, however, that Wiener was referring to the Stalinist attack on science when he stated:

> Science is a way of life which can only flourish when men are free to have faith. A faith which we follow upon orders imposed from outside is no faith, and a community which puts its dependence upon such a pseudo-faith is ultimately bound to ruin itself because of the paralysis which the lack of a healthily growing science imposes upon us.[31]

A rapid succession of scholarly conferences made cybernetics a scientific orientation of universal appeal and the major propelling force of the modern revolution in science and technology. In November 1963 L. F. Il'ichev, in a paper read before the General Assembly of the Academy, referred to cybernetics as an inexhaustible fountain of theoretical ideas, enriching both science and dialectical materialism:

> The concept of information was known and was widely utilized before the emergence of cybernetics. But, after having become one of the basic notions of cybernetics, it quickly became filled with new content. So enriched, it spilled over the limits of cybernetics and began to acquire much broader significance and application. As it turned out, it came to express a new dimension of reality—an infinite, universal and general bond existing in all substances and phenomena, which could not be expressed adequately by other existing categories. In other words, the notion of information acquired not only a scientific, but also a philosophical—epistemological—meaning.[32]

Cybernetics spearheaded the computer revolution, and the computer revolution became the essence of the twentieth-century revolution in science and technology. Cybernetics connected mathematical logic, one of the most abstract branches of modern science, with electronic computers, one of the most practical devices of modern technology—and modern civilization. In 1964 the journal *Studies in Soviet Thought* published a list of over 200 Western translations of Soviet studies in cybernetics.[33] Soon it became generally accepted that cybernetics had not only benefited *from* Pavlovian studies but had also *contributed to* elevating them to previously unscaled heights of experimental refinement. The theoretical domains of the new interdisciplinary study covered the mathematical problems of cybernetics, information theory, technical cybernetics (concerned with the problems of control in technical systems), reliability theory, signs theory, and a long series of problems of a philosophical nature. Cybernetics found a place in economics, chemistry, energetics, transportation, engineering, many branches of biology, psychology, medicine, and jurisprudence. No other creation of the modern scientific mind had acted so swiftly in reshaping the course of scientific development in the Soviet Union. A philosopher described the new situation in science by pointing out that just as the splitting of the atom created inexhaustible substitutes for man's physical power, cybernetics created inexhaustible substitutes for mental powers engaged in studying and controlling the development of society, technical systems, and living nature.[34] This dual conquest, he said, was the essence of the twentieth-century revolution in science and technology.

Molecular biology, a new science that triumphed in 1953 when J. D. Watson and E. H. C. Crick shook the scientific world with their model of the inner structure of the DNA molecule (a model that accounted for the chief genetic and biochemical characteristics of hereditary material), stood in direct opposition to Lysenko's theory, which recognized work "in the field" rather than "in the laboratory" as the most potent tool in the study of heredity and which categorically rejected the existence of localized corporeal centers of heredity. The Institute of Biophysics, founded in 1952, sought to explain the physical processes involved in the work of muscles, the diffraction of light by the lens of animal (and human) eyes, the acoustics of the ear, and similar problems; all efforts to use the tools of modern physics and chemistry to unravel the genetic code of heredity were beyond the pale of recognized research. The enthusiasm with which the Crick-Watson discovery was received in the world of scholarship outside the USSR made a profound impression on Soviet scientists. The early signs of the erosion of Lysenko's academic power encouraged biologists to accept the challenge of molecular biology. Disregarding Lysenkoist taboos and doctrines, the Academy proceeded quickly to establish laboratory centers for new research and to create favorable conditions for a full employment of the tools of modern physics in the study of biological phenomena.

V. A. Engel'gardt, a leading biochemist and a full member of the Academy, viewed the triumph of molecular biology as a triumph of the unity of biology, physics, and chemistry and as a sure sign of the forthcoming revolution in life sciences.[35] At a time when Lysenko was making a desperate effort to consolidate and propagandize the biologistic slant of his theory of evolution, N. N. Semenov announced that the ongoing transformation of biology from "a descriptive and qualitative science" to a science guided by physics and chemistry promised to produce "a scientific and technical revolution of proportions equal to those wrought by nuclear physics."[36] As if to accommodate Lysenko, the Academy, at this time, showed an inclination to build most of the research base for the new biology and related fields away from Moscow—in the Siberian Department and in a special center in Pushchino. In a few years, however, new research facilities began to appear in Moscow as well.

Until the end of the 1930s mathematical logic did not find much encouragement in the Soviet Union, particularly in the age of Stalin. Despite this situation, the country had made noted contributions in this field. The work of M. I. Schönfinkel, A. N. Kolmogorov, A. A. Markov, and P. S. Novikov was recognized in and outside the Soviet Union.[37] Discouraged by Soviet ideology, mathematical logic received help from an unexpected source: the indifference of Soviet philosophers toward logic and toward the puzzling areas where logic and mathematics meet. Mathematical logic developed in isolation from philosophy, a situation that worked to its advantage.

In 1956 a group of scholars and engineers published an article in *Sovetskaia Rossiia* lamenting the fact that the country did not have a single institute of logic, nor a single journal devoted exclusively or primarily to logic.[38] In the same year, a special session of the Academy of Sciences dealt with the problems of automation in industry and with the role of mathematical logic in constructing the instrumental base of automation. It did not take long for the combination of the post-Stalinist thaw and the mass production of digital computers to catapult mathematical logic to the heights of academic respectability. In 1948 S. A. Ianovskaia noted the imperative need for the sparse ranks of Soviet experts in mathematical logic to turn their discipline toward the practical goals of building socialism and toward the theoretical goals of fighting the idealism of bourgeois mathematics.[39] As the 1950s came to a close, mathematical logic ceased to be suspected of total detachment from the realities of modern technology; on the contrary, it was now recognized not only as a key instrument in the study of the growing complexity of scientific cognition but also as a basis for computer technology.[40] The next step was to consult history in order to legitimate and bolster the newly recognized discipline by showing the impressive depth of its historical roots.[41] Mathematical logic was now proclaimed the primary contributor to "the beginning of a new epoch" in the evolution of scientific thought.[42] According to A. A. Markov, mathematical logic could justly be called "logic at the present stage of its development."[43]

By the mid-1960s there was no scientific research interest in the world that did not attract receptive attention in the Soviet Union. John von Neumann's game theory, previously considered a typical product of the moral cynicism and intellectual decadence that prevailed in capitalist scholarship, now received unqualified praise as the epitome of the daring and free spirit of modern mathematics. The unlimited flow of scientific knowledge and hypotheses created in foreign laboratories produced problems of a new kind. While in the days of Stalin many theoretical ideas were rejected without the benefit of careful scientific scrutiny, now there prevailed a strong tendency to take many Western ideas and research hints seriously even when they appeared to be tenuous or unrealistic. Alert scholars, led by L. A. Artsimovich, fought the new tendency to replace the excessive criticism of Western developments in modern science—a tool of the Stalinist war on cosmopolitanism—by a stance that was devoid of all criticism.

THE SCIENTIFIC AND TECHNOLOGICAL REVOLUTION

One side of the Stalinist equation emphasized the rigid ideological control of science, an effort that violated the moral code of science and required extensive institutional repairs in the 1950s and 1960s. The second side emphasized the need for a rapid growth of institutions and manpower engaged in scientific research, a policy that committed the Soviet government to virtually unlimited financial investments in science. While the first side of the equation produced monstrous attacks on the professional identity and intellectual standards of science and gave universally unfavorable publicity to Stalinist cultural policies, the second side made the Soviet Union a giant in modern science and helped to produce the scientific and technological revolution of the twentieth century. This revolution, as viewed by its Soviet interpreters, was a synthesis of a revolution in science that began to unfold at the beginning of the twentieth century—with the first triumphs of the quantum and relativity theories—and a revolution in technology, heralded by the technical conquest of atomic energy and by the advancement of electronic computers.[44]

The revolution in science was the main catalyst of the revolution in technology. As recounted by M. V. Keldysh, president of the Academy since 1961, the revolution in science was most decisive in four major areas of activity. First, it brought to prominence various disciplines known under the general name of microphysics. Physicists faced the task of discovering new particles and reducing the bewildering number of discovered—and yet to be discovered—particles to a small number of basic units. In order to achieve these goals, physicists worked constantly on improving the efficiency of nuclear accelerators. The work on controlling thermonuclear fusion received high priority. Second, the revolution in science added new importance to space exploration and to the astrophysical study of the structure of matter. The study of non-stationary aspects of the universe held the keys to understanding the transformation of matter. Third, it made cybernetics—and computer technology—the backbone of the new civilization. And fourth, the revolution produced a total refocusing of biological research, with studies of life centering on molecular and cellular analysis. "Like physics, which at the beginning of the century made a minute study of inorganic matter its pivotal activity, biology has undertaken a deeper study of living matter, concentrating on the structure of the cell and 'the elementary composition of its numerous components.'"[45]

The new revolution upset the traditional interrelations of science and technology: while previously science was guided by and responded to the needs of technology, now technology was guided by and reflected the achievements of science.[46] Science became a "direct productive force," a vital part of the "socioeconomic base" of Soviet society. According to a student of new developments in science:

> The modern scientific and technological revolution is characterized by a particular organic unity of all branches of science, technology and production. Revolutions in science and technology are not new. A revolution in science ushered in modern experimental science and classical mechanics. A revolution in technology brought forth the industrial revolution at the end of the eighteenth and the beginning of the nineteenth century. No doubt there was a connection between these two revolutions. . . . However, the technological revolution of that time cannot be labeled "scientific-technological." . . . Technology and science grew, as a rule, in two separate spheres—one in the sphere of direct material production and the other in the sphere of nonmaterial culture. The revolution in science began to be a scientific and technological revolution only when the application of science became the most important factor in the development of material production, when science began to be transformed into a direct productive force and when material production became a technological application of science.[47]

The dawn of the new era came on the wings of remarkable achievements in harnessing nuclear energy, space exploration, and the science of macromolecular matter and was accompanied by a romantic faith in the unlimited promises of electronic computers. The new revolution made it possible to transform industrial plants into scientific laboratories and to change scientific laboratories into elaborate technical systems resembling modern industrial plants. It deeply affected the institutional makeup of science: it accelerated the process of creating larger research organizations, made group research an institutional norm, and encouraged new styles of collaboration among experts in various sciences and branches of technology.[48] All these developments, in turn, produced pressing demands for broad changes in the organization of political, economic, and administrative authority on various levels of the national system of scientific institutions. According to a social analyst: "We have named the modern progress in technology the scientific and technological revolution . . . because . . . it has made science, technology, and education the main spheres of human activity."[49]

In the eighteenth and the nineteenth century, a revolution in industrial technology was the primary cause of the growing adaptability of science to the needs of industrial production. In the twentieth century, a revolution in the theoretical and experimental base of natural science was the primary cause of the growing adaptability of industrial technology to the massive breakthroughs in science.[50] In the earlier period, the growth of industry directed the growth of science; in the modern period, the growth of science directed the growth of industry. What was the effect in an earlier period became the cause in the twentieth century. According to S. R. Mikulinskii, a leading student of the organizational and sociocultural dynamics of science:

> The scientific and technological revolution is not simply the application of new types of energy, new materials and computers, and comprehensive automation of production and management; it is a drastic transformation of the entire technical basis, the technological mode of production, the forms in which production is organized and managed, and man's attitude to the production process. The scientific and technological revolution is a drastic qualitative transformation of the productive forces of society on the basis of the development of science into the leading factor of technical progress and the development of social production. It changes the entire face of social production, the conditions, character and content of work, the structure of productive forces and social division of labor and, through the latter, affects the social structure of society. Thus, the scientific and technological revolution covers not only the sphere of science and technology, but also the sphere of production, which is why it has such a strong impact on all aspects of present-day society, including everyday life, culture, psychology and the interaction between nature and society.[51]

Whatever the scientific underpinnings of the new revolution, Soviet scholars made a solid contribution to it. In 1953 they unleashed the awesome power of thermonuclear fusion, and in the mid-1950s they established the Tokomak project, which quickly became "the front-runner of all controlled fusion concepts."[52] On June 27, 1954, the Soviet Union put into operation the first atomic power station, located in Obninsk. The station had a usable power output of 5,000 kilowatts. The first atomic icebreaker soon followed. At this same time, Soviet scientists and engineers built a 10-GeV proton synchrotron in Dubna.[53] This was only the first step in the construction of increasingly more powerful and sophisticated accelerators, all contributing to the creation of a distinctive pattern of national nuclear technology. In 1968 a 70-GeV proton synchrotron was completed in Serpukhov. In 1954 two Soviet scientists— N. G. Basov and A. M. Prokhorov—presented startling ideas on the possibility of amplifying electromagnetic waves, a discovery that led to the construction of masers and lasers.

The first sputnik, which orbited the earth on October 4, 1957, brought man closer to the cosmos and the cosmos closer to man, made interplanetary travel a realistic possibility, and widened the horizons of science beyond previously contemplated bounds. This feat, together with the orbiting of the first manned sputnik on April 12, 1961, attested not only to the remarkable achievements of Soviet science but also to the high level of Soviet technology. The era of space exploration, in turn, produced a wide range of new research topics, such as interplanetary plasma, cosmic vacuum, and solar radiation outside terrestrial limits. It also gave birth to such new sciences as space biology, space medicine, and space physiology, as well as to the astronomy of ultraviolet rays, gamma-astronomy, and X-ray astronomy. The study of the planet Earth benefited from such new branches of scientific knowledge as space meteorology, space mineralogy, and space geography.

The notable achievements of Soviet science and technology were part of a general outburst in scientific creativity that affected most disciplines and, in its total impact, opened radically new perspectives on the development of the national economy, social relations, and cultural values. International recognition of the achievements of Soviet science reached unprecedented heights, as attested to by the decision of various American scholarly associations to publish, in English translation, over fifty Soviet scientific journals, cover to cover.

The inauguration of the new era in science called for a fresh look at the socioeconomic correlates of the alliance of science and technology, which produced an endless stream of sociological studies, philosophical discussions, and literary elaborations. Most philosophers and scientists agreed that basic research—or "pure science"—was the true fountain of the scientific-technological revolution. New scientific discoveries produced more advanced technological processes and new branches of industry. Nuclear physics, for example, created a new industrial source of power. Solid-state physics and high-pressure physics led to the industrial production of semiconductors, artificial diamonds, and new types of heat-resistant materials. The ongoing revolution accelerated the processes of transforming scientific ideas into industrial products and of establishing research centers in industry. It brought together sciences previously isolated from each other. (Cybernetics, for example, brought neurophysiology, structural linguistics, and the mathematical theory of communication into an enormously fruitful process of cooperation that helped to resolve the theoretical and practical problems of automation.) The revolution in science laid the foundations for a more effective organization of the managerial system of the national economy and for more reliable techniques of effecting directed social and cultural change. New science added an intellectual quality to industrial labor: it made the industrial worker a master of technical knowledge and production processes rather than a mere appendage of machinery.[54]

Behind realistic assessments of the broad canvass of sociocultural and economic effects of the scientific and technological revolution loomed a vast and enticing area of ideological coloration. The ideologues produced an immense literature on the comparative superiority of the Soviet system in realizing maximum benefits from the modern explosion of scientific thought and its alliance with technology. In this literature ideological cliches often ran far ahead of factual analysis and contributed to an unrealistic assessment of mounting efforts to keep up with the challenges and potentials of the ongoing revolution. In most writings, the realities of today were adorned with lavish expectations for tomorrow. Almost all writers on the subject rejected the idea that, in a socialist society of the Soviet type, the growing automation of industrial production could produce an even greater subordination of man to the impersonal power of control mechanisms, as well as an even more debilitating alienation than Marx wrote about. Conditioned to idealize the contributions of the ongoing revolution to humanity, most writers overlooked the ominous threats to the expression of individuality in scientific creativity and, in general, to the moral code of science.

The widely heralded optimism did not go completely unchallenged.

Criticism of the less commendable results of the revolution was recorded. Limited in scope, it came primarily from individual members of the Academy of Sciences, whose critical points usually referred to the realities of their institution. Limiting his discussion to the dilemmas of the national commitment to science, P. L. Kapitsa used several occasions to make a plea for checking the forces unleashed by the scientific and technological revolutions, which threatened both academic autonomy and the humanistic base of the scientific vocation.[55] At one point he asserted that science lost freedom when it became a "productive force."[56]

In his comments at a session of the Academy of Sciences in 1965, Kapitsa emphasized three problems that, in his opinion, required immediate and thorough scrutiny. The first problem, which he identified as the moral problem, dealt with the acute need for alleviating the rigidity of structured impersonality in the performance of research duties. He thought that the authorities should encourage the existence of informal "clubs of scientists" as mechanisms for mobilizing the spontaneous exchange of ideas and serving as important instruments for placing "unplanned" creative thought into the lines of scientific communication. Without these informal mechanisms, he said, there could be no true scientific community, the only rightful certifier of new contributions to science and the only proper guardian of the scientific legacy. He wrote in an article published in 1965:

> The creation of a healthy and progressive community of scientists is an enormous task to which we give far too little attention. This task is more difficult than the training of selected young talent or the establishment of large institutes. . . . The community of scientists alone can objectively judge the achievements of science. . . . Only an advanced scientific community can fully appraise the intellectual power of a scientific discovery independently of its direct practical significance.[57]

The scientific community, as Kapitsa saw it, drew its major strength from personal contact between scholars. An international phenomenon, it depended to a large measure on face-to-face contact between scientists from different countries. Lack of contact was the main reason for the slow recognition of Russia as an active contributor to scientific thought. Kapitsa acknowledged the growing scope of personal encounters of Soviet scientists with their foreign peers, but he contended that this area of activity needed much improvement. Zhores Medvedev had argued in favor of legal safeguards enabling Soviet scientists to have "personal relations with citizens of other countries." These safeguards, he said, would mark a major step in transforming Soviet society "from a semi-closed to an open system."[58]

Obviously influenced by Kapitsa's pronouncements, Soviet students of human relations in scientific institutes had come to recognize so-called informal organizations as an important factor bearing on scientific activity.[59] Informal groups were a source of relations mitigating the rigidity and impersonality of prescribed behavior in laboratories and structured research projects. They were potent vehicles for reacting to unpopular decisions emanating from various links in the formal organization. Above everything else, these informal groups were a source of spontaneous ideas unencumbered by the stringency of paradigmatic unity in various branches of science and ideology. In the age of Stalin, informal groups provided dynamic and relatively safe havens that nurtured scientific interests forbidden on ideological grounds. Although the extraordinary sociological potency of these groups was readily recognized, no systematic study of their anatomy and dynamics had been published. A much stronger interest existed in informal groupings as a potential source of challenge to formal authority than as a positive force for humanizing and enriching modes of scientific communication.

The mores of the scientific community demanded much more than the existence of "clubs of scientists," international face-to-face contact, spontaneous, unrestrained discussion and informal groups. Scientists also required that they be guaranteed the right—and that they be encouraged—to criticize aspects of national science policy that they considered inadequate or misdirected. In 1965, afraid that the momentum in the development of Soviet science was ebbing and that the scientific productivity of the Soviet Union was beginning to fall even more behind that of the United States, Kapitsa noted that the best way to reverse this trend was to encourage unrestricted criticism of "our shortcomings and errors" and free expression of recommendations for improvement.[60]

The second problem—Kapitsa called it the problem of cadres—referred to the search for the most rational employment of scientific manpower. He opposed the preponderance of rash efforts to cope with the growing momentum of the scientific and technological revolution by building research centers before adequately trained personnel were available and by spreading talent too thinly over the rapidly growing and occasionally fashionable institutes and laboratories. Instead of dispersing the talent, he said, the country would do much better by concentrating on disciplines that constituted strong national traditions in science. L. A. Artsimovich, who elaborated these arguments, mentioned solid-state physics and radio physics—but not nuclear physics—as the areas of the most original contributions of Soviet physicists.[61] Neither Artsimovich nor Kapitsa argued that the Soviet Union should give up on the areas that were not supported by strong traditions: what they wanted was a gradual, careful, and systematic building of new traditions to replace uncontrolled and indiscriminate adoption of every lead, tested or untested, generated by foreign scholarship. Right or wrong, they reasoned that the real strength of Soviet science could be measured most reliably by its contributions to the universal pool of knowledge.

Seemingly unlimited financial investment in science made it possible for the authorities to heed Artsimovich's criticism without retreating from strategic positions in the rapidly widening scope of scholarship. The scientific community continued to build upon strong national traditions, but it also kept a watchful eye on every new research lead in the full spectrum of the sciences. On the one hand, it stuck closely to the cultural uniqueness of Soviet science—created, to give a few examples, by P. L. Chebyshev's engagement in probability and number theories, I. M. Sechenov's and I. P. Pavlov's neurophysiological theories of behavior, A. M. Butlerov's structural chemistry, and V. V. Dokuchaev's soil science. On the other hand, it never deviated from the idea that the building of a new society required a national commitment to the full arc of scientific endeavor. "Industrialization, socialist construction, and cultural revolution" demanded an "unbroken front" of Soviet science—a national involvement in "all branches and orientations" of scientific inquiry.[62] If this rule was occasionally violated, as in the case of genetics, the protection of science from unscientific admixtures was given as the explanation. This kind of excuse, however, continued to live mainly in the persistent and frightful memory of Stalin's war on heterodox thought.

The emphasis on the "unbroken front" of national science—on the versatility of national involvement in scientific activity—required a reexamination and reassessment of practical and sentimental attachment to the strong Russian traditions in science. A critical look into the previously unchallenged hegemony of I. P. Pavlov's neurophysiological legacy brought home the necessity for a careful balancing of national strengths in science with the need for an "unbroken front" of scientific activity.

A representative group of leading scientists now admitted publicly that the Pavlovian "bias" had caused much asymmetry in research in physiology and related fields, and that the time had come for major restitutions. Until the end of the 1960s, for example, the official Soviet classification of scientific professions did not recognize psychologists as a distinct group. This was an obvious legacy of Pavlov's repeated assertion that the psychological profession, as a scientific profession, was covered by neurophysiologists engaged in laboratory work on conditioned reflexes. E. M. Kreps stated in 1970 that the prolonged dominance of Pavlov's thought was responsible not only for the low interest in physiological research not directly related to the neurophysiology of conditioned reflexes but also for insufficient concern with theory. He concluded that more versatile and comprehensive psychological studies were needed in three areas: health, culture, and society.[63] Kreps added that even within their limited context, some pivotal ideas of Pavlovian neurophysiology needed thorough reexamination. A growing number of scholars claimed that the time had come to abandon the practice of disproportionate representation of Pavlovian experts in the general areas of physiology, psychology, and psychiatry.

The third problem, which Kapitsa called the fiscal problem, dealt with the established practice of financing the scientific activity of the Academy. Continuing a personal practice that he started two decades earlier, Kapitsa noted that the government met the financial needs of rapidly expanding research, but that it did not grant research institutes sufficient latitude in distributing available funds.[64] Implicit in his argument was the idea that tight external controls over financial matters prevented institute directors from undertaking the necessary steps to maintain high-level standards in selecting and retaining research personnel and to meet the challenges of "unexpected" and "unplanned" turns in authorized research. These controls tied the hands of institute directors in recruiting necessary—and in dismissing unnecessary—personnel and in using funds for technical readjustments in laboratories to meet the challenge of "unexpected" research leads. This practice, he said, made many leading scientists reluctant to work as institute administrators. No doubt Kapitsa agreed with the statement made by I. P. Pavlov in the early 1930s that the bureaucratic methods of financing research were totally incompatible with the intellectual autonomy of scientific activity, and that the function of the government should be to provide adequate funds for science but not to administer and control their distribution among and within various components of research centers.[65]

Kapitsa's plea did not go unheeded: in 1967 the government granted institute directors an unprecedented degree of independence in assigning available funds, commanding the flow and distribution of personnel, and determining the sums allocated for equipment.[66] A few years later, R. W. Davies and R. Amann noted that "in the budget grant to every Academy institute a definite part is allocated to entirely free research."[67] Artsimovich argued that the health of the scientific community did not depend solely on the greater autonomy of institute directors; the scientific community, he said, must introduce appropriate measures to prevent abuses of government funds by persons in positions of authority who placed self-aggrandizement over the interests of science. He also pleaded for a more vigilant role on the part of the scientific community in resisting various efforts to impose the undue and harmful controls of some sciences over others. For example, Artsimovich resented the unmistakable signs of unchecked growth—e.g., the "imperialism of physics"—among some members of the family of sciences. He could not understand why the national investment in astronomy constituted only "a few percentage points" of the investment in nuclear physics.[68] And he voiced categorical disapproval of the fiscal treatment of astronomy as a mere extension of physics. In all his lamentations, however, Artsimovich never disagreed with the complaint of A. N. Kosygin, at that time deputy chairman of the Council of Ministers, that the Academy provided a haven for research projects that had neither theoretical nor practical significance.[69]

In his appeal for a more systematic and rational approach to the national science policy, Artsimovich stated that the question as to whether scientific research should follow self-generated theoretical and methodological leads or the practical needs of society, as defined by the government and included into five-year and annual plans for economic and cultural development, was obscured by serious secondary questions. He took note of extensive—and often expensive—research that was guided neither by the internal logic of science nor by the external needs of society but was instituted mainly as a response to "international competition for scientific achievement." Such competition, Artismovich said, allowed foreign countries to dictate the research and development priorities in the Soviet Union and to impose unnecessary financial bur-

dens on the Academy and the Soviet scientific community.[70] He added
that "irrational factors" of this kind had found their way into national
science policy and were a force to be reckoned with.[71]

INSTITUTIONAL GROWTH

The post-Stalinist thaw and the swift growth of new disciplines—
mainly at the crossroads of established sciences—accounted for the
broad refashioning, reorienting, and expanding of the organizational
base of the Academy. In 1955 the Presidium of the Academy appointed
a long list of "brigades" of leading scholars to study the general dynamics
of modern science and to make recommendations for institutional adaptations. Almost immediately, the authorities began the intricate process
of recasting the complex network of research components. In some
cases—typified by the Institute of Biological Physics, the Institute of
Plant Physiology, and the I. P. Pavlov Institute of Physiology—the major
change consisted of reinstating research specialties that had lost scientific
status during the conflict won by Lysenkoism at the end of the Stalin
era. In other cases, it consisted of creating special research bodies—
mainly institutes—to meet the needs of new scientific fields. In quick
succession, the Academy organized research in cybernetics, molecular
biology, bioorganic chemistry, quantum chemistry, mathematical economics, empirical sociology, and many other newly carved research
areas.

The most extensive changes in the anatomy of the Academy came as
a result of organizational adjustments to new developments in physics.
Institutional responses to processes of fission within the bulging body
of physics led to the creation of institutes specializing in such new
branches of knowledge as the physics of ocean water, the physics of
metals, and the physics of atmosphere. There was also a rapid expansion
of previously established institutes in biophysics, chemical physics,
physical chemistry, electrochemistry, radioactive elements, astrophysics, physical crystallography, and acoustics. New journals specialized in
atomic energy, acoustics, crystallography, solid-state physics, optics,
spectroscopy, geomagnetism, physics of metals, radio physics, and
physical equipment.[72] Research equipment produced by physics laboratories found extensive application in many disciplines, some far
beyond the traditional boundaries of physics. This equipment also
helped to modernize and redirect work in most branches of physiology,

biochemistry, and microbiology. Organic analysis alone changed many of its traditional research patterns through a growing reliance on techniques supplied by infrared spectroscopy, nuclear magnetic resonance, mass spectrometry, and polarography.

The expansion and repatterning of the research base acquired a new dimension in the creation of modern scientific centers removed from Moscow and Leningrad. There were several major reasons for the new policy. World War II, which forced the Academy to move most institutes from Moscow and Leningrad, demonstrated the dire consequences of overcentralized research. The war showed clearly that it was much easier to move scientists than their laboratories and that it was exceedingly difficult, and practically impossible, to organize advanced research in communities without strong traditions in scientific activity. The policy of organizing elaborate research facilities away from the traditional centers received strong support from influential leaders of the Academy who felt that relative isolation from firmly rooted scientific traditions and vested academic interests would be helpful to exploratory work in new disciplines. They felt, for example, that the need for accelerated work in biology, using the tools of modern physics and chemistry, could not be carried out effectively in the Moscow institutes, which were staffed by a large number of Lysenkoists, even though the power of Lysenkoism was rapidly waning.

In 1957 the academic authorities announced ambitious plans for a large science community in Novosibirsk that would serve both as an agency to guide and coordinate the application of the most advanced research techniques in the study of natural riches of Siberia and as a center for exploratory work in selected branches of new science. In 1958 the construction of the new community—Akademgorodok (Academy City) or, officially, the Siberian Department of the Academy of Sciences of the USSR—was well under way. By 1961, with a population of over 6,000 (including family members), Akademgorodok was a thriving scientific center. In 1968, of the 25,827 persons employed by the department, 5,555 were classified as scientists.[73] It was widely acclaimed as the leading center in a number of the most promising disciplines or scientific orientations of modern vintage and as a breeding ground for new, creative ideas in organizing research, facilitating the flow of scientific ideas to industry, and training future leaders of science. Petr Kapitsa greeted Akademgorodok as a particularly attractive place of work for promising young scholars who needed more independence than they would have

had working in central institutes, which were dominated by older scholars with set ideas.

The Siberian Department occupied a unique position in the academic
system. Unlike other academic departments, it was governed by its own
general assembly and presidium, and its administration was subordinate
to both the USSR Academy and the government of the RSFSR. It was
funded by the RSFSR, which was without a republic academy of its
own.[74]

Although the Siberian Department was presented as a new center of
fundamental research in all the basic sciences, it was clear from the
beginning that its major function was to work in the areas that required
radically new perspectives, more independence from the strongholds of
academic traditions, and a decisive dismantling of the barriers built
under Stalin. The Institute of Cytology and Genetics undertook extensive
work in "the physicochemical and structural foundations of heredity"
and in "the genetic foundations of evolution and selection." This institute helped set the stage for a revival of research in genetics and for the
tying of classical genetics to the rapid advances in molecular biology.
The Institute of Mathematics (with the Computer Center) concentrated
on the new branches of mathematics, such as the theory of linear programming and mathematical logic. Such luminaries of Soviet mathematics as M. A. Lav'rentev, L. V. Kantorovich, S. L. Sobolev, and A. I.
Mal'tsev helped the Siberian mathematicians give their discipline many
new forms of expression, a broader base of operations, and a more
flexible philosophical orientation. The institute was engaged in pioneering work on mathematical methods for such widely varied branches of
knowledge as sociology, economics, geology, and chemical technology.
The Institute of Physics worked on new types of accelerators and on the
regulation of thermonuclear reactions.

The department became deeply involved in a search for new styles of
organized research and new institutional links between academic laboratories and industrial plants. It also became an important national
center for designing and constructing complex laboratory tools for modern scientific research. From the outset, its aim was not only to become
a workshop of new orientations in science but also to explore ways of
establishing new types of interdisciplinary research and new modes of
cooperation between institutes.[75] A growing number of practically oriented institutes appeared in the immediate vicinity of the Siberian
Department. Financed by individual ministries, these institutes worked

on techniques for a speedy transmission of new scientific ideas from academic institutes to the economy. The Novosibirsk center quickly emerged as the publication headquarters of a long string of high-quality scientific journals—including the internationally reputed *Siberian Journal of Mathematics*. To add another dimension to the unity of the scientific community, Novosibirsk became a center for learned societies—voluntary bodies, such as the Mathematical Learned Society—organized for the purpose of providing additional media for scientific communication and for intensifying personal contact between scientists working in the same or related fields.

The Siberian Department took on yet another function of vital significance: the search for new ways of selecting and training young scientists.[76] Since the early 1950s, the Academy of Sciences vastly expanded the recruitment of promising young scholars to meet the challenges of the scientific and technological revolution. While in 1940 it added 84 new university graduates to its research staff, during the next fifteen years it added an average of 1,357 young scientists annually.[77] The number of "scientific workers" employed by the Academy grew from 4,200 in 1940 to close to 35,000 in 1970. Particularly during the 1960s, the Academy was concerned not only with the sharply rising need for expert cadres but with the quality of scientific training as well. It instituted the policy of filling vacant positions by competitive methods and of making it obligatory—and financially rewarding—for leading scholars to play a more prominent role in training the future elite of Soviet science. Despite various measures, the Academy was not in a position to apply uniformly high standards in all its research components, particularly in the new areas where biology merged with physics and chemistry, which were areas devastated by Lysenkoist hegemony in the Department of Biology. The same applied to the fast-growing interest in modern computers, which demanded intense involvement in previously underemphasized or nonexistent branches of mathematics, such as information theory and theory of algorithms.

The role of the Siberian Department in raising the standards of scientific training was particularly innovative. The department founded a modern university dedicated to improving the techniques of selecting new students: in mathematics, for example, it relied on "mathematical olympiads," annual academic-achievement contests involving Siberian high school graduates. The university made student participation in activities of experimentally oriented research institutes a key part of the required curriculum. Attached to the university was a special boarding school for high school students in mathematics and physics, who received special tutoring in current developments in the sciences of their choice and who were provided with continuing opportunities to communicate directly with the established leaders of modern scientific thought.[78] The obvious purpose of this school was to prepare talented students for university studies.[79] The total educational process was based on the principle that the "revolutionary" developments in new science did not represent a break with tradition but a stepped-up and more diversified building upon it. It also concentrated on preparing students for work in vital disciplines—such as mathematical logic, quantum chemistry, molecular biology, and mathematical economics—that previously had received insufficient attention.

In the evolution of the Siberian Department, two trends were unique. In the first place, the department became the core of the Novosibirsk Scientific Center, which combined the academic research units with selected sectoral institutes financed by various economic ministries and with newly founded design and development bureaus. It also included the local branch of the Lenin Academy of Agricultural Sciences, which undertook an expeditious and eminently successful effort to eliminate institutional and philosophical vestiges of Lysenkoism. The center represented an effort to coordinate the full spectrum of scientific activities—from the engagement in high theory to the production of industrial prototypes. In the second place, the Siberian Department embarked on an ambitious plan to establish its branches in the major cities of Siberia—or to expand the branches it inherited from the Academy of Sciences. These branches, in turn, became the core units of local research systems, modeled on the Novosibirsk Center. All these actions were part of a new plan to achieve the maximum regional integration of institutions working under different administrative jurisdictions and covering all aspects of research and development.

The history of the new scientific centers actually began in 1969 when the Presidium of the Academy of Sciences decided to establish the Ural Scientific Center in Sverdlovsk and the Far Eastern Scientific Center, entrusted with the task of coordinating the work of the local branches of the Soviet Academy, selected sectoral institutes, and research units of institutions of higher education. New centers were dedicated not only to the study of regional problems but also to a deep involvement in basic research and to offering high-quality training to aspirants for higher degrees.[80] The Ural Center concentrated on various branches of mathematics, solid-state physics, polymer chemistry, high-temperature electrochemistry, and biophysics; the Far Eastern Center became an important base for the study of marine biology, biologically active natural substances, vulcanology, and computer methods in the study of the archeology, ethnography, and languages of the Soviet Far East.[81] The growth of the new complexes—particularly of the Far Eastern Scientific Center—proved to be a slow process. Academic authorities decided on a gradual building of local talent rather than an accelerated migration of established scientists from central institutions.

The new centers encountered two serious difficulties: the extremely slow construction of modern laboratory facilities and the lack of experience and technical skills in coordinating the work of research units scattered over wide areas.[82] Although the government was not stingy in providing funds for modern laboratory equipment, Soviet factories continued to be sluggish in making deliveries of more sophisticated scientific equipment.[83] While recognizing these difficulties, the academic authorities greeted the new centers as an important step in concentrating talent and modern research equipment in strategic places. They also welcomed the new moves to achieve a deconcentration of scientific activities. According to an Academy spokesman, the future of Soviet science was in regional competition and cross-fertilization rather than in rigid controls by traditional centers.[84] Regionalism was greeted as a powerful weapon in the war against monopolies in science. The purpose of regional centers was to strike a balance between overconcentration and excessive deconcentration of scientific talent and modern research facilities.

The search for new types of regional research organizations proceeded also in previously unexplored directions. The most noteworthy new creation was the North Caucasian Scientific Center, administered by the regional system of higher education rather than by the Academy or various ministries.[85] Heralded as "a new form of organization of scientific research in the Soviet Union," the center, anchored at Rostov-on-Don State University, was a unique experiment designed to raise the intensity and to widen the scope of scientific work in provincial institutions of higher education. The new center coordinated the general lines of research carried out by 50 institutions of higher educations and 200 research organizations affiliated with government agencies and universities.[86] The most practical purpose of the new center was to organize interdisciplinary and mobile research teams, in order to study specific problems of regional economic significance, and to integrate students into active research undertakings. The burgeoning system of regional scientific centers opened new vistas in the national search for improved methods of bringing high science—science at its most advanced frontiers—into direct contact with the productive forces of individual economic regions. Its goal was to raise optimum returns on the financial investment in science and to give new stimulus to scientific research in institutions of higher education.

Part of the expansion of the Academy's institutional network came through the establishment of individual institutes in cities away from Moscow and Leningrad. For example, in 1969 the Academy established the Sverdlovsk Branch of the Mathematical Institute, which concentrated on the mathematical theory of control processes, computational mathematics, and engineering and economic mathematics. In the same year, it founded the Institute of Chemistry in Gor'kii to work on the chemistry of heterocyclic organic compounds. The expansion of the Academy also came through absorption of previously established institutes known for research excellence in highly specialized fields. In 1970 it absorbed the well-established Biological Research Institute of Kazan University, distinguished by its original work in plant physiology and biochemistry, particularly in the study of the chemical structure and function of water in plants. The Academy was particularly interested in absorbing advanced specializing institutes in the areas of modern science that were both new and inadequately represented in the central research organizations. The establishment of academic institutes or research centers in urban communities away from Moscow and Leningrad was highly selective. The result was that entire provinces were left outside the geographical limits of academic institutions. The Saratov, Kuibyshev, Volgograd, Penza, Ulianovsk, and Astrakhan provinces were among the regions without a single research body within the institutional framework of the Academy.[87]

Since the early 1960s the Academy has pursued yet another—and rather spectacular—line of institutional expansion: the creation of specialized "scientific cities" near Moscow and Leningrad, particularly the former, as complexes of research institutes and laboratories working in limited clusters of closely related sciences. The Scientific Center for Biological Research in Pushchino, opened in 1967, consisted of the Institute of Biological Physics (transplanted from Moscow), the Institute of Protein, the Institute of Biochemistry and Physiology of Microorganisms, and elaborate workshops for designing and producing laboratory equipment for experimental research in biology.[88] Although it was meant to be a "closed system," the center provided research experience for advanced graduate students from Moscow University. A special council on education explored the most effective techniques of mobilizing local talent in the challenging task of transmitting the avalanches of new biological ideas to organized groups of secondary-school students. Relying primarily on generalized lectures by known scholars, this venture was undertaken for the purpose of expanding the base of cultural conditioning of future candidates for advanced training in the new branches of biology.[89] All the institutes concentrated in Pushchino shared one common interest: the application of physicochemical methods to the study of the phenomena of life. There was also a burgeoning interest in the study of living nature as a source of suggestive ideas on the sensitivity of electronic instruments. The lively interest in bionics marked an experimental step in the direction of using biological models in physics.

The Noginsk Center united a number of institutes working in various areas where physics and chemistry meet; it was particularly noted for research in the kinetics of chemical reactions and the chemistry of phosphorus-containing compounds. The Krasnaia Pakhra Center worked on earth magnetism, high-pressure physics, and spectroscopy. The Leningrad Institute of Nuclear Physics in Gachina quickly emerged as one of the country's most elaborate centers for research in high-energy physics, neutron physics, and molecular and radiational biophysics. In a way these specialized scientific centers were results of a compromise between the government's pressure for geographical deconcentration of science and the unwillingness of scientists to live in communities removed from the cultural advantages provided by the great metropolitan areas.

The accelerated growth of the Academy at the periphery and of academic cities near Moscow and Leningrad paralleled the explosive growth of central institutions and scientific personnel. In 1965 L. A. Artsimovich, academic secretary of the Department of General and Applied Physics, noted regretfully that the Academy was still a "Moscow institution" rather than an "all-Union institution," because "more than one-half" of its manpower was located at the center.[90] Acknowledging implicitly the substantial expansion of the network of academic institutions at the periphery, Artsimovich urged an acceleration of the process that was

already moving at high speed and that saw the Novosibirsk Akademgorodok grow during the next decade to a scientific center of fifty huge
research institutes.[91]

Several other lines of development helped the authorities to carry out
the new policy of transforming the Academy into a true national institution. The authorities made a concerted effort to broaden the geographical base of recruitment of junior scientific staff. During the 1950s over
80 percent of the young scholars employed by the Academy were graduates of Moscow and Leningrad schools of higher education; during the
1960s (and early 1970s) the number was reduced to a little over 60
percent.[92] Similar efforts were made to phase out the practice of favoring
Moscow scholars in electing new regular and corresponding members
of the Academy. The new trend reached a peak in 1970 when thirteen
of the twenty-nine newly elected academicians—and forty-one of the
seventy-two newly elected corresponding members—came from scientific institutions outside Moscow.

The wide geographical distribution of academic institutions brought
high science closer to regional economic resources, fresh reserves of
talent, and national cultures. It also created a broader base for instituting
new modes of research organization, more effective interaction of
research centers and institutions of higher education, and closer relations
between scientific bodies and economic enterprises.[93] The institutional
matrix of Soviet science acquired unprecedented fluidity and enormous
proportions. In efforts to widen the general pool of science talent,
research institutions—including the Academy—encountered three
deeply rooted and historically conditioned obstacles.

In the first place, Soviet census data showed clearly that engagement
in science, and higher education in general, varied enormously from
one republic to another. In 1970, for example, there were 51 scientists
in Armenia, 48 scientists in the Russian republic, and 39.5 scientists in
Georgia for each 10,000 inhabitants; in the same year, for each 10,000
inhabitants there were 24.2 scientists in Belorussia, 20.8 scientists in the
Kazakh republic, 20 scientists in the Kirghiz republic, and 16.9 scientists
in the Turkmen republic.[94] The new policy was intended not only to
help raise the participation of individual national groups in the country's
general effort in science but also to provide quality research facilities for
particularly gifted individuals regardless of their ethnic origin. The policy
worked on two fronts: it brought science to the talent of various national
groups, and it assured smoother flow of ethnic talent to the center. It

worked against the repetition of the 1952 experience when the Academy chose to fill only 58 percent of the vacancies for doctoral aspirants rather than admit promising young students from republic academies.[95] The new geography of research institutions took science to the periphery without reducing the overpowering position of Moscow in the national distribution of scientific manpower: in 1970 one out of four Soviet scientists worked in Moscow.[96]

In the second place, the participation of women in the national scientific effort was considerably below ideal expectations and required a broad reexamination and an institutional restructuring. Although this problem was more pronounced at the periphery, it was very serious at the center as well. The Soviet Union was the first modern country to make a concerted effort to attract women to scientific scholarship. At the end of the 1950s women constituted 59 percent of the total number of persons with higher or specialized secondary education; they made up 49 percent of the total student body of the institutions of higher education. In 1970 they made up 38.8 percent of the nation's total research personnel, or 43 percent of "scientific workers" employed by the Soviet Academy of Sciences.

The ratio of women on higher rungs of the academic ladder, however, was much less favorable. This was particularly true for the positions of full and corresponding members of the central and republic academies, the highest positions on the scale of professional ranks in science. In 1970, in all academies, women held 94 of these coveted positions;[97] yet the total number of full and corresponding members in the central Academy alone was 693. The law of the land required that women make up no less than 40 percent of the staff of a research establishment. But in academic achievement women did considerably less well than men. At the end of the 1960s 35 percent of the total number of male scientists and 19 percent of the total number of female scientists had higher degrees.[98] Women constituted only 25 percent of the nation's graduate students. While they commanded a decisive majority among the junior research associates, men held 70 percent of the administrative positions. Statistics also showed that the higher the level of professional qualification required by individual sciences, the lower the proportion of women.[99] In the Siberian Department 80 percent of young specialists in mathematics and physics were men; among the young chemists, however, 50.8 percent were women. The experts readily admitted that the "double loading" of women—the splitting of time between "research" and "domestic" duties—was an important factor in holding them back from professional advancement. At least one writer observed that "the system of awarding degrees and promotion of personnel" deserved careful reexamination.[100] Despite this demographic disproportion, Soviet women had a larger representation in the scientific community than in any other country.

In the third place, rural inhabitants, who in the 1960s made up close to 40 percent of the total population, represented a formidable potential source of scientific manpower that was tapped far below the level of urban, particularly metropolitan, population.[101] This applied to the rural community in the center of the country almost as much as at the periphery. In 1968 only 11 percent of Moscow University's freshmen came from rural communities. There were no precise data on the social origin of students on the national level. Representative local and regional studies agreed, however, that while two-thirds of all high school graduates from families of the urban intelligentsia enrolled in the institutions of higher education, only one-third of high school graduates from the families of urban workers and employees and one-tenth from rural families extended their education beyond the secondary level. According to Vladimir Kantorovich, "the road to the intellectual professions has been made easier for the hereditary intelligentsia."[102] Inadequate scholastic preparation placed the graduates of rural high schools at a disadvantageous competitive position in college entrance examinations. A Moscow University registration officer noted that rural high school graduates had developed a "psychological block" against entrance examinations and that an inordinately high number of promising students shied away from them. The low level of rural education received wide publicity thanks primarily to proliferating studies in "concrete sociology," a significant product of the post-Stalin thaw. It was revealed that rural schools were in short supply of auxiliary teaching materials. *Pravda* reported in January 1968 that 40 percent of all schools in the Cheliabinsk region did not have a physics laboratory, and less than 20 percent had a chemistry laboratory.

Despite compounded flaws in the mobilization of human resources, the number of "scientific workers" in the Soviet Union increased from 223,000 in 1955 to 927,700 in 1970.[103] The number of scientists doubled during the thirteen-year period from 1947 to 1960; it doubled again during the seven-year period from 1960 to 1966.[104] Nevertheless, the gigantic structure of expanding science was not without problems: judged by the distribution of higher degrees, the rapid growth of the ranks of scientists produced a lowering of qualifications, measured by the proportion of higher degrees in the total makeup of certified qualifications.[105] From 1950 to 1967, for example, the number of specialists with higher degrees declined from 33.1 percent to 24.3 percent of total scientific manpower.[106] Lower admission and training standards and the hurried creation of graduate programs and new research centers worked against the hope that the increase in the number of scientists would produce more true pioneers in science. A group of Rostov sociologists stated in 1968 that the time had come for shifting the emphasis from numbers of scientists added annually to research centers to the quality of professional training, a more certain route to a higher national output in scientific ideas.[107]

Many institutions of higher education entered the field of postgraduate training even though they had neither faculties nor laboratories to meet the challenge. They were mainly responsible for the proliferation of "doctors of science" and "candidates of science"—holders of the two standard advanced degrees—most of whom were underprepared for creative work in science. Pedagogical institutes were typical representatives of this category. A correspondent of *Literary Gazette* referred to these schools when he stated: "Offering watered-down programs in the basic natural sciences—in comparison with the universities—as well as working under considerably less favorable conditions for scientific research, pedagogical institutes, as a rule, cannot train candidates of science of the same quality as universities."[108] During the twenty-year period after 1954, two Moscow pedagogical institutes—or teachers colleges—produced over 5,000 holders of candidate degrees. While only 4–5 percent of these candidates received their degrees in education and psychology, the vast majority were "specialists" in various natural sciences. Schools of this kind not only lowered standards of education but also contributed to an unbalanced distribution of higher degrees among specialists in various scientific fields. Two Moscow schools alone produced more "experts" in "molecular acoustics" than the national academic market demanded; at the same time, many other fields in physics were in short supply of competent personnel. The demand for a rapid growth of scientific manpower was so intense and unyielding that it caused a breakdown in the planning mechanism of the national system of higher education, a situation that began to be remedied at the end of the 1960s. This was one of many critical situations that helped legitimate and accelerate the extensive national investment in the science of science, particularly in its concern with the forecasting of scientific developments.

During the 1960s the Academy expanded its programs of postgraduate studies leading to the two advanced degrees, candidate of science (whose standards were somewhere between the M.A. and Ph.D. degrees in the United States) and doctor of science. From 1960 to 1970 it produced close to 5,000 candidates of science and 2,000 doctors of science.[109] Because of its enormous prestige and high position in the hierarchy of scientific institutions, it continued to attract the most promising young scholars; this helped it to meet the growing pressure for higher degrees, at the same time preventing a serious erosion of academic standards. In 1963 the Academy institutionalized the practice of probationary employment of carefully selected young scholars who earned their higher degrees in other institutions. This was one of several ways of choosing new members of the Academy's research staff. The Soviet Union shared one problem with the United States: an alarming number of qualified persons who ceased to be involved in publication-oriented research as soon as they received higher degrees.

Neither the general proliferation of higher degrees nor the rapid expansion of the institutional base of science had helped in the national effort to raise the productivity of the scientific community to Western standards. The problem of productivity, as pointed out by R. Amann, J. Cooper, R. W. Davies, and their coworkers at the Center for Russian and East European Studies at the University of Birmingham, was the chief factor explaining the persistent gap between Soviet and Western technology. This gap was manifested at both the prototype level and the level of diffusion of new technology.[110] Gigantic and accelerated technological advances in the West and Japan during the 1950s and 1960s were also among the most important reasons for the difficulties encountered by Soviet scientists and engineers in their effort to close the technological gap. In 1966 P. L. Kapitsa noted that the productivity of Soviet scientists, determined by the number of publications per individual engaged in research, was only one-half the labor productivity of their American peers.[111] As a first step in the right direction, he suggested that appropriate measures be introduced to ensure a continuous transfer of unproductive scientists to other forms of employment, or to the less demanding sectoral laboratories in industry. Heeding Kapitsa's observations, which were not free of rhetorical exaggeration, the government undertook a series of actions specifically designed to place stronger emphasis on postgraduate training, to establish obligatory annual norms for weeding out incompetent personnel, and to create fresh pools of tested scientific forces ready to take the positions vacated by scientists transferred to other employment. At the same time, it sought to improve methods of selecting students for training in scientific research—a problem previously attended to by intuitive measures.[112]

Specialists began to work on more precise indices for the measurement of "productivity" in basic research, as distinct from applied research directly related to economic activities. They were also involved in an effort to determine optimum sizes of scientific cadres in individual categories of research centers.[113] V. G. Afanas'ev expressed a popular view when he stated that the large research centers provided the most efficient institutional response to the needs of modern science. Large centers, he said, were more flexible and more efficient in meeting the needs of rapidly developing science and in setting up temporary research teams to deal with theoretical and experimental knots requiring urgent attention. According to Afanas'ev, these centers were in a good position to be in command of material resources and professional manpower necessary for setting up independent design bureaus and for producing unique types of research equipment. They offered more favorable conditions for discussion and exchange of ideas. And they made it possible to set up modern automatic information-retrieval systems.[114]

L. A. Artsimovich, academic secretary of the Department of Physical and Mathematical Sciences, agreed with Afanas'ev in principle. He warned, however, that in order to perform the central role in organized research with requisite efficiency, institutes must be ready to fight off constant threats generated by various centrifugal forces. Referring to the Academy, he noted in 1961 that

> the structure of our institutes does not meet the needs of modern science; most institutes are organized on the model of early feudal estates. They are made up of tenuously connected laboratories, each striving to separate itself from other units and to become a small state with its own permanent realm and inventory. These conditions make it exceedingly difficult to move from one research topic to another and to solve new problems.[115]

This isolation, according to Artsimovich, produced a narrow and rigid specialization and led to a rapid "ossification" of large scientific institutions. Traditionally, academic institutes were static, adhering strictly to rigid principles of specialization in science; during the 1960s the tendency was to build more fluid and interdisciplinary institutes.[116]

Artsimovich did not deny the functional superiority of large institutes; he merely wanted to point out the types of centrifugal forces encouraged by institutional giants. Huge institutes continued to be the main line of the Academy's involvement in science, and they continued to serve as models for a reorganization of sectoral research centers, most of which were financed and administered by economic ministries. In 1969 the government admitted that most of the small sectoral research centers were antiquated and inefficient; their small size forced them to handle limited numbers of research topics, creating institutional isolation and uneconomical operation. Another argument against small sectoral research units was that they worked on minute and random themes outside the mainstream of modern science. Among their many woes was the difficulty in retaining highly competent personnel.[117] Nevertheless, the development of the sectoral segment of organized research followed its own course: while a large number of these centers grew rapidly and became functionally diversified, others continued to be strictly specialized and relatively small. Although the Academy of Sciences served as the most authoritative source of ideas for a rational and efficient organization of the major categories of research centers, it did little to raise the quality of the sectoral component of the national system of scientific institutions. Academic leaders, including Petr Kapitsa, did not hide their feeling that one of the functions of the sectoral system was to absorb scientists who did not make the grade as members of the Academy research staff.

THE RECONSTITUTION OF AUTHORITY

With the explosion of scientific thought in the age of space exploration, nuclear technology, and digital computers, and with the vast expansion of the national engagement in science, came the need for a more precise and concrete determination of the place of the Academy in the overall system of scientific institutions. In the age of Stalin, the country did not have a unified system of administering science. The authorities concentrated much more on the internal organization of the Academy than on creating functional links between this institution and the other components of the national system of scientific organizations. The place of the Academy in this system was defined only in the most general terms. Soon after 1955 the place of the Academy in the national network of scientific institutions became the topic of heated debates in high academic circles. Highly placed experts in "technical sciences" and the government preferred direct participation of the Academy in meeting the technological needs of the national economy and defense. Most leading scientists, in contrast, wanted the Academy to be involved primarily in basic research not necessarily related to the practical needs of the day. They contended that the ongoing revolution in science made it mandatory for the Academy to concentrate on new developments in scientific theory and methodology and, in order to achieve this goal more effectively, to withdraw from the areas of practical research. The division was clear and deep seated: while the engineers and the government looked at technical institutes as the very heart of the Academy, the leading scientists looked at them as a misplaced burden hindering the Academy's efforts to widen the horizons of scientific thought.

In 1957 A. N. Nesmeianov issued a comprehensive report on the achievements of the Academy of Sciences. He noted that mathematics, theoretical mechanics, and physics were particularly strong sciences in the Academy. But he went out of his way to state that he was not happy with advances on every front. Research in astronomy suffered, he said, because of a pronounced lack of modern equipment. Chemistry was weak in large areas of more recent development. Experimental biology was underdeveloped, and the same applied to animal physiology, a science with strong roots in national tradition. Geology, too, was below the level of its national tradition; it did not use modern methods in the study of the earth's crust and was lost in endless minutiae. Some institutes of the Technical Department were not clearly separated from the sectoral institutes in industry and did not establish working relations with the physics and chemistry institutes in the Academy. In general, they had shown much difficulty in involving themselves in the "specifics of academic work." Some social sciences, too, were far below par.[118] Nesmeianov's rhetorical dramatization of the weaknesses on various fronts of academic science was done for what he and a number of leading academicians considered a good cause: it underscored the pressing need for the Academy to become prepared to keep up with—and to add to— the flood of ideas unleashed by the modern revolution in science. The emphasis on basic research was also needed to make up for the crippling effects of isolation during the period of violent anticosmopolitanism and rampaging Lysenkoism. One way to enable the Academy to meet the rising demands for exploratory work in basic theory was to free it from heavy commitments to applied science.

Khrushchev and the government did not share these views. They thought that the rapid expansion of the Academy's research facilities should allow for a proportionate expansion of practical research. Khrushchev, who would have welcomed the Academy's extensive involvement in "metallurgy and coal industry," was nevertheless willing to listen to the leaders of the academic community. Representing the top level of the academic group, N. N. Semenov, a recent Nobel laureate, developed an interesting and rather moderate argument. Industrial production, he said, drew help from two sources: technical experience and science, each with its own "logic of development."[119] Industry accumulated technical knowledge and built upon it. Science, in contrast, often made discoveries in areas where industry had no previous experience and no immediate needs. Each should be given full opportunity to develop its own internal resources, and, for this purpose, each should be treated as an independent unit. This did not preclude their drawing on each other's resources; it merely shifted the emphasis from converging points to distinctive potentialities of the two vast areas of social activity. His argument implied that the country should have two distinct and separate systems of research: one devoted to basic research and the other concerned with the immediate technical needs of industry.

In Semenov's view, the Academy of Sciences should be the center and the chief coordinator of fundamental research. It should concentrate on the structure of matter as the most effective method of enhancing man's control over natural forces, on the chemistry of synthetic substances, and, in general, on the basic theory of modern science. A science, he said, must look at the technical needs of society through the prism of its own internal logic of development and theoretical and experimental strengths. The Academy must drop the institutes of the Technical Department; it must combine the seven remaining departments into three huge units covering the experimental sciences (mechanics, physics, chemistry, and experimental biology), the geological and geographical sciences, and the social sciences. This type of organization, Semenov said, should be more efficient in meeting the challenges of the interdisciplinary orientation of the most active areas of modern research. G. I. Naan, a distinguished member of the Estonian Academy of Sciences, pointed out that a sharp separation of "science" from "practice" did not violate the principles of dialectical materialism: "In order to contribute to practice, to practical life, science must be separated from it as much as possible; it must rise above it and scale the heights of abstract thought. The more decisive the separation and the more abstract the abstractions, the more revolutionary the technical contributions of science."[120]

As expected, the Semenov plan evoked two kinds of reaction. The leaders of the Technical Department, led by I. P. Bardin and A. A. Blagonravov, argued vigorously in favor of preserving and expanding the close ties of the Academy with national economic plans. Heartened by the succession of revolutionary achievements in physics, molecular biology, and computer-oriented branches of mathematics, the theorists united their ranks in an effort to emancipate the Academy from too close an involvement in engineering activities—a creation of Stalinist science policy. Each group worked in earnest to bolster its arguments and to solidify its ranks. The engineers argued that pure science and applied science had become so closely intertwined and interdependent that their separation would remove the Academy from the main avenues of modern scientific activity and technological advancement. In response, the "pure scientists" pointed out that some key technical institutes worked in full isolation from the Academy's centers of basic research. A. N. Nesmeianov, president of the Academy, became the most eloquent defender of "pure science" or "basic research." According to Alexander Korol, Nesmeianov found the time opportune to emphasize "the need to increase the 'store of Soviet science,' to give attention 'to theoretical science and scientific research,' and viewed fundamental research as the key function of the Academy system." On the eve of the Twentieth Congress of the Communist party, held in 1956, Nesmeianov "again emphasized basic research, pointing out the need 'to begin the intensive development of theoretical science'—what he called 'the science of the foundation of physical-mathematical, chemical, and biological complexes of sciences.' 'Herein,' he said, 'is the chief task of science and above all of the Academy of Sciences.'"[121]

Before all the arguments were heard, the authorities adopted the basic principles of the Semenov plan. In 1961 the government accepted the idea that the Technical Department must go.[122] In rapid succession, the numerous institutes of this department were either dismembered or were subordinated to individual ministries or other government departments. A few institutes dealing with automation and other problems of modern technology were absorbed by the other departments of the

Academy. Within a year the Academy had lost one-half of its institutes and one-third of its scientific personnel.[123]

While the Academy retreated from research activities concerned with purely technical problems, it moved rapidly into the domains of newly developed theory. It was particularly attracted to areas where various branches of biology depended on the methods of modern physics and chemistry, and where the influence of cybernetics and the general systems theory was particularly promising. It was at this time that the Academy became deeply involved in a practical problem of a new kind: the interdisciplinary approach to the biosphere, part of the national effort to resolve the critical problem of protecting nature from mounting human abuses. Ecological problems became a special interest for an ever-widening aggregate of sciences, from old-fashioned geology and soil science to the burgeoning molecular biology. Especially notable was the institutional response of the Academy to a strong interest in the biosphere as the substantive link between genetics, evolutionary biology, physiology, and sociobiology.

In 1961 the government established the State Committee of the Council of Ministers for the Coordination of Scientific Research, which was assigned the task of supervising the national organization of major research projects, as stated in official government documents, and of coordinating the work of such systems of scientific institutions as the Soviet Academy of Sciences, the republic academies of sciences, sectoral research units, and institutions of higher education. Prior to the establishment of this committee, the huge national network of research units was controlled by 170 federal and republic ministries and offices, each concerned with a specific area of activity.[124] Neither the Scientific and Technical State Committee of the Council of Ministers of the USSR, established in July 1957, nor its successor, the State Scientific and Economic Council of the Council of Ministers of the USSR, established in the spring of 1959, were vested with sufficient authority to become a serious factor in the search for a system of centralized administration of the national effort in science.[125] The new committee synchronized science and technology, drafted plans for new research and development units, and worked on reducing unnecessary parallelism in scientific research. Furthermore, it was authorized to organize a national council to study the basic trends in the development of science. Roland Amann has noted that "one of the main tasks of leading industrial managers and distinguished scientists who were members of the committee was to define long-term goals for national research and development, and thus to identify and give strong backing to important projects which were regarded as crucial to the economic and social well-being of the country."[126] The Academy, in contrast, concentrated on the fundamental questions of modern scientific theory and on coordinating the theoretical work conducted by republic academies and universities.[127] The change prompted Helgard Wienert to note that although the Academy of Sciences was a Soviet institution only tenuously resembling the Imperial Academy, "it has completed the full circle and has come back to its original occupation—emphasis on theoretical research."[128]

The dual administrative system satisfied both the government and the Academy: it satisfied the government by giving a clearly defined institutional recognition to practical orientation in science, and it satisfied the Academy by recognizing theoretical exploration as its main function. Nevertheless, individual members of the Academy did not treat the new reform completely as a blessing. Some academicians, led by Artsimovich, expressed the fear that the abolition of the Technical Department did not remove the Academy's heavy involvement in technical problems but that it merely transferred this duty to institutes in other academic departments. At the beginning, however, it was clear that the new system of centralized authority required extensive readjustments to achieve a fuller synchronization of component parts, a more precise division of labor, and a more consistent philosophy.[129]

In 1963 the government issued a new order specifying in more detail the role of both the State Committee and the Academy of Sciences in what was referred to as the unified science policy. In more specific terms, the functions of the State Committee were: to determine the future lines of scientific and technological concentration; to coordinate the work of scientific institutions related to the key technological processes; to supervise the preparation of plans for current research and development; to plan the future territorial expansion of research centers; to exercise general supervision over research activities on all levels of the institutional hierarchy; to determine the main lines of capital investment in science; to devise methods for raising the level of scientific productivity; to plan the establishment of intersectoral research centers; to prepare and approve plans for the training of future scientists; and to supervise and improve mechanisms facilitating the flow of scientific information.[130] The State Committee was also directed to assist the government in drawing up and implementing plans for a unified national science policy, to organize a national network of scientific and technical information agencies, and to promote scientific and technical relations with the outside world.[131] According to one authority: "The creation of the State Committee means in essence the creation of a single central state organ entrusted with the task of administering applied research and development on a national scale."[132]

In 1965 the committee was renamed the State Committee of the Council of Ministers for Science and Technology (GKNT), and its functions were adjusted to recent changes in the administration of the national economy, now organized on an industrial rather than on a regional basis. The government decision made it appear that the State Committee was the highest administrative-coordinating agency in the country and that the Academy of Sciences was its main arm in the field of basic research. The interpreters of the new regulation quickly became accustomed to treating the two agencies as coequal—one in charge of pure research and the other working in the general area of the flow of scientific knowledge to the national economy in the broadest sense. By official definition, both organizations worked for the national economy: one by enriching basic science and the other by maintaining and expanding the ties between science and technology. The Academy became the main source of designs for new centers of scientific inquiry and for involving the periphery in basic research. Its role in coordinating and supervising the work of republic academies became much more concrete and involved. All this, however, did not mean that the two organizations were in full charge of scientific research; their basic function was to serve as coordinating boards in planning the general lines of development in the multiple fields of scientific activity. Their task was to assure the world of science of a planned, efficient, and economical development. Forecasting future developments in both science and technology became an important part of their activity.

The State Committee was the modern successor of the Scientific and Technical Section of the 1920s. It did not finance science, but one of its major tasks was to design methods for a rational distribution of financial resources over the main branches of scientific effort and to establish procedures for the measurement of economic returns from the investment in science. It had the authority to terminate individual research activities considered to be unnecessary duplications of research already in progress or to have insufficient grounding in the present state of science and technology.[133] The committee was the central agency collating information on—and planning broad actions in—the four main areas of science policy: research priorities, growth of human and material resources of scientific research, rational utilization of existing scientific forces, and flow of scientific knowledge to the national economy.[134] The Academy, meanwhile, worked on the strategy of scientific research: it evaluated the potentialities of various scientific orientations and methods of inquiry, searched for new institutional responses to the needs of rapidly advancing science, and maintained relations with the scientific world outside the USSR.[135]

As early as 1945, the Soviet Academy of Sciences established a special council to coordinate its research activities with those of the republic academies. In 1963, when the government gave the Academy added authority as a coordinating center, each republic had a national academy. Now for the first time, the Soviet Academy acquired the authority to make decisions mandatory for the republic academies, which were otherwise financed and administered by the councils of ministers of the republics.[136] This authority, however, was of the most generalized nature: it consisted primarily of synchronizing research plans of republic academies, giving directives for planning long-range developments of research units, designing uniform procedures for the training, recruitment, and distribution of expert personnel, and recommending candidates for full and corresponding memberships in these academies. A scholar recommended by the Soviet Academy for full or corresponding membership in a republic academy was not guaranteed election, particularly in cases where there were more recommendations than vacancies; however, a candidate not recommended by the Soviet Academy had little chance to be elected.

Having become the chief coordinator and supervisor of basic science in the country, the Soviet Academy of Sciences acquired another duty of vast proportions: it became the pivotal agency of centralized control over research activities in universities and other institutions of higher education. On the fiftieth anniversary of the October Revolution, the country had 862 institutions of higher education, with a total enrollment of 4 million students; this total included 65 universities, with an enrollment of 600,000 students. In the pre-Soviet period, the universities were the true centers of scientific research. During the Soviet period the emphasis on research rapidly shifted to special institutions, topped by the system of the academies of sciences. The shift found clearest expression in statistical data on the distribution of scientific manpower: in 1970

63.3 percent of all scientists worked in research organizations unaffiliated with higher education. This tells only part of the story: at least as important is the fact that although university professors took part in research, their participation was appreciably below their professional capability. The school system under the Ministry of Higher and Specialized Secondary Education had a total of 52 research institutes, 900 laboratories of various kinds, 19 computer centers, 13 botanical gardens, and 11 astronomical observatories.[137] Some institutes—such as the P. K. Shternberg State Astronomical Institute at Moscow University, the Institute of Solid-State Physics at Tomsk University, and the Institute of Radio Physics at Gor'kii University—were well known and highly esteemed organizations. The basic problem was that two-thirds of research activity in the institutions of higher education belonged to the realm of applied science. Universities were encouraged to work on contracts with research units of the various ministries, always attuned to the concrete and immediate needs of the national economy.

During the late 1960s the government began to think seriously of bolstering basic research in the institutions of higher education and of making the Soviet Academy of Sciences the central planning and coordinating agency entrusted with carrying out the new policy.[138] The Academy acquired the right to decide which institutions of higher education should be granted the status of university. In performing this task it relied solely on the quality and magnitude of basic research in institutions under consideration, but this rule was not always adhered to. During the five-year period from 1966 to 1970, the country acquired sixty new institutions of higher education, including nine universities. The Academy was concerned with raising the standards of science training in universities as much as with involving professors and graduate students in basic research. The advocates of these reforms argued convincingly that the active participation of the professorial staff in basic research was the best guarantee for improving the quality of scientific training. The new policy was designed to make universities fountains of new scientific knowledge as much as workshops of practical knowledge.

Concrete steps were taken to organize research projects by combined teams of university professors and Academy scientists and to make the academic laboratories accessible to both professors and promising graduate students. Every year after 1962, 1,000 instructors were relieved of teaching duties for up to two-year periods to complete their doctoral dissertations on topics related to the basic theory of individual sciences.[139] Generally, however, universities and other institutions of higher education had far to go to reach the level of scholarly excellence. A. D. Aleksandrov, rector of Leningrad University and a member of the Academy, described the plight of universities in 1961 when he stated that the excessive concentration of talent in the Academy of the USSR and republic academies had made it exceedingly difficult for the universities to fill positions in advanced research. He noted also that "by taking the best talent from the universities, the Academy has undermined the source of its own renewal."[140]

The experts now agreed that the universities—along with the academic institutes—were the only scientific institutions with strong potential for exploratory work in theory.[141] But the process of freeing universities from primary involvement in practical research was a rather slow process, thwarted by compounded uncertainties and legal ambiguities in the organizational web of higher education. At the end of the 1960s the experts conceded that "static organization," archaic laboratories, and limited resources had prevented most universities from keeping up with the dynamism of modern science.[142] Noting that most university and college professors were not engaged in research, V. G. Afanas'ev explained that this was a result of heavy teaching loads, underequipped laboratories, outmoded buildings, and grave shortages of technical personnel.[143] But as the 1960s came to a close, the attention of planning authorities became strongly attracted to the urgency of transforming the latent resources of the universities into a vital fountain of scientific knowledge. The authorities approached this task with the full knowledge of its immense complexities and unsteady turns in recent university tradition. Solutions were sought only in experimental measures of local significance and in administrative readjustments of limited consequence. The institutions of higher education, however, were most successful in expanding their programs in short-term postgraduate courses for various lines of employed specialists not interested in advanced degrees.[144]

Despite their important and diverse activities, the State Committee and the Soviet Academy were not the key links in the organization of decision-making authority in the vast empire of science. For the real loci of supreme authority, one must look to GOSPLAN or to the federal ministries. It was true that in carrying out their functions, the two organizations had considerable authority in issuing administrative orders to scientific and technical agencies and in dispensing special funds. But it was equally true that their function was to help the executive branch of the government in its effort to make Soviet science a consistent and fully centralized system of complementary institutional components, to balance basic research and practical science on a national scale, to evolve the most efficient organizational forms in scientific and technical knowledge, to anticipate the main lines in the future development of science, and to build additional reserves of scientific manpower. The Soviet Academy and the State Committee worked less on running the national system of science on a day-to-day basis than on formulating new functional principles and organizational models for giving this system more unity, efficiency, and adaptability to rapidly changing science and technology. Among the many key functions of the Academy, the tasks of planning the future development of the institutional base of basic research, creating a balanced distribution of research activities, and maintaining a symmetrical relationship between basic and applied science were of paramount significance.

All these developments placed the Academy in a new situation of complex relations, crisscrossing lines of communication, and vacillating policies. It had ceased to be the national center of technical research or applied science; but it did not make theoretical and experimental exploration in modern science its exclusive preoccupation. It continued to be the chief forum to which the government could turn for technical advice. But it also became a "ministry of science" (an oratorical exaggeration of Artsimovich's making), heavily involved in guiding, coordinating, and administering a huge empire of scientific institutions scattered throughout the country and dominated by its own regional units, republic academies, and universities. Responding to the internal criticism of the new burdens imposed on the Academy, Keldysh sided with the group of academicians who claimed that the demands for transforming the Academy into a "science monastery," deaf to the practical needs of the day, were both ill advised and most unrealistic.[145] In the meantime, P. L. Kapitsa led a small but vociferous group of academicians who, while recognizing the merits of national planning in the organization of research, demanded broader areas of academic autonomy. Academy institutes, Kapitsa said, must be disencumbered of as much external control and administrative activity as possible in order to grow "successfully and freely." He suggested that the plan retain only "its broad organizing principle" and that individual institutes be granted the right to work out operational details.[146]

The Academy became more firmly integrated into the system of Soviet government; but it also acquired unprecedented strength in influencing the government's decisions on the development of national science policy. The signs of autonomous action appeared in many areas of academic activity. Institutes, the nerve centers of basic research, acquired limited independence in the disbursement of funds and in the distribution and procurement of expert personnel. The budget of the Academy now contained a percentage of unassigned funds that could be freely channeled into unanticipated developments in individual sciences. The position of the chief scientific secretary, previously known as permanent secretary, was brought closer to the academic community.

The experience of the Stalinist era had shown clearly that the chief scientific secretary, as a de facto representative of the government, infringed on the rights of the president, as a representative of the scientific community. Three such secretaries—N. P. Gorbunov, N. G. Bruevich, and A. V. Topchiev—were forced on the Academy by nonacademic authorities. All were minor scholars and all were elected full members of the Academy by the most irregular procedures—that is, by a total disregard for the selection and election steps specified in the charter of the Academy. In order to avoid similar "elections" in the future, the Academy—obviously through government permission—decided in 1962 that only the members of the Presidium of the Academy, normally academicians of long standing and high esteem in the scientific community, qualified for nomination to the position of chief scientific secretary. This ruling eliminated the probability of future reliance on quick "elections" of outsiders to fill this key administrative position. At the same time, it made the position clearly subordinate to that of president. Thus, the position of president was now not only clarified but also strengthened. Zhores Medvedev went so far as to assert that "the position of the president of the Academy of Sciences of the USSR is more influential than the position of minister of the government, and the change of the Academy president is a more serious affair for Party leadership than the change of a minister in most branches of industry."[147] The same ruling specified that the position of chief scientific secretary be open for elections every fourth year.[148] The new measure gave the Academy only one chief administrator and spokesman; it established an area of autonomy in the vast ocean of government regulations.

In 1963 the Academy acquired a new charter.[149] Its basic task was to recognize the new alignment of research bodies that formed the institutional core of the Academy. The number of departments increased from eight to fifteen. Departments were now grouped into three huge sections administered directly by the Presidium of the Academy and covering the "physicotechnical and mathematical sciences," the "chemical-technological and biological sciences," and the "social sciences."[150] The aim of the new division was to improve administrative efficiency rather than to expand the participation of academic groups in decision-making processes. According to M. V. Keldysh, the aim of the new division was to avoid unnecessary scattering of scientists working in the same and closely related fields. He pointed out that the previous division was not based on precise and logical principles of professional classifications: he cited the work on semiconductors, which was previously carried out not only in different institutes but in different departments.[151] Implicit in his argument was the idea that at least some functions carried out by the defunct Department of Technical Sciences did not easily fit into the logic of the Academy's division of scientific labor and its primary dedication to basic research.

The new charter attended to the growing awareness of the need for a broader participation of academic personnel in managerial decision making. To broaden the base of the involvement of research personnel in administrative activities, a new network of institutional units was created. Among these units, scientific councils were basic and most popular.[152] Organized for interdisciplinary exploration in areas that promised important results of a practical nature or key leads for new developments in science, these bodies passed through three phases of development.[153] The first phase, 1957–1961, was dominated by a search for organizational principles and functional orientation. The second phase, 1961–1963, featured a rapid growth of scientific councils and intensive experimentation with a limited number of organizational forms. The third phase, after 1963, was marked by an active role of scientific councils in the administrative reorganization of the Academy and in institutional adjustments to new developments in science.[154] The Scientific Council on the Physiology of Man and Higher Animals, for example, worked on the adjustment of Pavlov's scientific legacy to current developments in cybernetics, biophysics, and biochemistry.

Scientific councils were distinguished by four major characteristics. First, they were subordinated either to the Presidium of the Academy or to individual departments: the main purpose of this administrative design was to make them independent of institute administrations. Second, their decisions were the result of group discussion and voting. Third, they had no executive power: their duty was to supply the Presidium or the individual departments either with suggestions for organizational changes or with pools of information on requested subjects. Fourth, they were strictly interinstitutional: they included not only representatives of various academic institutes but also delegates from various central ministries and other institutions. At the end of the 1960s 58 percent of associates of scientific councils in the Department of Physiology came from nonacademic institutions.[155] In an age of rapid and radical changes in science, the primary task of scientific councils was to enhance the institutional flexibility of the Academy—to help the Academy in its constant search for more rational modes of carrying out its multiple functions. Despite high expectations, councils were beset with many difficulties: they were too big and too unwieldy to act firmly and expeditiously; they were plagued by disruptive effects of overlapping personnel; and they were easily manipulated by skillful persons with administrative ambitions. These and many other difficulties, however, did not prevent the councils, as advisory bodies, from playing a notable role in reshaping the organizational and functional structure of the Academy and in directing the development of science in the Soviet Union.

There was one basic difference between institutes and scientific councils: while institutes worked on advancing individual sciences and were strictly compartmentalized, scientific councils worked on finding answers to "complex problems" requiring interdisciplinary cooperation. At this time more than ever before, there was a clear recognition of two major types of interdisciplinary research: one engaging various sciences in a search for solutions to modern technological problems, and the other crossing the boundaries of individual disciplines in search of theoretical elaborations dictated by the internal logic of evolving science. Both types of interdisciplinary research were concerned with scientific theory: one with translating pure theory into technical advances and the other staying close to explorations in pure theory. In this dual endeavor, the scientific community faced minor and fragmented opposition to its traditional claim that depth of theory was the best indicator of the scope of the utility of science: the deeper the theoretical abstractions, the more sweeping the range of applied knowledge.

In 1970 there were close to 200 scientific councils in the Academy, each working as a distinct administrative unit. Referred to as "mass organizations," they were clearly separated from the main body of Academy administration. Aside from the unquestionable practical value, the councils, as a new source of "collective management," were important sounding boards for criticism of particular policies of the Academy and a valuable pool of suggestions for administrative improvements. Their main purpose, however, was to open a line of "administrative" positions to younger scholars who lacked sufficient seniority—or other qualifications—to occupy higher positions in the formal organization of the Academy. They were expected to help allay the smoldering conflict between the old scholars, who rose to high administrative positions after they had passed the zenith of their scholarly productivity, and the younger scholars, who, at the peak of productivity, were normally underrepresented in academic councils vested with authority. The slow process of earning higher degrees contributed to making positions of authority inaccessible to younger scholars; in 1966, for example, a scholar affiliated with the Academy received the degree of candidate of science at an average age of thirty-five and the degree of doctor of science at an average age of forty-eight.[156] Without an advanced degree, a person could not expect to become a serious candidate for decision-making positions.

The inclusion of a "mass organization" into the managerial structure of the Academy did not pose a serious challenge to the principle of monocratic management on the institute level. Directors of institutes, the research units forming the main line of the Academy's involvement in scientific work, were elected by secret ballot in the general assemblies of the department with which individual institutes were affiliated.[157] The Communist party units operating on the institute level played a strong role in the selection of candidates for directors. Once elected, directors could not be challenged by the constituency. Each institute had an academic council (uchenyi sovet) consisting of a specified number of persons—including party and trade-union representatives—selected by the director to serve as an advisory board. The government preferred this arrangement—known as one-man management (edinonachalie)—for it allowed a more precise definition of administrative authority and more effective government supervision of the Academy's performance. Liberal academicians, including Kapitsa, preferred this system, for, they thought, a strong institute director was in a position to modify unpopular orders emanating from the higher echelons of academic and government administration on the basis of suggestions made by his staff.

The liberalization of academic administration was not a special concession to the scientific community's perennial search for wider areas of
academic autonomy. It was part of a general deconcentration of authority, inaugurated by the Khrushchev regime, within all phases of economic and cultural activity. The apparent growth of autonomous action
in the national system of administrative links did not affect the role of
political considerations in managing the multiple activities of the state.
Many administrative innovations that gave the periphery and the lower
echelons a stronger voice in academic affairs changed the techniques of
management without affecting the strategy of the national science policy.
The changes in science management were a response to the availability
of more sophisticated administrative techniques, but they did not challenge the primacy of political considerations and procedures in strategic
decision making.[158]
The academic search for additional parameters of autonomy was
enhanced by the increasing role of experimentation with so-called matrix
models advanced by Western corporations. By instituting a bifurcated
managerial system that recognized distinct "functional" and "project"
lines of organization, these models attenuated the rigidity of monocracy.
Scientific councils, for example, were intended to expand the horizontal
layers of management, which cut across vertical lines of organized
authority on institute and department levels. The matrix system consisted of superimposing a secondary—that is, a horizontally organized—
authority on the primary authority structure, organized vertically and
empowered with ultimate decision making.[159] The Academy provided
an excellent environment for such a meshing of the two branches of
management. But the new experiment was stifled by the spirit of tradition, which made the flexibility of academic organization a modification of, but not a challenge to, the monocratic system of management.
Academic administration continued to be a typical expression of Soviet
principles of highly concentrated decision making and highly ramified
forms of supervision, enhanced by the ubiquitous and expanding presence of Communist party organizations and their affiliates.

VI
SCIENCE AND CHANGING VALUES

THE SCIENCE OF SCIENCE: A DEFENSE OF
BROADER AUTONOMY FOR SCHOLARSHIP

The scientific and technological revolution received much more publicity in the Soviet Union than in the West. Intertwined with the liberalizing policies of the post-Stalinist thaw, it covered a much broader area
of human activity and generated emotional responses of incomparable
intensity. The changes it wrought evoked avalanches of comment in all
the social sciences, in philosophy, in newspaper editorials, in party
propaganda, and in belles-lettres. Most writings concentrated on the
joyous recountings of its blessings. Only isolated critics raised the issue
of rapidly growing threats to balanced relations between human society
and the biosphere. More serious, but always circumspect, scholarly literature made it clear that the scientific community required new and
wider parameters of independent action in order to maximize the beneficial effects of the revolution. Most discussions of institutional adjustments to the ongoing revolution contained implicit arguments in favor
of new avenues of academic autonomy.

The enormous proportions and prodigious expansion of the network
of scientific institutions in the post-Stalin period pressed for a thorough
and systematic study of the "strategy of science," in an effort to determine national goals and research priorities in science. The rapid growth
also brought about the need for a more careful and reliable study of the
"potential of science," including such problems as the training of scientists and keeping up with the growing need for sophisticated research
instruments. And, finally, the expansion required a broad and contin-

uous study of the "effectiveness of science"—a search for a more rational utilization of the existing scientific manpower and a more expeditious transfer of new discoveries to industrial production. These and related needs and pressures became the main topics of a specialized discipline called the science of science (*naukovedenie*, in preference to *nauka o nauke*), strongly influenced by pioneering work in Poland and the United States.[1] G. M. Dobrov defined the science of science as the "theoretical basis of the science policy."[2] In inaugurating the new science, S. R. Mikulinskii and N. I. Rodnyi stated in 1966:

> It is becoming increasingly more difficult, and almost impossible without the help of a specialized and professional study, to view the entire field of modern science, to unveil its core problems and their interrelations at a given moment and to fathom its internal logic and development perspectives. Without special study, it is impossible to establish the rational foundations of the organization and planning of science. It has become clear that science is more than the sum total of knowledge—a totality of fruit produced by the tree of knowledge—for it is also the tree itself. In order to produce more abundant and better fruit, and to achieve this goal with reduced labor, it is also necessary to study the tree itself—to examine its internal structure and dynamics, its functional systems and elements and its adaptability to the external world.[3]

The science of science, it was stated explicitly, was the product of the enormous expansion of the institutional base of science and of revolutionary developments in science and society. It came on the wings of momentous progress in modern science. From the beginning, the new discipline took on many tasks: it studied the logic of scientific inquiry as the only gateway to the discovery of regularities and predictabilities in the evolution of science; it examined the "internal" and "external" environments of science and their interaction; it laid the theoretical— and mathematical—foundations for forecasting scientific developments; it sought to unravel the varying tempos in the development of science, including the nature of crises and stagnations; it illuminated the formal structure of scientific theory and the methodological and theoretical interrelations of various sciences; and it scrutinized the organization of science and the web of government controls of scientific institutions.[4]

One of the most general functions of the science of science was to help the government in the difficult task of formulating and integrating the major principles of the national science policy. Among the most acute problems deserving careful and thorough study, Mikulinskii mentioned the structure and distribution of scientific cadres, the rational utilization of scientific talent, the efficiency of the communications system in the scientific community, the harnessing of the tools of the general systems theory in expediting the storing and dissemination of scientific information, the continued need for the systematization of knowledge, and the bewildering organizational intricacies and psychology of collective research. Mikulinskii and his associates were successful in convincing the authorities that solutions to these and similar problems could be accomplished through a combined professional study of the internal logic, socioeconomic correlates, and historical perspectives of evolving science. "The purpose of the science of science is to create a theoretically founded branch of knowledge concerned with the organization, planning and management of science, that is, to promulgate a system of measures consonant with the objective logic of science growth, to establish optimum tempos of scientific research and to increase research productivity."[5] The duties of the new discipline included also work on developing a mathematical apparatus for forecasting developments in science and making recommendations for institutional adaptations to anticipated changes. One of its major tasks was to discover general regularities in the evolution of science.

Experts working in the new discipline concentrated on setting up criteria for planning future developments in the organization of research that would harmonize the economic needs of the country with the internal logic of science growth.[6] Supported by such veterans as P. L. Kapitsa and N. N. Semenov, and encouraged by both the Academy of Sciences and the government, the rapidly expanding ranks of experts in the new discipline inaugurated an era of the versatile study of science as a special and widely ramified subject of "self-examination." Such special areas as the sociology of science, the philosophy of science, the logic of scientific inquiry, the psychology of scientific creativity, and the economics of science became vital fountains of information flowing into *naukovedenie.*

Prior to 1970, however, the new discipline possessed two distinct meanings. Some scholars viewed *naukovedenie* as a generic term for a number of separate disciplines—such as the philosophy of science, the history of science, and the sociology of science—united by science as their common subject of inquiry. Other scholars viewed it as a unique and fully integrated discipline treating historical, philosophical, sociological, economic, and psychological aspects of science as interrelated manifestations of more general regularities in the growth of science.[7] *Naukovedenie* included also scientometrics *(naukometriia),* which was concerned with applying elaborate quantitative methods to the study of the dynamics of science growth.[8] Although the Academy of Sciences did not have a monopoly on the science of science, it played the major role in carving out its subject matter, in shaping its theoretical matrix, and in setting up its empirical research and broader philosophical meaning.

The science of science owed its existence to the critical need for an understanding of general trends in the development of science and technology, as well as the interaction of economic, social, historical, logical, psychological, and organizational aspects of this development. It concentrated on the theoretical foundations of the organization, planning, and administration of science and on concrete measures called for by the logic of scientific evolution. It became the basic reservoir of information needed by the architects of the national science policy.[9] But the most distinctive and in many respects most impressive role of the science of science was its function as a source of intellectual directions and modern attitudes. The science of science gave order and broader meaning to the inordinately acute and critical public interest in the interrelations of science, as a body of knowledge and a system of beliefs, with the dominant modes of esthetic expression, with moral norms, with the twists and turns in modern philosophical inquiry, and with the creative impulse of the search for individual expression.

The sociology of science, the most rapidly developing branch of the science of science, attracted the attention of the general reader interested in the humanistic side of the modern scientific equation. A lively and vigorous area of intellectual concern over one of the major dilemmas of modern civilization, the sociology of science dealt extensively with three general problems: the place of science in modern society, "human relations" in scientific institutions, and the types and structures of social hierarchies in the pursuit of science.[10] It was closely related to the theory of management as well as to the psychology and philosophy of science. With a few notable exceptions—such as the work of Dobrov—most of the research in this area was empirically oriented, without the benefit of elaborate and integrated theory.

In the multitude of issues tackled by the sociologists of science, the question of science as a major source of the increasing impersonality of human existence and as a potential threat to the survival of civilization came up for consideration. But it did not receive the careful and dispassionate scrutiny it required from both a practical and a humanistic point of view. Blinded by narrow patriotism, or fettered by censorship, the sociologists clung tenaciously to the stereotyped view of their country as a unity of cultural values and social institutions providing automatic safeguards against moral abuses of science. As Linda Lubrano has pointed out, they operated from a position that socialism provided "the most compatible form of social structure for the advancement of science."[11] Only the rebellious voice of a concerned physicist—A. D. Sakharov—expressed a stern warning that the Soviet Union, no less than capitalist countries, had a long way to go in its obligations toward universal peace as the primary safeguard against the destructive potentials of science. He pleaded for a broad and open discussion—"without pressure of fear and prejudice"—of the danger of "a thoughtless bureaucratic use of the scientific and technological revolution in a divided world."[12]

Despite institutional restraints and constant reminders of the moral purity of Soviet society, the sociology of science—and the science of science in general—worked quietly and persistently on marshaling reasonable arguments that would tie the future progress of science to the strengths and the magnitude of academic autonomy. Quietly and unobtrusively, the sociology of science, carried out by more thoughtful experts, became the leading branch of the science of science in supplying articulated details depicting the acute need for a wider base of independent action within the institutional framework of science. It assumed the challenging task of removing the restriction on the ethos of science wrought by the successive waves of Stalinist onslaughts. The Stalinist ethos, based on the principles of technological supremacy, *partiinost'*, anticosmopolitanism, and Marxist philosophical orthodoxy, was purged of its most crippling and degrading rules of behavior.

The post-Stalinist thaw created relatively favorable conditions for an impartial reexamination of the history of national science and of science in general. But the historians of science were slow in divorcing themselves from the lingering vestiges of the war on cosmopolitanism. During the 1950s the Institute of the History of Natural Science and Technology published over fifty volumes of papers devoted almost exclusively to the history of Russian science. Although the search for Russian priorities in science and technology had lost most of its nationalist pathos, it continued to be very much in evidence. The institute published two general historical surveys of Russian science—the *History of Natural Science in Russia*, in three volumes, and the *History of the Academy of Sciences of the USSR*, in two volumes. Both surveys were monumental compilations of data, and both lacked analytical depth. Neither study made a particularly serious effort to cast the history of Russian science and scientific institutions within a Marxist theoretical framework. Both dealt with Russian science in complete isolation from the major developments of Western scientific thought.

In 1959 N. A. Figurovskii, director of the Institute, called for sweeping changes in the organization and methodology of historical studies of science.[13] He said that most of the work carried out under the auspices of the institute was compilatory rather than analytical and that its non-theoretical nature had made it particularly attractive to all kinds of amateurs and half-trained personnel. Worst of all, he added, numerous writers had distorted the history of Russian science by an inexcusable disregard of its close ties with the mainstreams of Western scientific thought. In no uncertain terms Figurovskii called for systematic and expeditious work toward removing anticosmopolitan injunctions from historical studies of science.

In 1962, after a careful retooling of its research endeavors, the institute announced its readiness to embark on a large-scale involvement in the study of the universal history of science.[14] At the same time, many signs indicated that it was ready to undertake a systematic search for a broader and more functional theoretical base for the history of science as a professional discipline. During the 1960s hundreds of biographical studies of the leaders of Western science, many of book length, came off the presses in rapid succession. The choice of great scientists depended more on the personal interests and tastes of biographers than on a generally conceived plan. For the first time, many Western luminaries, previously known only to a narrow circle of specialists, received extensive biographical treatment. John Napier, the inventor of logarithms, Evariste Galois, the founder of group theory in mathematics, and Alphonse de Candolle, the French botanist who wrote the first systematic history of science with a sociological orientation were among those accorded elaborate treatment. Such masters of twentieth-century science as J. J. Thomson, Ernest Rutherford, Max Planck, Niels Bohr, Max Born, Werner Heisenberg, and Albert Einstein were the heroes of a prodigious output in modern scientific biography.

Prior to 1960 Soviet scholars did not produce a systematic and comprehensive survey of the development of modern science or its individual branches. E. Ia. Kol'man noted that the country was still waiting for the first systematic and detailed study of the evolution of Einstein's theories, written by a physicist in the spirit of dialectical materialism.[15]

He could have added that equally absent was such a study of quantum mechanics. The plethora of unsettled issues in the relationship between dialectical materialism and the revolutionary changes in modern science—particularly in physics and genetics—worked against the creation of an intellectual atmosphere favoring such an undertaking. At the beginning of the 1960s the situation changed dramatically. At that time the institute embarked on publishing general surveys of the history of individual sciences. Although the interest in world science became an enterprise of large scope, the history of national science continued to be attended to with considerable vigor. Now, however, strong effort was made to treat Russian scientific contributions in their close relations with Western thought. The production of wild and undocumented claims, a dominant feature of Stalin's nationalist campaign, gave way in most cases to a careful concern with adequately documented history.

All this, however, amounted to only a small portion of the tasks taken up by historians of science in the period of the thaw. The major task consisted of rewriting the history of Soviet science after the October Revolution—of integrating the disciplines, scientific theories, and scientists victimized by the waves of ideological purges back into the annals of Soviet scientific thought. The historian did not acquire the task of rehabilitating condemned aspects of the scientific past; his job, rather, was to incorporate the relevant results of rehabilitation orders issued by government authorities into the annals of science and to assemble historical and scientific information justifying the new interpretation. As a rule, the historian of science refrained from analyzing the intolerable conditions that led to the outlawing of entire disciplines or from dissecting the operations and motivations of the Stalinist architects of organized attacks on unwanted sciences and scientists. This job was left to the select group of scholars—such as N. N. Semenov and P. L. Kapitsa—who occupied high positions in the academic hierarchy and who enjoyed great prestige in the scientific community. Leading the list of rehabilitated victims of Stalinism was the geneticist Nikolai Vavilov. The Academy republished his collected works, sponsored numerous symposia in his honor, issued a volume of glowing reminiscences by friends and admirers, made public his correspondence with leading scientists, initiated prizes honoring his name, and made him one of the most popular subjects of biographical studies.

The rehabilitation of fallen heroes was not limited to the victims of

Stalinism. Some victims of Lenin's biting criticism were reevaluated and even honored, though on a limited scale. For a long time, A. A. Bogdanov, the main villain in Lenin's *Materialism and Empirio-Criticism*, was a target of the most scornful attacks. Everything he said and wrote was rejected as revisionist folly. During the 1960s Bogdanov's intellectual legacy began to evoke favorable comments, though on a rather limited scale. Students of the historical roots of cybernetics and organization theory, very important elements in the ongoing revolution in science and technology, showed no hesitancy in placing Bogdanov among the modern pioneers of these theories.[16] They were careful, however, to separate Bogdanov's general organization theory—tektology—from his neopositivist flirtations.

The rehabilitation of victims of previous regimes went hand in hand with reevaluations of the heroes of "socialist science." Many of these heroes ceased to be worshiped and presented as syntheses of great achievement in science and exemplary dedication to socialist ideals. The soil scientist V. R. Vil'iams was among the most notable examples of this process of reevaluation. Slowly and quietly, his name disappeared from the honor rolls of generally recognized pioneers in national soil science. In 1948 he was featured prominently in *Men of Russian Science*, an ambitous biographical survey of prominent scientists who had made or had begun to make their scholarly contributions prior to 1917. The 1963 edition of the same book had no article on Vil'iams. Generally, however, his name was not dropped from professional literature; but it had lost most of the luster given to it by Stalinist propagandists. O. Iu. Shmidt underwent similar metamorphosis. Several years after Stalin's death, he ceased to be an exalted hero of Soviet science. His planetary theory was no longer regarded as an unchallengeable national theory. Numerous critics pointed out that it lacked solid empirical backing, contradicted the internal logic of theoretical developments in modern astronomy, and relied on Lysenkoist metaphors in explaining the role of the environment in cosmological processes.[17] Shmidt received credit for interesting ideas on the axial motion of the earth and the role of gravitational pull in the formation of planets, but he was removed from the pantheon of national scientific luminaries.

The demotion of Lysenko, once the greatest and most worshiped hero among "socialist scientists," was a more complicated process, for it proceeded from several directions. The biologists started the attack on the sanctity of the Lysenkoist intellectual empire. At first the attack was indirect and rather quiet: the most common procedure was to omit Lysenko from discussions centered on contemporary developments in biology. For example, in 1955 N. V. Tsitsin, a full member of the Academy, published an article in *Voprosy filosofii* in which he rejected genetics, criticized Darwin, and disregarded Lysenko as an active contributor either to the Michurinist school or to biology in general.[18] Soon after this, open attacks on Lysenko became common. The biologists now concentrated on Lysenko's failure to base his theory on incontrovertible empirical data. Lysenko simply had claimed more than his evidence could sustain. They also complained of Lysenko's heavy reliance on "administrative pressure" in silencing opposition. Although the philosophers were generally among the last groups to abandon their blind loyalty to Lysenko, the philosopher B. M. Kedrov was among the first serious critics of the major assumptions that formed the operating base of the philosophers of Michurinism. In 1955 Kedrov criticized I. I. Prezent, Lysenko's philosophical aide, for heavy reliance on unsupportable arguments in his attack on Darwin's notions of the struggle for existence and continuity in the transformation of species.[19] Kedrov's main argument was that Prezent, in his effort to refute Darwin's ideas, relied exclusively on a corrupted version of the Marxist philosophical legacy.

Soon the burgeoning criticism received strong help from the country's leaders in science working outside biology, who changed both the nature and the intensity of the attack. N. N. Semenov, the leader of the new offensive, placed major emphasis on Lysenko's violation of the moral code of science. He accused Lysenko of giving both inadequate and deliberately distorted information on the methods of his investigations and the results he claimed to have achieved. Lysenko built a personal empire by relying on unlimited fabrications of "scientific" claims, by misinterpreting the state of biological theory, by withholding information from avenues of academic communication, by depending on political authorities to give his theories the validity of established science, and by charging the defenders of opposing theories of unpatriotic deeds against Soviet society. Encouraged by Semenov, the biologists expanded their attack: now they too criticized the methods of Lysenko's subterfuge.

Party scholars were the last to recognize the evils of Lysenkoism. They were the last to stop worshiping the heroics of Lysenko's dedication to the practical needs of socialist agriculture and to the theoretical needs of dialectical materialism. These scholars became particularly active only

after Lysenko was deposed. In 1969 I. T. Frolov, representing the authoritative party view, came out with what was meant to be the final verdict
on Lysenkoism. Frolov admitted that Lysenko had committed gross
errors in scholarship and that his work was injurious to the Soviet scientific community. He contended that Lysenkoist errors could have been
committed "at a certain stage of the development of socialism," but that
they were incompatible with Soviet socialism in general and could not
be repeated in the future. His basic argument was that Lysenko's errors
were in theoretical judgment rather than in deliberate digressions from
the moral code of science. The birth of Lysenkoism, he maintained, was
part of an untempered and undisciplined dedication to the cause of
communism rather than a planned search for self-aggrandizement. Having exonerated Lysenko from calculated attacks on the moral fabric of
science and the scientific community, Frolov suggested that the most
violent features of the Lysenko affair be erased from the collective memory.[20] Historians of science were now advised to view Lysenko as a minor
actor on the stage of modern biology, notable more for his errors than
for his contributions. He erred primarily in making "agrobiology" the
central theoretical discipline in biology, in treating Darwinian thought
and modern genetics as mutually exclusive bodies of theory, and in
staking his fortunes on the idea of the full incompatibility of "formal
genetics"—which included all traditions in biology rooted in Mendelian
thought—and Marxist dialectics. Frolov's intent was to refute Jacques
Monod's claim that Lysenko was essentially correct when he stated that
modern genetics was a total denial of the dialectical method as a tool of
science.[21]

   To the efforts to correct the historical record and to expand and diversify the study of the annals of science and scientific institutions, Academy experts added an acute interest in the theoretical foundations of
the history of science as an academic discipline. Of particular relevance
here was the question: Did historical materialism, in its traditional formulation, provide an adequate explanation of the growth of science?
Could the development of scientific thought in individual countries,
and in general, be explained in terms of socioeconomic causation or determinism? The atmosphere of the 1960s produced an intense interest
in the inner logic of the development of science. External causation was
not dismissed, but its undivided reign was subjected to serious
challenge.

   B. M. Kedrov, the new director of the Institute of the History of Natural

Science and Technology, deserved much credit for bringing forth the arguments in favor of "inner logic" as a propelling force of scientific development. Modern history, he said, offers many examples showing that, in the search for a solution to specific practical problems, science depends not on the urgency of external pressures but on the level of its internal development. Einstein's contributions to physics were not so much an adaptation of science to socioeconomic needs at the beginning of the twentieth century as they were a grand and creative synthesis of certain unique intellectual strands in the development of modern science, including the rise of the very impractical non-Euclidean geometries (particularly Riemann's), the Michelson-Morely experiments designed to verify the existence of ether, Lorentz's transformations, and the discovery of the physical and chemical nature of radiation. Kedrov stated explicitly that historical materialism could not explain why the theory of chemical structures and the periodic law of elements were discovered in the Russia of the 1860s and not in the economically more advanced Western countries.[22]

To understand why A. M. Butlerov made the first serious move in laying the foundations of structural chemistry and why D. I. Mendeleev formulated the periodic law of elements, Kedrov believed it was necessary to lay bare the complicated mechanisms of the inner logic of the development of modern chemistry. Butlerov and Mendeleev made their discoveries not because Russian society asked for them but because chemistry was ready for them. Once chemistry—or any other science— is ready, new discoveries come from persons of astute mind and advanced training, regardless of practical motivations. Once science is ready for it, he theorized, the forward step can be made simultaneously by scholars working independently of one another and under different socioeconomic conditions. The mathematician-philosopher A. D. Aleksandrov gave the idea of "internal logic" a clear and forceful expression:

> Guided by the internal regularity in the growth of scientific thought, the scholars concern themselves primarily with problems generated by their disciplines. Science cannot skip any steps in the internal logic of development. Material conditions of human existence and economic interests may stimulate or slow down the development of science, depending on the tasks expected to be fulfilled. But the solution of these problems becomes possible only when science has achieved the necessary level of development. Moreover, the most significant achievements of modern technology have come from scientific research guided by a quest for pure knowledge rather than by practical goals.[23]

Kedrov did not deny the great influence of "external" factors—of social forces outside the world of science—on the course of scientific thought. What he showed, however, was that the scientist must stay close to the lines of research dictated by the internal logic of science development. He asserted categorically that a full understanding of the internal momentum in the development of science was necessary before an adequate study of "external" influences could be undertaken. In advancing the new orientation, Kedrov and his colleagues recognized their debt to the ideas advanced by Thomas Kuhn in *The Structure of Scientific Revolutions*. There was much in Kuhn's book with which they disagreed: for example, they took issue with his excessive emphasis on discontinuities in the paradigmatic growth of science and with his antiaccumulationist orientation. Nevertheless, they were ready to admit that Kuhn's was the only serious, consistent, and thorough effort to give the history of science a theoretical base and that no future work in the history of science could disregard Kuhn's views.[24] They appreciated Kuhn's analysis of the internal logic of science growth and of the part played by antinomies in the evolution of science. Most of all, they favored his emphasis on the active role of the scientific community in determining the course of scientific development—and in serving as an intermediary between "external" and "internal" conditions of science development.

The new views stood in sharp contrast to the well-known thesis advanced by Boris Gessen in his paper devoted to a Marxist interpretation of Newton's *Principia*. Gessen saw direct ties between Newton's laws of mechanics and the current need for faster and better ships, more reliable clocks, more extensive canal networks, and more efficient water-pumping and ventilating equipment. In their efforts to reconstruct the history of the twentieth-century revolution in science and technology, Soviet scholars did not turn to Gessen for explanatory clues and analytical guides. Indeed, their thinking produced a serious challenge to Gessen's theoretical model. They now took advantage of growing evidence showing that, at least in some notable cases, science produced epochal discoveries by following its own impulses rather than by responding to the needs of economic production. Entire new industries were created by "sudden" and "unforeseen" advances in science. New historians avoided a major clash with historical materialism by portraying science as an organic component of the newly regeneralized notion of the "forces of production"—by moving science from social "superstructure" to social "base." The new interpretation was welcomed, for it gave a strong boost to the scientific community's perennial search for a recognition of internal propulsion as a basic force of science development. As part of the "base," science became the cause rather than the effect of social development, and this resulted in a stronger claim for the independence and relative self-sufficiency of science. The "independence of science could only be translated into the "autonomy" of the scientific community. The emphasis on science as a "force of production" corresponded to the emphasis on academic autonomy as a legitimate claim of the scientific community. The new orientation attracted widespread attention in the scientific community. Loren Graham has observed: "The emphasis of the Soviet writers of the 1930's was on the impact of society on science. Soviet scientists after the war, however, began to speak more and more frequently of the 'inner logic' or 'self-flow' of science, and of the impact this relatively autonomous science had upon society, instead of society upon science."[25] It was not until the mid-1960s that the new view received minute and explicit treatment. The arguments in favor of the paramount historical role of the "self-flow" of science were also the arguments in favor of widening the base of academic autonomy and strengthening the Academy as a scientific community based on intellectual and moral unity.

Soviet scholars interested in building the history of science upon a theoretical base went beyond the obvious effort to reconcile the "internalist" and the "externalist" orientations. They also asked questions that did not fall within the traditional framework of the two orientations. The question that attracted them most dealt with the role of the scientific community in integrating new discoveries into established systems of knowledge. The scientific community was viewed neither as a mechanism of the inner logic of the growth of science nor as an environment external to science: occupying a strategic position, it was wedged between, and only partially coincided with, the internal and external determining factors. A new scientific idea might be rejected not because it violated the logic of science (the internal impulse) and not because the larger society (the external factor) was unreceptive to it, but because the scientific community, sometimes influenced by irrational forces, chose not to absorb it. When confronted with the choice between the corpuscular and the wave theories of light, the scientific community accepted the former, not because it was more penetrating and promising but because the authority of Newton was behind it. For example, the scientific community at first rejected Caspar Wolff's theory of epigenesis because Albrecht Haller, its bitter critic, occupied a more prestigious position in the scientific community.

The new interest in theory led Soviet scholars to appreciate the history of the scientific community as a vital component of the history of science. Some experts emphasized the need for a study of the scientific community as the sole builder and guardian of the scientific legacy. But this type of study was as rich in design as it was rudimentary in concrete application. It was important as part of a quiet effort to clarify the role of the scientific community in setting up the criteria for the acceptance of new knowledge. The real question was not whether the scientific community occasionally relied on "illogical" or "incongruous" criteria in ratifying new knowledge, but whether the scientific community alone had the right to accept and reject scientific innovations. Without this autonomy, the scientific community could not advance the internal unity that was needed for cooperative effort toward common professional goals or for the mobilization of science toward solving problems of social relevance. The message was simple and forthright: without autonomy in ratifying knowledge, there could be no science. The subtleties and articulations of the historians of science were manifestations of a sustained war on the ghost of Stalinism.

Despite the tormenting difficulties of the past, science was gaining on every front. No less a figure than M. V. Keldysh, president of the Academy, announced in 1966—in an article published in *Kommunist*—that science had become an active ideological force, a creator rather than a mere reflection of ideological values. The natural sciences, he said, exerted influence on "both ideology and social governance."[26] He added that the social sciences, previously considered the main funnels of ideology, had begun to influence "material production"; they had become a factor of technological consequence. These arguments, too, served the scientific community in its continuous effort to broaden the institutional base and legal guarantees of academic autonomy and to legitimate its increasing participation in shaping the national science policy.

Among the many branches of the science of science, the psychology of scientific creativity occupied a unique and most challenging position. It concentrated on the intricate relations between science and the scientist. In searching for personal attributes most favorable to scientific innovation, experts found current theories on the formation of personality particulary suggestive. They relied on the work of S. L. Rubinshtein and A. N. Leont'ev, the country's most eminent social psychologists, who recognized two distinct categories of processes in the formation of personality: the processes whereby the individual absorbs (or internalizes), and learns to conform to, the values and norms of his society; and the processes generating idiosyncratic behavior, which contains residual sources of nonconformism, the mechanism of sociocultural change. The personality of a scientist is no exception to this rule: it combines a "conforming" and a "nonconforming" component, the former working for the unity of scientific knowledge and the latter for its diversity.

In the age of Stalin, the ideology of group research emphasized strict conformity to established science and scientific theory as an essential prerequisite for fruitful cooperation within the framework of the scientific community. The unity of science was a specific manifestation of the inner unity of society. During the 1960s the experts did not deny the enormous potential of unified group effort as an efficient adaptation to the growing complexity of modern research projects and as an efficient style of interdisciplinary investigation. But they also argued that the individuality of the scientist—the source of skepticism and challenge to the established authority—is the most genuine fountain of new scientific ideas. The function of the scientific community and of society in general is to create conditions that favor a working unity in, or a conformity to, the established paradigm and style of work, and a nonconformity (or "creative individuality") in the interpretation and evaluation of the inventory of science.

In his noted *Sociology of Personality*, I. S. Kon stated that the "conformist mentality" is antithetical to the spirit of scientific research—and that the "poverty of ideas" and "inflexibility of thought" go hand in hand with conformism.[27] Without nonconformism as a social norm there can be no creative individuality, and without creative individuality there could be no scientific advancement. The science of science, it was now emphasized, must not be limited to the search for a more perfect order in the pursuit of science; it must also provide room for "disorder"—for "creative individuality"—the true source of "scientific creativity": after all, "where everything is connected and determined, there is no room for creative work." Without the lacunae of "disorder"—of challenge to the existing order—there could be "no new hypotheses, nothing to verify by experiments and nothing to transfer to production."[28]

By recognizing "creative individuality" in science, the science of science recognized "individuality" as the basic social force. The emphasis on the combination of "nonconformism" and "creative individuality" was an expression of both the relaxed atmosphere in the world of scholarship and the general groping for a broader institutional base for academic autonomy. This groping marked a return to the early decades of the twentieth century when V. I. Vernadskii made creative individuality a common value of advancing science and living democracy. But the notion of "creative individuality" was advanced more as an expression of the ideals of the scientific community than as a measure of real achievements, even though there were some of major consequence. Despite its many limitations, the overt recognition of the cultural value of nonconformism was among the brightest spots on the horizon of Soviet science during the 1960s.

SCIENCE AND PHILOSOPHY

As soon as the general thaw began to take a firmer hold of the scientific community, several large-scale conferences undertook a broad and critical reexamination of "the philosophical foundations" of modern science. The members of the Academy of Sciences set the stage and provided the main speakers at all the national conferences. What happened at these conferences was premeditated by the members of the Academy engaged in a war against the lingering vestiges of Stalinism. As expected, the discussion on the relations of physics to dialectical materialism attracted most attention and cleared the way for the emancipation of the scientific community from the rigidity of Stalinist thought.

Soon after Stalin's death, a rebellion against the Stalinist critics of the so-called idealistic foundations of quantum mechanics and the theory of relativity began to grow in breadth and intensity. In 1954 *Questions of Philosophy* sponsored a running discussion focused on a philosophical— or Marxist—reinterpretation of the theory of relativity. After all the comments were in, the journal wrote:

> An analysis of numerous addresses, some of which appeared in our journal, shows that a vast majority of commentators consider the theory of relativity one of the greatest achievements of physics. Even though the fruitful debate had not produced unanimity, it has made it clear that the physicists are in agreement on the major issues. The existing differences are mainly of a terminological nature.[29]

Most commentators agreed in disavowing the Stalinist philosophers who treated the theory of relativity as pseudoscience grounded in a misdirected philosophy. Stalinist defenders of Marxist philosophical orthodoxy were now identified as purveyors of nihilism totally alien to dialectical materialism. According to Stalinist philosophers, since a "wrong" philosophy can only produce a "wrong" science, Einstein's flirtations with neopositivism could not but distort his views in science. According to the new view, however, scientific ideas of a scholar are not necessarily related to his work in science; Einstein's neopositivist leanings were not necessarily reflected in his scientific thought. The primary duty of the philosopher is to separate experimentally verified achievements of science from "false" philosophical structures built upon them.[30]

In 1957 the new philosophical orientation received a powerful boost when *Questions of Philosophy* published A. D. Aleksandrov's article "Dialectics and Science," which attacked the leading critics of Einstein's theories—including A. A. Maksimov and I. V. Kuznetsov—openly charging them with incompetence in both modern physics and Marxist dialectics.[31] The culminating point of the new course came in October 1958: the occasion was the First All-Union Conference on the Philosophical Problems of Natural Science. The Academy of Sciences was the chief organizer of the conference and supplied the key speakers.[32] Presented as "the first general and creative discussion of the philosophical problems of modern science by Soviet philosophers and scientists," the conference heard a series of major papers and numerous comments, all showing clearly that the hegemony of Stalinist philosophers was coming to an end. The conference brought about a dramatic curtailment of philosophical interference with the work of the scientific community. The major speeches had a clearly stated leitmotif: the time had come to remove philosophical barriers that had misdirected science and had made it, in some cases at least, work against itself. The astronomer V. A. Ambartsumian made the main plea on behalf of the scientific community. The time had come, he said, for philosophers "to stop placing barriers on the paths to new knowledge. . . . Whenever we showed courage in asking new questions and whenever science turned to problems that were not already answered and were extremely complex and challenging, we heard grumblings from some of our philosophers that all that could lead to idealism!" In order to meet the basic task of helping science in its search for truth, philosophers must become thoroughly familiar with "the principal questions of natural science."[33] Ambartsumian's argument was both clear and acrid: the philosophers, in their war on the most challenging ideas of modern science, operated from positions of ignorance, doing a disservice to both science and philosophy.

The voices of Lysenkoism were also heard at the conference, but they seemed feeble and anachronistic. And the same applied to the scattered opposition to cybernetics and molecular biology.[34] By the early 1960s the relationship of physics to philosophy had been thoroughly reexamined and recodified, and the results served as a model for other conferences recasting the positions of Marxist philosophy in biological theory, neurophysiology, and many other branches of modern science. At the end of the decade, B. M. Kedrov was ready to admit that the philosophy of Engels, which presented "a dialectical-materialistic interpretation of nineteenth-century natural science," did not provide sufficient basis for a philosophical analysis of the revolutionary advances in twentieth-century science.[35]

Making a radical departure from the Stalinist paradigm, the new orientation called for an end to the treatment of Western philosophies of science solely as different verbalizations of "idealism." The trouble with Stalinist philosophers, wrote A. D. Aleksandrov in 1957, was, in part, that they were not able "to separate the idealistic interpretation of the theory of relativity from its content," or "to find an adequate dialectical-materialistic approach to this content."[36] They identified the theory of relativity as "reactionary Einsteinianism" without trying to establish the points of substantive contact between Einstein's theories and Marxist philosophy of science. The post-Stalinist orientation, articulated by Aleksandrov and Fok, encouraged exploratory work on similarities between Soviet and Western thought as related to the theoretical foundations of modern science.

Careful and often tortuous public debate brought the theoretical ideas of Einstein, Heisenberg, Bohr, Born, and de Broglie closer to the philosophical pulse of dialectical materialism. The theory of relativity was now viewed as a confirmation of such postulates of dialectical materialism as the unity of space and time as "forms of matter," the inseparability of matter and motion, and the reciprocity of "absolute" and "relative" views of nature.[37] Einstein's "materialism" acquired three distinguishing features: it was firmly attached to an objective epistemology—a theory of knowledge that recognized physical reality as an entity external to and fully independent of human cognition; it had a clear historicist orientation, for it recognized "time" as an essential attribute of nature; and it recognized causality as the key explanatory principle in science. The basic purpose of these and similar maneuvers was to reduce the differences between Western and Soviet philosophers to a level of secondary importance and thereby to ensure a smoother flow of scientific ideas in both directions. Their function was to free the scientific community of unnecessary and annoying philosophical intrusions. To establish more intimate contact with Western scholarship, the Academy spearheaded an ambitious effort to publish—or, in some cases, republish—the major works of modern physicists in Russian translation. In 1966–1967 the Academy published four volumes of Einstein's works and comments.

The reinterpretation of the general theoretical views of Western scientists encouraged a similar reinterpretation of the views of Western philosophers. Leibniz, Kant, Charles Peirce, and Bertrand Russell, once viewed as advocates of extreme idealism and agnosticism in the philosophy of science, now became important links in the development of the logic of scientific inquiry and in the epistemological foundations of scientific knowledge. Favorable references to previously much-maligned philosophers such as Carnap, Reichenbach, Popper, and Wittgenstein appeared at increasing frequency. Attacks on the "idealism" of Western science and philosophical tradition did not come to an end; however, these attacks lost much of the belligerency bequeathed by the age of Stalin.

The new orientation demanded a shift in emphasis from the interpretation of modern science in the light of dialectical materialism to the interpretation of dialectical materialism in the light of modern science. In the age of Stalin, the philosophers rejected the key notions of modern genetics, the principles of indeterminacy and complementarity in physics, the basic views of relativistic cosmology, and the mathematical base of quantum chemistry. In the 1960s the new philosophers made these scientific ideas the major topics of dialectical elaboration; they stretched Marxist philosophy to make it a home base for previously heavily scorned scientific ideas. Stalinist philosophers dealt with the effects of ideology on science much more than with the effects of science on ideology. It was not until 1966 that M. V. Keldysh, president of the Academy of Sciences, found it timely to acknowledge the paramount influence of science on ideology—and, by implication, to stress the need for making philosophy a discipline concerned with the dynamics of secular thought rather than with the regularities of sacred prescription. This development helped science in reacquiring some of the key attributes of intellectual and institutional autonomy that it had lost during the age of Stalin.[38]

During the 1960s the philosophers outgrew the habit of challenging the theoretical foundations of modern science. L. F. Il'ichev, in a long presentation before the General Assembly of the Academy in November 1963, was the last philosopher during the 1960s to chastise errant scientists. He lauded the fortitude and vision of the physicist M. A. Markov for philosophical statements on the general theory of quantum mechanics, made in 1947, that exposed him to vicious attacks by Zhdanovian critics of cosmopolitanism; but he could not resist the temptation to note publicly that some of Markov's key arguments were erroneous.[39] He behaved like a typical Stalinist philosopher when he told the linguists to ignore both Stalin's and Marr's linguistic theories and to focus their research on "the history of language in connection with the development of society," the theory of signs, semantics, cybernetic nature of language, style, "the culture of words," and dialects.[40]

M. E. Omel'ianovskii replaced A. A. Maksimov and M. B. Mitin as the chief articulator of the new Marxist philosophy of science. He tried to make the adjustment to the new situation as gradual as possible. At first he stressed the growing opposition of Western physicists to the "idealistic" leanings of the Copenhagen school. This opposition, he noted, came not only from such established scholars of the older generation as Louis de Broglie but also from younger scholars typified by David Bohm and Jean-Pierre Vigier.[41] Then Omel'ianovskii looked into the growing dissension within the Copenhagen school, accentuated by Max Born's search for a middle course between Western "positivism" and Soviet "materialism."[42] He argued that the leaders of Western physics had changed their philosophical views and had come closer to the positions of dialectical materialism.[43]

During the 1950s Omel'ianovskii fought the efforts of the leaders of Soviet physics to treat the principles of indeterminacy and complementarity as other than pure expressions of "physical idealism."[44] During the 1960s, however, he not only accepted these principles but tried to integrate them into the Marxist conceptualization of causality and into the mainstream of the Marxist philosophy of science. Particularly in the principle of complementarity he saw an incontrovertible proof of the work of dialectics in nature.[45] The criticism of the "decadence" of Western idealism in physics was still very much in evidence, but now it was softened by a decisive effort to avoid building elaborate arguments against philosophical "aberrations" in physical theory.

The new orientation manifested yet another radical feature: it encouraged diversity in the interpretation of the high theory of modern science. The war on cosmopolitanism was a war on theoretical and philosophical disunity. In contrast, the 1960s blossomed with a diversity of theoretical views. Einstein's general theory of relativity, for example, was now subject to three major interpretations. V. L. Ginzburg and Ia. B. Zel'dovich, representing a majority view, regarded the general theory of relativity as essentially complete in the form presented by Einstein.[46] V. A. Fok and A. Z. Petrov accepted the general theory of relativity as a major step in the advancement of modern physics, but they insisted on the need for a major overhauling of its structure. Fok, in particular, emphasized the local character of the equivalence of acceleration and gravitation and claimed that the covariant form of equations was not related to the notion of physical relativity. He also contended that the rejection of the uniformity of space-time in the Riemannian sense meant a shrinking rather than an expanding of relativity.[47] D. D. Ivanenko represented the physicists who argued in favor of a nonrelativistic interpretation of Einstein's principles, which were accepted as a starting point. All orientations were united by the acceptance of Einstein's conceptualization of the curvature of space-time in terms of Riemann's non-Euclidean geometry.[48]

Those present at the 1958 conference on the philosophical questions of natural science heard three different interpretations of quantum mechanics: V. A. Fok and A. D. Aleksandrov, representing the largest group of physicists, expressed views very close to those of the Copenhagen school, which considered quantum mechanics a theory of the individual microobject; D. I. Blokhintsev defended his own version of a statistical interpretation, the so-called ensemble theory; and D. D. Ivanenko and Ia. P. Terletskii showed some sympathy for the new "causal theory" of quantum mechanics advanced in the West by David Bohm and Louis de Broglie.[49]

Another example of disagreement was provided by the diversity of views on the general theory of sociology. In this newly revived and much-emphasized field, there were four major orientations. Older philosophers, mainly the defenders of Stalinist thought, were willing to accept sociology only as an empirical subsidiary to historical materialism—as a discipline without its own body of theory. The second group argued that historical materialism should supply a universal theory of society that is historicist in its basic orientation, and that sociology should concentrate on middle-range theory—the theory of social subsystems— that is structuralist in its basic interest. According to this view, the two theories should be essentially independent of each other: sociology should be an empirical discipline with its own source of empirical data and body of theory.[50] The third group viewed sociology as a combination of philosophy and science. As a philosophy, it coincides with historical materialism; as a science, it consists of an empirical study of specific historical complexes. Historical materialism and sociology were thus placed in a reciprocal relationship: historical materialism supplies sociology with the universal abstractions about the nature and dynamics of human society; sociology supplies historical materialism with empirical generalizations.[51] The fourth orientation, most consistently articulated by E. S. Markarian, envisaged both historical materialism and sociology as parts of a general systems approach to the study of human society, a most-effective synthesis of historicist and structuralist orientations.[52] The "official" Marxist position in sociology was best described by L. F. Il'ichev in 1963 when he stated that all social sciences must recognize two general rules: they must avoid "dogmatic and schematic" interpretations of historical materialism, and they must avoid "empiricism" that allows no room for the general theory of historical materialism.[53]

While retreating from some areas, the philosophy of science broadened its activities in others. Indeed, it underwent a process of rapid growth and diversification. A mere leafing through the issues of *Voprosy filosofii*, which directed philosophical discussion, showed that philosophers were deeply involved in a search for common denominators of contradictory theories and for universal principles of science. They sought to explain the growing number of formal logics, new conceptualizations of structural levels in nature, society, and technical systems—a task that demanded functional and acausal explanations—and conflicting views on the "styles of thought" in the natural sciences. The journal did not allow its readers to forget that dialectical materialism was essentially a "Weltanschauung philosophy" that explained the unity of Soviet cultural values, political ideals, and the general theory of modern science. Now, however, philosophers were more defensive than offensive, more subtle in their expressions, and more fluid in their convictions.

Marxist philosophers showed a particularly strong interest in explaining and codifying the fundamental principles of the unity of scientific knowledge as a reflection of the unity of nature. In this endeavor they helped create a general picture of science and nature that differed from prevalent Western views more in rhetoric than in the interpretation of the mechanics of the guiding principles of scientific research. The emphasis was clearly on bringing the traditional principles of dialectical materialism, as a philosophy of science, in tune with the grand principles of scientific explanation advanced by modern physics, various branches of biology, and cybernetics and its offshoots.

The law of causality occupied the central position in the dynamics of philosophical adjustment to the demands of new science. At no time were Marxist philosophers willing to part with the law of causality as the key indicator of the unity of nature and the most general principle of scientific explanation. Under Stalinism they fought against the principles of indeterminacy and complementarity on the ground that these principles worked against the universal and absolute validity of the law of causality. During the 1960s the picture changed radically. In physics the principle of indeterminacy became the principle of statistical determinacy, denoting a specific kind of causality operating on the microphysical level. The principle of complementarity, formulated by Bohr to make up for assumed inapplicability of causality on the microphysical level, was now treated as a specific dialectical expression of the work of causality in nature. Marxist philosophers devoted much attention to purposefulness, or teleonomy, as a unique form of causality operating in the living world. To the "primary causes" of the Newtonian mechanistic world they now added a modified form of "final causes," a teleology shorn of metaphysical impurities.

The principle of causality underwent yet another modification: in the confrontation of Marxist historicism with the structuralism of modern science, Soviet philosophers found it expedient to tone down and modify the traditional claim of the universality of causal explanation. They did not abandon the idea that a natural or social phenomenon could be explained scientifically only by placing it into a historical context—that science works by establishing cause-effect sequences in nature and society. Historicism continued to be the supreme and inviolable law of scientific explanation, and the consistent and all-pervasive emphasis on historicism meant the recognition of the universality of change and its "objective necessity, laws, and specific forms." It also meant an emphasis on concrete historical analysis of phenomena and sequences of phenomena from which historical "regularities"—and scientific laws—are distilled. Above everything else, it meant the recognition of the possibility of scientific forecasting of future developments in nature and society.

All this did not prevent philosophers from recognizing modern structuralism, an orientation that shifts emphasis from cause-effect sequences to functional interrelations—from historical instability to the structural stability of natural and social systems and subsystems. The philosophers resolved the problem simply by making structuralism a special ramification of historicism—by recognizing relatively stable elements within the context of historical change. They criticized Jacques Monod not because he emphasized the philosophical and scientific primacy of structural principles but because he went out of his way to "denigrate" the importance of a historical orientation in the study of life. They now conceded that historicism also covered the necessary connection between coexisting phenomena or events.[54]

During the 1960s philosophical workshops and colloquia in scientific institutions grew at a rapid pace. The main task of these bodies was to involve as large a segment of the scientific community as possible in carefully organized philosophical debates and to achieve a new synthesis of modern scientific theories and dialectical materialism. Above everything else, they were expected to bring an end to the alarming alienation of scientists from philosophical issues—and, as a Soviet writer put it, to halt the unwelcome growth of "philosophical nihilism," a view of philosophy and science as mutually exclusive orders of knowledge. It was readily conceded that the antiscientific campaign of the Stalinist philosophers was the main source of this "nihilism." Scientists looked at the philosophers of the 1930s and 1940s as a reactionary force that lent support to all kinds of pseudoscientific ideas. The Stalinist philosophers caused unwarranted confusion and gross distortions in the theoretical studies of modern physics; they fought against cybernetics; and they wrote enough pseudoscientific literature against genetics "to fill a railroad car."[55] Worst of all, they influenced a part of the scientific community to sacrifice the best interests of individual disciplines to the fleeting interests of ideology. P. L. Kapitsa noted in 1962 that the philosophers' treatment of natural science was a bizarre display of ignorance and dogmatism.[56] His views were seconded by V. N. Sukachev, who charged openly, in early 1965, that the ignorance of philosophers was responsible for allowing Lysenko to "discredit Soviet science" while corrupting dialectical materialism.[57] Kapitsa obviously had in mind the lingering breed of Stalinist philosophers like S. T. Meliukhin, who resisted the sobering moves of the post-Stalinist thaw and who refused to give up his traditional right to criticize "idealistic" aberrations in the theoretical thought of leading Soviet scientists.[58]

An unusually strong interest in logic was the most exciting and innovative development in Soviet scientific and philosophical thought during the 1960s. It was also one of the most significant indicators of change in the views of the scholarly community on the interrelations of science and philosophy. Moved by the spirit of the time, the new logic invited a scientific analysis of philosophical propositions no less than a philosophical scrutiny of modern science. It invited an introduction of exact methods into philosophy, or, in the words of the mathematician A. A. Markov, "a mathematization of logic."[59] The proponents of the new orientation were involved in two major activities: they adduced powerful arguments in favor of making "the logic and methodology of scientific inquiry" a key branch of philosophy; and they endeavored to show that the new species of logic were too broad and too complex to be accommodated by dialectical materialism.

A. A. Zinov'ev, a Soviet pioneer of new approaches to logic, noted that the time had come to stop treating the deep involvement in logical analysis and logical construction as an exclusive trademark of neopositivist philosophy, the archenemy of dialectical materialism.[60] In 1959 I. S. Narskii, professor of philosophy at Moscow University, made "a clear distinction between the 'serious scientific contributions' of Carnap and others to formal logic and their neopositivist philosophy."[61] The new "philosopher-logicians" worked steadfastly but unobtrusively in showing that the work of Reichenbach, Carnap, Russell, and many other Western "idealists" could be used profitably in constructing the philosophical foundations of a modern logic of scientific inquiry. This development marked a turning point in the history of Soviet philosophical thought. A decade later, two leading logicians, in a joint article, noted that Soviet scholars had overcome the bias of treating the "application of logical and other exact methods to the analysis of knowledge as a distinctive feature of logical positivism."[62] In numerous and challenging logical studies, Zinov'ev went so far as to ignore dialectical materialism altogether as a possible base for the complex structure of modern logic. In this regard, he represented not the community of philosophers in general but a small, articulate, and important minority. In an unprecedented move, the journal *Voprosy filosofii* chose to publish several articles written by him.

A sizable group of logicians did not steer clear of dialectical materialism; however, they worked toward making Marxist philosophy more tolerant of the formal-logical involvement of neopositivist philosophies. These scholars deserve much credit for the emergence of a conciliatory attitude toward the leading Western orientations in the foundations of mathematics—Bertrand Russell's logicism, David Hilbert's formalism, and L. E. J. Brouwer's intuitionism—which were previously rejected in the Soviet Union because they did not consider the practical and historical underpinnings of mathematical structures and because they were linked with "idealistic" philosophies. The Soviet orientation in constructive mathematics, based on "constructive logic," was now viewed as complementing rather than negating the chief Western orientations.[63]

The philosophers of the Stalin era were concerned exclusively with translating Leninist epistemology into ideological dictates. The new philosophers undertook an additional task: they worked toward making epistemological propositions parts of elaborate logical analysis. A symposium sponsored by *Voprosy filosofii* in 1970 concentrated on "quantum-relativistic logic," the "logic of microphysics," and "the logic of the 'possible.'"[64] A strong and rapidly growing group of promising experts in logic accepted Wittgenstein's dictum that the function of philosophy is not to fashion new ideas in science but to clarify propositions by means of logical analysis. A relatively sparse breed of new philosophers agreed with Bertrand Russell that "it is not the business of philosophy to make discoveries, but to assess the merit of different ways of talking about what is admitted on all hands."[65]

The lively interest in V. I. Vernadskii, which began to gather momentum during the 1960s, was also related to this recasting of the relationship between dialectical materialism and science. Searching through Vernadskii's published and unpublished scholarly legacy, modern commentators found support for the mounting efforts to emancipate the logic of scientific inquiry from epistemological detours and, in general, to narrow the scope of philosophical control over the construction of scientific theory.

The new philosophers saw the future of their discipline in the independence of logic from epistemology, and of logic and epistemology from dialectics. In carrying out this task, they added strong, but implicit, arguments in favor of reducing the impact of externally imposed ideological limitations on the functional autonomy of the scientific community. Epistemological analysis of the determinism-indeterminism controversy led inevitably to the heart of Marxist ideology and accentuated the feud between dialectical materialism and "bourgeois" idealism. A logical analysis of determinism and indeterminism, however, took most of the heat from the feud. It did this by recognizing that the term "cause," looked at from a logical point of view, had many gradations in meaning rather than a monolithic definition.[66] Thus, logical analysis softened the conflict between determinism and indeterminism and between dialectical materialism and "bourgeois" idealism, and it worked against a categorical rejection of the ideas built into such influential Western philosophies as neopositivism and neo-Kantianism.

In the past, Marxist philosophers were unanimous in their rejection of the neopositivist claim that "description" rather than "explanation" was the chief goal of science; their argument was that the emphasis on "description" meant an effort to reduce science to a lower—that is, empirical—level of systematic search for knowledge. The thaw—and the general attacks on philosophers—made it possible for the biologist A. A. Liubishchev to make the following statement:

> A significant group of progressive scientists (R. Kirchoff, E. Mach and H. Poincaré, for example) contended that the task of science was not explanation, but correct, brief and full description, which makes prediction possible. Many persons find the word "description" unsettling, for it appears to suggest the return to a primitive—"descriptive"—level of science. In actuality, however, this return is purely dialectical, operating from a higher base.[67]

Instead of trying to force dialectics on the new branches of formal logic, a group of traditionalists resorted to creating a new logic, known as "dialectical logic." The basic task of the new discipline was to preserve and advance Marxist interests in the development of modern logical strategies in the study of the dynamics of nature and society. It quickly became clear that the purpose of "dialectical logic" was not to supplant but to coexist with other kinds of logic. The experts in this field made no effort to interfere with their colleagues working in the other branches of logic, perhaps because their discipline was built on shaky foundations and was not united.[68] Some experts treated it as the most general science of the laws governing the processes of knowledge, including formal logics.[69] Others viewed it as a study of the role of antinomial thought in the construction of the rational base of scientific cognition, or as a formal inquiry into the empirical moorings of logical processes. Experts in dialectical logic made it their primary duty to issue constant reminders—and to produce illustrative material—in favor of the view of logic as a system of historically evolved, culturally molded, and socio-economically determined rules of human cognition. Dialectical logic was presented as a study of "motion, development, and progress," as categories of Marxist dialectics.

D. D. Comey recognized three distinct emphases in the efforts of Soviet scholars to describe the role of dialectical logic: as a "content-oriented" logic, it was clearly separated from logic as a science of the "forms of thought"; as "general logic," it dealt primarily with the historical aspects of logical forms; and as a "logic and methodology of science," it covered the sequences and dynamics of logical processes involved in the creation of scientific theories.[70] A nonmilitant discipline, "dialectical logic" sought to establish its domain at a safe distance from the multitudinous systems of logic concerned with the formal principles of abstract thought, in general, and with the unique modes of modern scientific thinking, in particular. In doing this it helped to create a lively philosophical enclave removed from the highways of heavy ideological traffic, as well as to reduce external interference with scholarship. The defenders of dialectical logic now complained about the increasing occurrence of "hidden or open" attacks on their discipline.[71]

The new philosophy met with opposition from the old Stalinist philosophers, who were exceedingly slow in recognizing philosophical concerns removed from the ideological front. In November 1969 the Presidium of the Academy held a special meeting devoted to the evaluation of the work of the Institute of Philosophy. Dominated by M. B. Mitin and F. V. Konstantinov, the tested champions of Stalinist philosophy, the meeting ignored logic altogether and emphasized the need for continued elaboration of materialistic dialectics and ideological warfare against idealistic philosophies in the West.[72] Mitin and Konstantinov continued their campaign against the new interest in logic, which they interpreted as an escape from the acute issues in Marxist philosophy. On March 10, 1970, at a meeting sponsored by the Department of Philosophy and Law—called to discuss a newly published book on Lenin's contributions to logic by P. N. Kopnin—Mitin and Konstantinov used strong words to express their resentment of the new drift in philosophy from the concrete ingredients of the Marxist Weltanschauung to the most abstract constructions of logic. They received vociferous support from G. V. Platonov, the diehard philosopher of Lysenkoism.[73] But it was a sign of the times that Mitin and his group could not reverse the trend that gave logic a prominent, though not a leading, place in Soviet philosophical thought.

In the past, Marxist philosophers considered Francis Bacon a forerunner of modern materialism because he underscored the utilitarian virtues of the search for scientific knowledge and because he declared war on all kinds of "idols" that challenged the sovereignty of reason. Modern Marxist theorists accepted these arguments, but they added a new one: Bacon, as a good "materialist," made it clear that logic must assume many forms in order to meet the specific needs of individual sciences and bodies of knowledge. It was Bacon, we were now reminded, who said that logic must come not only from the nature of reason but also from the nature of things.[74]

Under Stalin—particularly during the war on cosmopolitanism—the philosophers presented a united front, not only in the interpretation of key theoretical principles but also in the classification of topics according to their ideological importance. They recognized and were guided by a unified list of philosophical problems arranged according to the scope and intensity of their relevance to the issues at hand. All dealt extensively with the epistemological foundations of scientific knowledge. During the 1960s many central issues were still alive and duly emphasized, but there was also a noticeable development of specialized interests on a regional level. Ukrainian philosophers, for example, showed a particularly strong interest in the logic of scientific inquiry. Philosophers of the Kirghiz republic became noted for their work in linguistics and communication in a broader philosophical context. Novosibirsk scholars concentrated on the meaning and role of determinism in natural science. Armenian philosophers led the country in the search for a general theory of culture as a synthesis of the basic principles of historical materialism and the leading Western orientations in sociology, ethnography, anthropological philosophy, and the general systems theory—an effort that demanded broad philosophical considerations.[75]

The full unity of science and philosophy was a Stalinist ideal. During the 1960s there was a pronounced, but not forcefully expressed, tendency to recognize the existence of basic disagreements in the guiding principles of the philosophical and scientific universes of inquiry. Philosophy adhered closely to an ontological position anchored in materialism. It claimed that everything in the universe—and in human society and culture—had, or was derived from, a material substratum. Science, in contrast, rejected the idea of ontological reductionism—of reducing all scientific explanations to a material base. It recognized various levels of emergent reality, irreducible to more basic levels of matter. Philosophers were satisfied with an emphasis on the neurophysiological base of human behavior and culture. Scientists, in contrast, worked on the assumption that the very essence of psychological and cultural reality was irreducible to neurophysiological mechanisms and processes.

On the epistemological level, philosophers recognized objectivity as the basic attribute of scientific knowledge. According to this position, scientific knowledge is viewed as a true reflection of natural phenomena, existing independently of the human mind. Under the pressure of the grand theories of quantum mechanics and relativity, physicists came around to viewing scientific knowledge as a combination of an objective reflection of external reality and a subjective reliance on instruments of observation, whether they be macrophysical laboratory setups or Einsteinian "frames of reference." Observer, as a "subjective" category, became an organic part of the strategy of theory construction in modern physics. Reluctantly and after considerable delay, Soviet philosophers of science admitted that man, disregarded by the logic and the method of Newtonian science, was an inextricable part of the theoretical structures of modern physics. One scholar noted: in epistemology there is an absolute difference between the material and the ideal; in the natural sciences, the difference is not absolute.[76] The two interpenetrate and depend on each other.

The Academy played a dual role in the development of philosophy at this time. It illuminated the major routes leading to new ventures in speculative thought and to new areas of philosophical discourse, particularly in the puzzling world of ideas opened by modern science. But the Academy also continued to be the stronghold of conservatism—the last citadel of Stalinist orthodoxy. In the Academy, younger persons without high academic titles were generally the leaders of liberalizing moves in philosophy; the older and titled philosophers were the last—though subdued—defenders of Stalinist narrowness and dogmatism and the leading guardians of the interests of dialectical materialism amid the outburst of new philosophical thought. Despite notable relaxations on the philosophical front, dialectical materialism continued to be a limiting factor in relation to science. There was considerable latitude in matching dialectical materialism with the high theory of modern science; however, there was little latitude in formulating the fundamental principles of Marxist philosophy. The suggestion, made by the Yugoslav philosopher P. Vranicki, that the time had come for the coexistence of various schools of Marxist philosophy, each with its own basic principles, was ridiculed as alien to the very spirit of Marxism.[77]

It would be incorrect to claim that the philosophers alone were guilty of efforts to place the leading theories of modern science into the Procrustean bed of Marxist orthodoxy. There were scientists, too, who followed a similar course. Typical among these scientists were those who tried to gain undue personal advantage by emphasizing the organic unity of their theoretical views with dialectical materialism. This practice usually enabled them to surge ahead of their peers whose theories were not so close to the axiomatic base of Marxist philosophy. A cosmologist defending the steady-state theory was clearly in imperfect competition with cosmologists who viewed the history of the universe as a constant succession of galactic ruptures. A molecular biologist espousing a historicist orientation was clearly in a more favorable position than a molecular biologist with strong structuralist leanings. The biochemist A. I. Oparin and the astronomer V. A. Ambartsumian, for example, two serious scholars, did not hesitate to capitalize on the patent proximity of their theories to the premises of dialectical materialism.

SCIENCE AND ETHICS

During the 1960s the relationship of science to ethics became a topic of public debate led by a select group of eminent members of the Academy. This issue had been widely discussed in Russia during the waning decades of the tsarist system. K. A. Timiriazev, a distinguished plant physiologist, popularizer of Darwin's theory of organic evolution, and commentator on the social and cultural attributes of science, upheld the nihilist view that science was the primary source of the moral code of modern society and the most reliable indicator of social progress. He fully endorsed M. Berthelot's assertion that since the seventeenth century, science was the only contributor to "the improvement of the material and moral conditions of social life."[78] The second view, by far the most popular in the Russian scientific community at the beginning of the twentieth century, expressed serious doubts about the notion of scientifically determined morality and about ethics as a science. V. I. Vernadskii was the most eloquent spokesman for this group. He thought that every effort to reduce morality to science was an intellectual by-product of the one-sided interpretation of the place of science in modern culture. One could not appreciate the power of science until he understood and acknowledged its intrinsic limitations and its complementary relations with moral, religious, philosophical, technical, and aesthetic modes of inquiry. He emphasized that most scientists, though influenced by the writing of the nihilists of the 1860s, gradually abandoned nihilist scientism and "materialism" and learned to live with a more modest view of science.[79]

The views represented by Timiriazev, fastened to a philosophy that attributed limitless intellectual and social powers to science, became an organic part of Soviet ideology. Marxist ideologues presented Soviet society as the first political community in which science became a crucial component of the social "base" and the first society in which morality became fully congruent with the intellectual inventory of science. "Moral truths," according to a typical Soviet writer, "are in no essential way different from scientific truths, as was claimed by Kant and modern positivists. . . . The concept of moral truth corresponds to the concept of truth in general and is a particular case of the latter."[80] "Morality and science," wrote another commentator, "shared a common object—the objective laws of history." Morality and science elevated man above the vagaries of everyday existence and placed him in the sphere of "historical reality" as an "objectively necessary reality."[81] Morality was viewed not only as a product of science but also as a proper subject of scientific inquiry. Moral norms acquired the authority of "principles with scientific foundations"; the true or false nature of moral norms could "be ascertained by means of a scientific analysis."[82]

During the 1960s the relationship of science to morality came up for a broad reappraisal. Unfolding gradually, the reappraisal received particularly strong impetus from a performance of Bertolt Brecht's drama *Galileo* in a Moscow theater in the 1960s.[83] Brecht did not adhere too closely to the documented history of the salient moments in the life story of his hero. Nor did he treat all the critical issues related to Galileo's struggle with the cultural forces inimical to the development of science. In a highly fictionalized account, punctuated by Galileo's philosophical discourses, the playwright concentrated on one problem: Galileo's confrontation with the moral obligations of a scientist to science and society.

Ridden with guilt for having "postponed" the beginning of the age of reason by his convenient "retreat" from the heliocentric theory, Brecht's Galileo was more than vindicated by his clear understanding and dramatic enunciation of the moral foundations of science and the nature of his "crime." Brecht, through his Galileo, demonstrated clearly that science and the moral code could not be placed into a causal—that is, a deterministic—relationship. Relying on involved and rather diffuse arguments, he indicated that it was too simple, sterile, and even dangerous to perceive rationalism, distilled from a scientific analysis of underlying conditions, as a reliable moral program for behavior. The power and compass of morality transcend the intellectual powers radiating from the rationality of science. Einstein expressed a Galilean thought when he stated—in his *éloge* to Marie Curie—that the moral qualities of leading personalities were of greater significance than their intellectual accomplishments. Galileo acknowledged the unique features of the moral code of science—for example, cultivated skepticism and unceasing challenge to authority—as a creation of the scientific community. But he also recognized the superior power of the moral law of humanity, which reaches far beyond the limits and rigidity of the logic of rationalism.

Concerned with the general sociological framework of the relationship of science to morality, the physicist E. Feinberg wrote in 1965—in the popular journal *Novyi mir*—that "no ethical criteria are immanent in science," that "there is no room for the ethical element in a scientific system." To think otherwise is to accept a "fetishization of logical thinking" and an application of it to domains outside its competence. All this, Feinberg argued, does not exclude the complementarity of science and ethics.[84] Two years later, A. N. Nesmeianov, former president of the Soviet Academy of Sciences, stated in *Literaturnaia gazeta:* "I do not see any connection between science and morality. . . . Morality varies not only from society to society but also from individual to individual. An honest merchant in bourgeois society could be considered an exploiter and a speculator in our society, but science . . . is the same everywhere."[85] Nesmeianov's statement did not go unchallenged, but the challenge came primarily from leading members of the Academy known for their efforts to harmonize the revolutionary developments in modern science with the philosophical postulates of dialectical materialism. A. D. Aleksandrov, one such academician, could only repeat the old dictum that "the unity of science and morality, the unity of the scientific explanation of the world and the moral demand for its change, is the alpha and omega of the Communist world view." Science, according to Aleksandrov, has an intellectual and a moral function: to understand the world and to change it. It is "as indispensable for morality as light is for vision."[86]

The mathematician P. S. Aleksandrov, another influential member of the Academy, wrote in *Literaturnaia gazeta* in support of Nesmeianov's statement. Like Nesmeianov, he was interested in the interrelationship between science and morality as a product of modern civilization rather than of Soviet ideology. According to him, science and morality are not only made of different cultural material but are also subject to different principles and tempos of development. Since the time of Rousseau, he said, science has advanced at an accelerated pace and has become a major component of modern culture; during the same period, however, man had stuck to many grand delusions and moral dilemmas. The synchrotron, lunar flights, and insulin had hardly improved the moral fabric of the world in which Leonhard Euler and Jean D'Alembert lived.[87] Aleksandrov made no effort to single out Soviet society as an exception to this sweeping and categorical assessment of the state of morality in the modern world. Neither Nesmeianov nor P. S. Aleksandrov actually thought that science and morality were unrelated. Both agreed with Henri Poincaré's view that although "there will never be scientific ethics in the strict sense of the word," science "can be an aid to ethics in an indirect manner."[88]

A writer summed up the "new" view in *Voprosy filosofii:*

> The level of knowledge possessed by individual persons as well as entire epochs does not correspond to the level of moral consciousness. . . . The differences in the "tempo" of development of knowledge and moral consciousness are strongly felt and cannot be resolved by a simple exaggeration of the power of scientific knowledge.[89]

Iu. Shreider used a different approach to affirm the same idea:

> To live in the world of exact sciences is in a way attractive and easy. In contrast to ordinary life, the world of science has a simple scale of values. But the simplicity of this scale leads easily to rigidly restricted consciousness, to isolation from the rest of the world [and] to a loss of human responsibility. There is something very infantile in the striving to achieve a simple scale of values, whatever the price.[90]

Science, it was now stressed, provided tools for the study of the structure and dynamics of moral norms, but all this did not make it the exclusive source and the architect of morality.[91] After all, morality emerged earlier than either philosophy or science. Finally, there was the towering figure of Einstein supporting the view that the scientific method did not define the goals of humanity; it merely provided some of the tools for their realization.[92]

The quiet renouncement of the intellectual imperialism of science— of a philosophy that sought to establish the primacy of science over all other modes of inquiry—could best be understood as part of the new emphasis on the autonomy of scientific inquiry. Nesmeianov and his colleagues were convinced that there could be no meaningful cultural autonomy for science without cultural autonomy of other modes of inquiry. The very fact that a scientist might stress the limited compass of science in the total picture of human values was an important index of the improved intellectual atmosphere in which he lived and worked in the 1960s. Although the search for a critical reassessment of social and cultural effects—and limitations—of science was rather sporadic and lacked a direct confrontation with the key problems of ethics, it paid handsome dividends. By separating ethics from science, it eliminated a wide and extremely sensitive area of ideological infringement on scientific activity. It also intensified the ongoing reexamination of the cultural matrix of science. It was, in short, part of a rapidly growing search for the affirmation of the inalienable right of scientists to define the domain and the limits of scientific inquiry.[93]

SCIENCE AND ART

By emphasizing the cognitive base of the unity of science and art, Soviet Marxist theory extended the criteria of science to the multiple forms of aesthetic expression.[94] It was thought that the function of art is to give added strength to the cognitive base of science. Art portrayed a reality that was both objective and scientific. For an artist to challenge science and the scientific world view was to commit sacrilege. The supremacy of science was enhanced when it was proclaimed to be "a direct force of production," a key element of the "base" of Soviet society, leaving art in the realm of derivative "superstructure." The basic function of art was to manifest and to prove the superiority of science in the total configuration of modern approaches to the universe.

At the beginning of the 1960s the interrelationship of science and art came up for a reexamination. The new debate was stimulated by several

developments in the interpretation of modern culture—its social underpinnings, historical roots, intellectual divisions, and moral dilemmas.
The reaction of Soviet writers to C. P. Snow's essay on "two cultures"—
a plea for reciprocity in the interrelations of science and the humanities—
was a factor of momentous consequence. The interest of literary historians in the scientific attitude of such giants of Russian literature as
Pushkin and Dostoevsky was another factor of considerable weight, as
was the intensity of current interest in the affinity of science with the
humanities as seen by the leaders of modern science. Heisenberg's warm
and affectionate comments on the scientific attributes of Goethe's poetic
mind, Bohr's discourse on science and art, Schrödinger's speculation
about "science, art, and play," and Louis de Broglie's views on "machine
and spirit" and on Bergsonian "intuition" provided ample evidence of
the basic need for a broader examination of the humanistic matrix of
science—and, in particular, for a critical and profound reassessment of
cultural bonds between art and science.

The new interest focused on Dostoevsky, who, in *Notes from the Under-
ground*, scoffed at science and made it a mere "20 percent" expression
of humanity. The time had come to take a deeper and more careful look
into the total literary work of the great writer. Einstein's remark to one
of his early biographers that he had learned more from Dostoevsky than
from Gauss served as a jumping-off point for the new discussion.[95] The
gigantic contributions of Einsteinian science were now universally
accepted in the Soviet Union, and Soviet writers found the time propitious to examine Einstein's humanistic views. In June 1965 the journal
Nauka i zhizn' published an article by B. G. Kuznetsov entitled "Einstein
and Dostoevsky," which produced two results: it reversed the traditional
view of Dostoevsky as a critic of scientific rationalism, and it reinforced
the view of the complementarity and interdependence of science and
art.[96] At the base of Kuznetsov's arguments was Claude Bernard's dictum
that while the artist found in science "a more stable foundation," the
scientist found in art "a more certain intuition."

The scientist Einstein and the writer Dostoevsky were united by a
search for harmony in the universe—for the unity of man and the cosmos. Both operated on the assumption that this unity could not be
comprehended as long as the human mind was limited to terrestrial
vision. To both, the harmony of the universe was not in the flat surfaces
of the earthly perspective but in the cosmic space-time curvature, a
curvature with no place for conditioned reflexes of human behavior. It

is common knowledge that Einstein developed the general theory of relativity by relying on the non-Euclidian geometry of Georg Riemann, a nineteenth-century German mathematician. What attracted less attention was that Dostoevsky, in *The Brothers Karamazov*, gave the first literary airing to the ideas generated by non-Euclidean geometries and that this was one of the reasons for Einstein's attraction to the great writer. In a thoughtful discussion with his brother Aliosha, Ivan Karamazov admitted that the greatest disappointment in his life was that his earth-locked mind prevented him from placing his God in a non-Euclidean framework. Ivan Karamazov was referring to the same idea that found its way into Einstein's thought—the idea that Euclidean geometry was only a small component of a much larger configuration of space-time. Einstein, like Dostoevsky, admitted readily that the path to the understanding of the harmony of the universe traversed successive domains of tortuous complexities and bewildering contradictions. To both, the strands of stereotypical experience—common sense—threw reality out of focus. To both, the reality of the universe was too complex to be placed into the monolithic system of unitary logic. Einstein echoed Dostoevsky's thoughts when he stated that logic did not precede mathematics but was an accommodation to it. In the search for the unity of the universe, Dostoevsky, an "antirationalist philosopher," acted as a "rationalist artist." In his search for harmony in the universe, Einstein depended on an aesthetic quality: the symmetry and perfection of physical theory.

The time had come to study not only the cognitive substratum of art but also "the aesthetics of scientific creativity."[97] The Commission on the Interrelations of Literature, Art and Science, established in 1963 in Leningrad, was given the task of exploring the inner dynamics of scientific and aesthetic modes of inquiry. In the same year a national congress of experts was held in Leningrad to discuss the multiple ramifications of artistic creativity—and to look into C. P. Snow's warnings about the growing rift between "two cultures" of modern society.[98] A similar congress, held in 1966, concentrated directly on "creativity and modern scientific progress." These and similar activities led to the proliferation of local organizations interested in the ever-fascinating problem of the overall unity of scientific and artistic "cognition." The literary legacy of the poet O. E. Mandel'shtam gave poetic support to the new search for the unity of science and art: "The needs of science concur happily with one of the basic laws of artistic activity. I have in mind the law of heterogeneity, which impels the artist to search for the unity of different sounds and of diverse and discordant images." Darwin's work, in Mandel'shtam's view, was a masterful expression of this unity: Darwin studied the most diverse forms of life, and yet he never lost track of the unity of nature.[99] Mandel'shtam denied poets a monopoly in depicting the beauty of nature: while poets were unable to see in the Russian earth anything more than an inscrutable monotony of endless grayness, the naturalist Peter Simon Pallas supplied unsurpassed descriptions of Russian earth rich in color, natural sculpture, and symbiotic relations among nature's infinitely varied components.

The new emphasis on the unity of science and art was not an emphasis on the identity of the two modes of inquiry but on their complementarity—on their distinct contributions to a unified view of the universe and man's place in it. Writers continued to emphasize the fundamental differences between the two approaches to the universe. According to one writer, science was originally a tool of man's practical adaptation to the external environment.[100] In his approach to the universe, the scientist rose above the limiting domain of personality; his universe became thoroughly impersonal. The artist looked at the universe through the prism of his personality; his universe was an outward projection of personal attributes. Again, in I. P. Pavlov's view, the artist, in contrast to the scientist, studied a universe presented as unfragmentable unity. The scientist, however, was a specialist. The artist, in the words of the poet A. Blok, was "a human being by profession."[101] "Individual sciences," stated another writer, "cannnot study man as a total, actively creative and universal being. This is the task of philosophy and art."[102]

The discussion of science vis-à-vis art went in many directions and on occasion led to "unexpected" results. A good example was provided by B. Runin's assertion that the more the scientist isolated his "biosocial I"—his sentiments, beliefs, and social-class loyalties—from the object of his study, the better were his chances of reaching the depths of scientific truth and becoming a recognized authority in his field.[103] In so many words, he repeated the utterances made by A. I. Herzen in 1843:

> Science is the kingdom of impersonality, a kingdom freed of passions, becalmed in profoundest self-cognition, illuminated by the all-pervading light of reason. . . . In science the truth is incarnated not in a material body, but in a logical organism, and is alive through the architectonics of dialectical development and not through the epic of the transient being. In science the law is the thought wrested and preserved from the tempest of existence and accidental disturbances.[104]

On the surface, Runin's statement—like that of Herzen—appears both generally true and rather obvious. But it was also a categorical challenge—although Runin did not spell it out—to the Leninist principle of *partiinost'* in science, which demanded from scholars a commitment to making their professional work a tool and an expression of ideological convictions and sentiment. It was one of many ways of expressing the growing sentiment in favor of taking politics out of science.

The principle of complementarity did not exhaust the relationship between art and science. In addition to providing two different approaches to the same universe, the two interpenetrated each other. From art the scientist learned the immense potentialities of intuition—the power of Einsteinian "pure ideas"—and the essential need for pictorial thinking at all levels of abstraction. According to I. Zabelin, Einstein's playing of the violin should not be discounted in the study of the genesis of the relativistic formulation of time. And, according to the same writer, without the poetry of the past there could not be the science of today. It was in this atmosphere that the mathematician P. S. Aleksandrov stated that, in order to instill a creative scientific spirit in his students, a professor must combine high-level expertise in his field and a "general cultural background," including the humanistic dimension.[105]

The defense of the complementarity of scientific and artistic modes of inquiry had two purposes. In the first place it was part of a mounting effort to crystallize a general strategy for the struggle against the ominous signs of growing disparity between humanistic and technical segments in modern culture. It was an affirmative answer to C. P. Snow's two-culture dilemma, and it was part of a burgeoning ideology that matched a scientific approach to the humanities with a humanistic approach to the sciences. In the second place its aim was to challenge the hegemony of science in the Marxist scheme of universes of inquiry. As in the relationship of science to ethics, the scientific community worked on the safe assumption that the doctrine of the intellectual hegemony of science—imposed on scholarship by an ideology external to it—was an invitation to make science the central weapon of ideological controls in the vast realm of intellectual and artistic activity. By protecting the autonomy of the arts, the scientific community sought to strengthen the ongoing search for broader autonomy in its own domain. This dedication was much more impressive as an expression of far-reaching ideals than as a summation of achieved reality.

As the Academy of Sciences approached the fiftieth year of its existence under Soviet rule, it was caught in a process of staggering expansion and wide-ranging efforts to find a strong home base for every branch
of modern scientific activity. It abandoned the Stalinist search for theoretical unity in science as a necessary condition for ideological unity of
society. It made tangible progress in emancipating science from corrup-
tive attacks by philosophers. But it also left the legacy of a growing
conflict between two antithetical forces of immense magnitude: the
growing rigidity and pervasiveness of administrative controls, which
made the Academy an integral part of central government, and the
intensified concern of the scientific community with guarantees of constitutional rights and professional autonomy. The endless discussions
of the relationship of science to philosophy, ethics, and arts achieved
one goal: they helped the scientific community to state its views on the
subtleties and multiple ramifications of cultural values and social conditions that, in its opinion, met the best interests of science as both a
system of knowledge and a social force of the first magnitude.

CONCLUSION

The Academy of Sciences—its institutes and its laboratories—
launched the Soviet Union into the age of atomic power, nuclear fusion,
space exploration, molecular biology, polymer chemistry, quantum elec-
tronics, and computers. It was the chief architect of grand designs that
took science, as a cultural activity, to every corner of the country. It
played the major role in coordinating the multi-institutional base of
science, in unveiling the hidden parameters of the national strength in
science, and in designing and maintaining elaborate channels for the
flow of scientific information.

The relationship of science, as defined by the scientific community,
to ideology, as articulated by Marxist theorists (primarily Marxist philosophers), made the Academy the center of a political drama that produced disruptive crises in the national dedication to science. This
relationship passed through four distinct stages. From 1917 to 1928 the
influence of Marxist ideologues on the scientific community was fragmentary and without clear direction. The reason for this was obvious:
the articulators of ideology did not agree among themselves on the key
issues involving the interrelations of dialectical materialism and modern
science. The Communist Academy, created for the explicit purpose of
making Marxist thought a unified system of philosophical propositions,
was deeply split on many major issues of modern science. The attitudes
of its members toward Einstein's theories ranged from total acceptance
to total rejection. The leading Communist theorists did not agree on a
Marxist interpretation of Heisenberg's uncertainty principle, Darwin's
evolutionary ideas, Freud's psychoanalysis, Pavlov's neurophysiology,

or many other ideas that challenged the skills and imagination of modern scientists. Ideological uncertainty, however, did not deter a mushrooming group of Marxist vigilantes—operating through such organizations as the Section of Scientific Workers, VARNITSO, and the Society of Militant Dialectical Materialists—from harassing scientists suspected of a less than favorable attitude toward the new system. Members of the Academy of Sciences, small in number and helped by scholarly detachment, survived the succession of crises with no major personal or professional damage. Members of the Communist Academy were too involved in intramural squabbles—and too unsettled in their own philosophical convictions—to launch a sustained attack on the leaders of non-Marxist scientists, concentrated primarily in the Academy of Sciences.

The two academies represented two distinct views on science as a cultural activity. The Academy of Sciences was dedicated to making Soviet science an active contributor to, and a beneficiary of, the world pool of scientific knowledge. The Communist Academy, in contrast, dreamed of setting the stage for the creation of Soviet science as a unique cultural activity. The unsettled situation among the Marxist theorists prevented this dream from finding concrete expression or developing in a systematic way. Lenin's forthright pronouncement that socialism could not triumph in the Soviet Union without the help of bourgeois science encouraged the search for a middle ground between the extremes of the two orientations.

The second stage (1929–1945) coincided with the first phase of Stalinist rule. It was dominated by massive efforts to make the unity of science and ideology a strategic component of national unity. Determined to control all aspects of Soviet society, Stalin spurred efforts to codify Soviet ideology, to eliminate the centrifugal forces characteristic of Marxist-Leninist philosophy during the 1920s, and to establish the full unity of official ideology, Marxist philosophy, and modern science. The effort to achieve the unity of science and ideology proceeded in two directions: purges of nonconforming scientists and a search for a unified Marxist philosophy of science.

The purges took place in two waves. During the late 1920s the Academy was purged of a substantial contingent of major scholars and research associates who were considered deeply committed to the ideology of the ancien régime. During the late 1930s the Academy lost many scholars identified—often erroneously—as members of Marxist groups

actively opposed to Stalinism. A national calamity, these purges produced incalculable losses in scientific talent. They staggered the Academy and opened its doors to external interference on a large scale. They
filled the academic air with a feeling of great uncertainty, misplaced
goals, and subdued pessimism. T. D. Lysenko exploited this atmosphere
for the purpose of creating a union of science and dialectical materialism
and making Stalinist patriotism a criterion in validating scientific theory,
thereby corrupting both science and Marxist theory. The terror also
produced an unanticipated result: the quiet but stubborn resistance of
a solid core of academicians to the attacks on the scientific community.

During this stage the Academy lost one of its most cherished traditional prerogatives: the right to be the sole judge in selecting and electing
its members. The watchful eye of ideological masters produced two
negative effects: it prevented the nomination or promotion of acknowledged scholars with questionable political records, and it forced the
nomination or promotion of ideological favorites without strong records
of scholarly achievement. In biology Lysenko was elected a full member
even though his biological "theory" was a backward step in the development of modern science; whereas S. S. Chetverikov, a true and widely
acknowledged pioneer in population genetics, and N. K. Kol'tsov, a
recognized ancestor in molecular biology, were prevented by ideological
barriers from reaching the peak of the academic hierarchy. In physics
V. F. Mitkevich, whose thinking did not evolve beyond nineteenth-
century mechanical philosophy, was elected a full member as payment
for his dedicated work on the ideological front. In contrast, Ia. I. Frenkel',
one of the most erudite Soviet nuclear physicists, was not promoted to
the rank of full member as obvious punishment for his unorthodox
philosophical views. By electing O. Iu. Shmidt in geophysics, V. R.
Vil'iams in soil science, and a score of members of the Technical Department, the Academy weakened its claim to represent the best in Soviet
science and the scientific community. M. B. Mitin, who reduced all his
writings to ideological slogans, became not only a full member of the
Academy but also the chief spokesman for Soviet philosophy. V. F.
Asmus, the most serious Soviet student of the history of philosophy but
a man averse to ideological cliches, was kept out of the Academy. For-
tunately for the Academy, and for Soviet science, the vast majority of
academicians and corresponding members were carefully selected on
the basis of scholarly achievement.

In the 1930s non-Marxist philosophers were completely silenced. Philosophical debate that did take place, however, played an important role in bringing about a unified and fully crystallized Marxist philosophy of science. The work of philosophers concentrated on establishing a full congruence of science and dialectical materialism. At last professional philosophers represented a united front; the only opposition came from scientists willing and courageous enough to discuss philosophical issues. The main battle was fought between the philosophers lodged in the Communist Academy (in 1936 transferred to the Academy of Sciences) and the leading physicists.[1] The search for an ideologically pure and philosophically elaborated view of science concentrated on physics, a discipline that, because of revolutionary new developments in the theory of matter, generated the strongest opposition to Leninist epistemology. The new campaign assumed gigantic proportions as the interests of the philosophers (to foster Leninism and to validate their own role in the conflict) clashed with the interests of the physicists (to protect their discipline from the attacks of those ignorant of its needs and dictates). The philosophers sought major revisions in quantum mechanics and the theory of relativity to make them fully congruent with Leninist epistemology. The physicists, for their part, wanted a freer interpretation of the Leninist legacy so that it would be adaptable to modern physics.

Both parties clung stubbornly to their positions. No unity of professional philosophers and theoretical scientists was achieved—not during the 1930s. Stalin was too busy with political purges to pay closer attention to philosophical subtleties. That both sides of the bitter feud could receive a full public airing indicated that the political authorities were not yet ready to take sides in this confrontation between science and ideology. World War II—and the intensified nationalism generated by it—produced relative tranquillity on the ideological front and a softening in the philosophers' views on the high theory of modern science. The increased search for academic autonomy went hand in hand with the search for reduced ideological controls and intensified contact with Western scholarship.

The third stage (1946–1953) coincided with the waning years of the Stalinist reign. This was a period of full unity in Marxist philosophy of science, a unity achieved by silencing the scientists who contended that dialectical materialism rather than science should undergo internal changes. This was the period of bitter and incredible attacks on Albert

Einstein, Werner Heisenberg, Niels Bohr, and other giants of modern physics. T. D. Lysenko assumed dictatorial powers in the Academy's gigantic Department of Biological Sciences, packing it with his followers, scuttling many ongoing research projects, and making biological journals tools of his own aggrandizement. The scientific validity of his theories, alien to the scientific community, was certified by the Central Committee of the Communist party and not by any consensus of scholarly opinion. Many disciplines underwent ideological purification that produced a merciless chastisement and demotion of scholars, bizarre claims for the superiority of national science, and massive institutional dislocations. I. P. Pavlov, whose general theoretical views were deeply embedded in a mechanistic orientation incompatible with Marxist thought, was now proclaimed a pure champion of dialectical materialism, and all branches of neurophysiology and psychology not anchored in his thought were dismissed as pseudoscientific. Since the identification of Pavlovian theory as a true expression of dialectical materialism was the result of a government edict rather than a consensus of scholars, it left little room for academic debate. Quantum chemistry, cybernetics, and genetics were made important only as telling indices of the decadence of Western scientific thought.

During the early phase of the Stalinist regime, individual scholars were punished not because their scientific ideas were unorthodox but because their real or alleged political associations were open to suspicion. In most cases the attacks on scientists were not attacks on science. During the later phase of Stalin's regime, scientists were attacked because their scientific ideas were deemed incompatible with dialectical materialism. The attacks and purges of the late 1940s and early 1950s were much tougher on science than on scientists. More scientists were executed or sent to prison during the earlier period; more sciences were abolished or distorted during the later phase.

The purges and attacks disrupted the intellectual momentum of Soviet scholarship and threatened to erode the moral fabric of Soviet science. A combination of circumstances, however, limited the scope and the depth of damage. Most sciences were not affected tangibly by the succession of ideological crusades and institutional disruptions. The government poured money into the Academy at a rapidly accelerating rate, even though the leading representatives of the academic community were involved in constant resistance to Stalin's ideological designs. Coming from many leading scientists and in many forms, resistance was a living force of great strength and moral determination. At a time when Lysenkoist hysteria had reached the peak of its destructive fury—in August 1948—V. S. Nemchinov defended genetics as part of "the golden fund of modern science." L. A. Orbeli displayed similar courage in criticizing the succession of ideological conferences that were planned in secret and thrust upon unsuspecting victims without providing sufficient time for them to prepare defense arguments and strategies.

There were other notable attempts to resist the Stalinist onslaught. In early 1952 V. N. Sukachev, editor of *Botanical Journal,* commissioned articles to reveal the fraudulent aspects of Lysenko's agrobiology. A. D. Landau chose to fight the ideological critics by ignoring their protestations, even their demands that he publicly renounce indiscretions in his statements on philosophical issues. V. A. Fok, G. I. Naan, and A. D. Aleksandrov defended Einstein's epoch-making contributions to science at a time when Stalinist philosophers emphasized the urgent need for a Soviet theory of relativity unburdened by heavy reliance on mathematical formalism and "idealistic" epistemology. Although the ideological regimentation of scholars, the disfiguration of the institutional makeup of the Academy, and the erosion of the moral foundations of science produced grave and deep-seated damages, these damages were, in the end, limited by yet another factor: the timely death of Stalin, which ushered in a period of moderation and concerted effort to undo the damages of Stalinist violence.

The post-Stalin era (1954–1970) coincided with two momentous developments: the post-Stalinist thaw, which produced a liberalizing trend, and the scientific and technological revolution of the mid-century decades. Step by step, the Stalinist union of science and ideology was dismantled, giving way to new developments that freed science from the real and potential damages of ideological interference. Philosophers lost their exclusive right to articulate the relationship of ideology to science, to criticize the grand theory of modern science, to chastise deviant scientists, and to denigrate the contributions of the giants of Western science. Dialectical materialism was opened to changes of major proportions.

In the days of Stalin, science had been called upon to bring the substance and the structure of its theory in tune with Soviet ideology. In the post-Stalin period, Marxist philosophy of science was recast to come closer to the spirit and the claims of modern science. For the first time, serious efforts were made to advance a branch of Marxist philosophy—

the logic of scientific inquiry—untrammeled by major ideological restrictions. New philosophers began to prefer topics far removed from acute theoretical disputes in the scientific community. Their job was not to judge the scientific validity of new theories but to translate the grand achievements of modern science into a scientific world view. The relics of the Stalinist era, typified by M. B. Mitin, were still very much in evidence, but now they received no help from shouting mobs that filled the galleries of academic halls, a practice exploited by Lysenko in his war on modern biology. They could no longer rely on patriotic slogans to sustain their philosophical claims. The ranks of philosophers were still saturated with vulgarizers of both science and philosophy (at the end of 1970, the Soviet Union boasted a total of 12,000 professional philosophers), but their impact on the academic community was of much less consequence. Only 10 percent of the books in philosophy were reviewed in journals.[2]

The philosophers of the Stalin era emphasized the unity of scientific theory, for without this unity, they thought, there could be no unity of science and ideology. The philosophers of the post-Stalin era emphasized the diversity of scientific theory—and the complementarity of opposite theories—as a reflection of the infinite complexity of natural and social reality and the inexhaustibility of the knowledge of the universe. Particularly at the frontiers of modern scientific development, Soviet scholars were guided by diverse orientations. There was no major theoretical orientation in modern physics, chemistry, astronomy, or biology that did not have representatives in the Soviet Union. This did not prevent Soviet scholars from advancing strong national orientations in science—such as the constructive orientation in the foundations of mathematics, the Biurakan theory of a nonstationary universe in cosmology, and the unique theoretical construction of neuropsychology (a synthesis of Pavlov's neurophysiology and Vygotskii's social psychology).

After the mid-1950s there was a pronounced change in the attitude toward Western philosophy. Criticism still appeared, but there was also a tendency to search for positive contributions. The Academy played the leading role in publishing selected works—sometimes in multivolume series—of such giants of Western philosophical thought as Leibniz, Berkeley, Hume, and Kant. It was Lenin who, in *Materialism and Empirio-Criticism*, had argued that there was nothing positive in neo-Kantianism or neopositivism that could be incorporated into dialectical materialism. Since the early days of the Soviet system, Marxist philosophers had waged an uncompromising war against these orientations. But in the late 1950s the negative attitude underwent considerable modification. Selected works of the leading representatives of these and related orientations began to appear in Russian translation, and a vigorous movement stood in clear opposition to the previous practice of total rejection of the contributions of these philosophers. Henri Poincaré, one of the main targets of Lenin's wrath, was now placed among the leading pioneers of modern science and the scientific world view. Bertrand Russell became a source of thoughtful citations depicting the involved and puzzling dynamics of the interaction of science and philosophy. His articulation of logicism as a distinct orientation in the foundations of mathematics was no longer totally dismissed as misrepresenting the relations between mathematics and logic. The posthumous publication of V. I. Vernadskii's essays bestowed new respectability on at least some of the ideas of Henri Bergson and Alfred North Whitehead.

Dramatic as it was, the reversal in the relationship of ideology to science—and the role of political authorities in the academic community—did not represent a structural break with the institutional limitations inherited from the Stalin era. The academic community continued to be handicapped in several areas of operation. The idea of autonomous scientific associations defending the professional interests of scientists and voicing the views of the scientific community on public policies continued to be a far cry from the stark reality of academic existence. Scientists were encouraged to discuss the moral dilemmas of science outside the Soviet Union, touching on such questions as the role of the scholarly community in fighting the threats of ecological despoilment, nuclear and germ warfare, and genetic engineering. But serious obstacles prevented Soviet scholars from raising the same questions about their homeland, a clear indication of censorship limitations. All serious criticism of real or potential abuses of science in the Soviet Union was disarmed by the a priori assumption that harmony reigned between Soviet socialism and the highest moral principles of science.

The Lysenko phenomenon—the most critical moral and professional dilemma faced by the Soviet scientific community—did not attract the postmortem attention it called for. The collapse of Lysenko's empire was not followed by a publicly aired inquiry into the scientific, ideological, and moral complexity of Lysenkoism and the conditions that made such a monstrous abuse of science possible. No public discussion of institutional changes necessary for preventing the rise of future Lysenkos was recorded. In 1968 I. T. Frolov, a Marxist theorist, published a detailed account of the ups and downs of genetic theory in his country and of the salient points in the history of the Lysenko empire.[3] In his opinion, Lysenko's errors were strictly in the realm of methodology and episte- mology, not in the realm of scientific ethics. Lysenko's faulty attitude toward the role of hypotheses in science, the relationship of theory to experiment, and the importance of conflict in the progress of scientific theory had led him, according to Frolov, to deny the fundamental com- patibility of dialectical materialism and genetics. That Lysenko built his "science" by depending on the fiat of political authorities rather than on the consensus of scholarly opinion and the weight of experimental verification did not engage Frolov's attention. Nor did he concern himself with Lysenko's regular practice of hiding the data on which he based his agrotechnical claims and of accusing his adversaries of unpatriotic deeds. Frolov's line of argumentation was followed by N. P. Dubinin a few years later. After that, Lysenkoism ceased to be a topic of public debate. Silence became the main tool in the effort to conceal the judgment of history that led to the demise of Lysenko.[4] It was part of an effort to forget that the nightmare of Lysenkoism had ever happened and to open a new leaf in the annals of Soviet science.[5]

The liberalizing atmosphere of the post-Stalinist thaw and the triumphs in space exploration and in harnessing nuclear energy helped scientists acquire a stronger and more articulate voice in shaping national science policy. These successes made it possible for the scientific com- munity to mount a search for wider domains of independent action in the pursuit of professional duties. Scholars marshaled elaborate but dif- fuse arguments in favor of the unqualified authority of the scientific community in fashioning the models of institutional response to the needs of science, in protecting and interpreting the values and norms embodied in the ethos of science, in certifying scientific knowledge, and in serving as the supreme guardian of the scientific legacy. The main lines of the struggle for wider areas of academic autonomy were drawn by Kapitsa, Artsimovich, Semenov, and other leading figures in the Academy and were kept alive by the quiet but persistent actions of the Academy's central administration. These efforts received the most extreme expression in the pronouncements of a growing group of polit- ical dissidents led and inspired by A. D. Sakharov, a full member of the Academy, who, apprehensive of inadequate constitutional safeguards against a recurrence of Stalinism, contended that nothing short of a fundamental restructuring of institutional ties between science and the society at large could make the Soviet scientific community a healthy social and cultural force. In the existing censorship setup these dissidents saw a major threat to the scientific community.

New lines of research in several disciplines produced subtle but formidable arguments in favor of a wider base of academic autonomy. Burgeoning activities in the new branches of logic helped to lift parts of the philosophy of science above the framework of dialectical materialism—and above the lines of ideological interference with the work of the scholarly community. By emphasizing the paramount role of the internal impulse in the development of science, the historians of science defended and elaborated the idea of the intellectual autonomy of science; they too tried to move their specialty a safe distance from ideological patronage. The same applied to psychologists, who emphasized the strength of nonconformism as a vital source of scientific creativity.

When the spokesmen for the scholarly community rejected M. Berthelot's notion of the close dependence of ethics on science, they were rejecting the ideologically dictated intellectual imperialism of science. They worked on the safe assumption that the larger the realm of science governed by a monolithic and all-pervasive ideology, the smaller the domain of academic autonomy. And for the same reason these spokesmen rejected the view of science as a primary, and art as a derivative (or superstructural), component of modern society. It is difficult to relate these new arguments, insights, and developments directly to the concrete achievements in the struggle for a wider base of the social individuality and relative independence of the scientific community. Nevertheless, it is abundantly clear that scientists received a new sense of internal unity and community identification and a new liveliness and determination in their search for self-assertion. Their real strength lay in making the advances in modern science the best argument for giving the scientific community broader rights in organizing its professional work and in interpreting its duties to both science and the society at large. Although the scientists and their spokesmen made a profound impact on the realities of Soviet science, their major achievement was in defining the ideal conditions for uninterrupted growth of scientific thought.

The post-Stalinist thaw produced tangible changes in the institutional and intellectual setting of the Academy. The crudest control mechanisms born in the Stalin era were either removed or shorn of instrumentalities that made them sources of arrogant attacks on the moral integrity of both science and the scientific community. The break with the Stalinist tradition, however, did not reach all the ramifications of the academic structure. While the criteria for selecting new members of the Academy had shown marked improvement, they continued to be marred by glaring imperfections. In some notable cases the Academy continued to give high positions to persons who were much better known for their ideological diligence than for their scholarly achievement. Such individuals made it impossible for the Academy to justify its claim to be the true forum of the best representatives of Soviet scholarship. The most notable representatives of this group—L. F. Il'ichev, P. N. Fedoseev, G. P. Frantsov, and A. M. Rumiantsev—were concerned with simplified generalities of dialectical and historical materialism, gracing patriotic occasions with florid speeches on the achievements of Soviet science, and rebuilding the bridges between science and philosophy destroyed by Stalinist ideologues in their ruthless war on some of the greatest achievements of modern science. They protected the high theory of contemporary science from attacks by philosophers, but they also maintained the supremacy of the branches of philosophy closest to ideology. Their academic duties were more in the realm of politics than in the realm of scholarship.

Il'ichev and his group were responsible for calculated efforts to preserve a place of honor in the Academy for M. B. Mitin, one of the most rigid and ubiquitous philosophers of Stalinism, who took on the responsibility of weaving Lysenko's "agrobiology" into dialectical materialism and Russian patriotism (as defined by Stalin) and who led Stalinist troops against the "agnosticism," "subjectivism," "relativism," and "decadence" of modern physics. In the post-Stalin era, Mitin fought a new generation of Soviet philosophers determined to take solid enclaves of their discipline outside the realm of ideological control. In July 1961 a united session of the Department of Economics, Philosophy and Law and the editorial board of *Voprosy filosofii* celebrated Mitin's sixtieth birthday. On that occasion P. N. Fedoseev, a new ideological leader in philosophy, delivered the main oration, in which he praised Mitin for major "contributions" to the formation and evolution of the Leninist phase of Marxist thought and to the application of the principles of dialectical materialism to modern science.[6]

In one important respect the Academy did not meet the challenge of its own history: it did not undertake a systematic, comprehensive, and critical analysis of its past, particularly during the Soviet period. It produced a long line of studies centered on its history, mainly in conjunction with important anniversary commemorations. But these writings were intended to add luster to festive occasions rather than provoke a realistic assessment of the twists and turns of academic history. Even when they were not part of anniversary occasions, most studies amounted to uncritical and laudatory reviews of achievements in individual fields of scholarly endeavor. If on occasion the Academy did assume a critical stance toward certain features of its past, it acted under the pressure of government authorities and in defense of interests not associated with the welfare of science as seen by the scientific community. By discouraging critical and objective studies of its history, the Academy deprived itself of a valuable opportunity to learn not only from the achievements but also from the errors of the past. Such studies could have prevented the rise of Lysenko, the foremost enemy of science and the creator of the blackest digression in the long—and for the most part admirable—history of the Academy of Sciences.

NOTES

In citing works in the notes, short titles have generally been used. The following abbreviations have been used for some titles.

AINT   *Arkhiv Instituta Istorii nauki i tekhniki*
BMOIP   *Biulleten' Moskovskogo Obshchestva ispytatelei prirody*
Bol   *Bol'shevik*
BSE   *Bol'shaia Sovetskaia Entsiklopediia*
BZh   *Botanicheskii zhurnal*
EiM   *Estestvoznanie i marksizm*
FE   *Filosofskaia entsiklopediia*
FN   *Filosofskie nauki*
FNiT   *Front nauki i tekhniki*
I-M   *Istorik-Marksist*
IRAN   *Izvestiia Rossiiskoi Akademii nauk*
LG   *Literaturnaia gazeta*
LP   *Leningradskaia pravda*
LRN:bio   I. V. Kuznetsov, ed., *Liudy russkoi nauki*, 2d ed. (biology volume)
LRN:geo   I. V. Kuznetsov, ed., *Liudy russkoi nauki*, 2d ed. (geology and geography volume)
LRN:mat   I. V. Kuznetsov, ed., *Liudy russkoi nauki*, 2d ed. (mathematics and physical science volume)
MiE   *Marksizm i estestvoznanie*
NiR   *Nauka i ee rabotniki*
NiZh   *Nauka i zhizn'*
NR   *Nauchnyi rabotnik*
NM   *Novyi mir*
ON   K. V. Ostrovitianov et al., eds. *Organizatsiia nauki v pervye gody sovetskoi vlasti (1917–1925): Sbornik dokumentov*, Leningrad, 1968.

OSN   B. E. Bykhovskii et al., eds., *Organizatsiia sovetskoi nauki v 1926–1932
           gg.:Sbornik dokumentov*, Leningrad, 1974
"Otchet"   "Otchet deiatel'nosti Imperatorskoi Akademii Nauk po Fiziko-mate-
           maticheskomu i Istoriko-filologicheskomu otdeleniiam"
Otchet-I   *Otchet deiatel'nosti Imperatorskoi Akademii nauk po Fiziko-matematiches-
           komu i Istoriko-filologicheskomu otdeleniiam*
Otchet-R   *Otchet Rossiiskoi Akademii nauk*
Otchet-S   *Otchet Akademii nauk SSSR*
PSZ   *Polnoe sobranie zakonov Rossiiskoi Imperii*
PZM   *Pod znamenem marksizma*
RBS   *Russkii Biograficheskii Slovar*
SR   *Slavic Review*
SRiN   *Sotsialisticheskaia rekonstruktsiia i nauka*
TIIE   *Trudy Instituta Istorii estestvoznaniia*
TIIE-T   *Trudy Instituta Istorii estestvoznaniia i tekhniki*
UFN·   *Uspekhi fizicheskikh nauk*
UMN   *Uspekhi matematicheskikh nauk*
UZMU   *Uchenye zapiski Moskovskogo Universiteta*
VAN   *Vestnik Akademii nauk SSSR*
VAR   *VARNITSO*
VF   *Voprosy filosofii*
VIE-T   *Voprosy istorii estestvoznaniia i tekhniki*
VIMK   *Vestnik Istorii mirovoi kul'tury*
VKA   *Vestnik Kommunisticheskoi Akademii*
VL   *Voprosy literatury*
VLU   *Vestnik Leningradskogo universiteta*
VMU   *Vestnik Moskovskogo universiteta*
ZAN   *Zapiski Imperatorskoi Akademii nauk*
ZhMNP   *Zhurnal Ministerstva Narodnogo prosveshcheniia*
ZhOB   *Zhurnal Obshchei biologii*

INTRODUCTION

1. T. S. Kuhn, *Structure of Scientific Revolutions*, p. 102.
2. R. K. Merton, *Sociology of Science*, pp. 268–269.
3. M. von Laue, *History of Physics*, p. 6.
4. A. Herzen, *Selected Philosophical Works*, p. 30.

ONE: ANCESTRY

1. P. P. Pekarskii, *Istoriia*, vol. 1, p. xxxvi.
2. P. P. Pekarskii, *Nauka*, vol. 2, p. 388.
3. C. Huygens, *Celestial Worlds*, p. 13.

4. For detailed information on borrowed words, see N. A. Smirnov, "Zapadnoe vliianie."

5. According to M. M. Bogoslovskii, Leibniz looked at Russia as "an endless country of Eastern barbarians," an ideal world for linguistic and ethnographic studies and a suitable area for political experimentation (*Petr I*, vol. 2, pp. 112, 324).

6. For details on relations between Leibniz and Peter I, see W. Guerrier, *Leibniz* and *Otnosheniia Leibnitsa*; and L. Richter, *Leibniz*. See also A. I. Andreev, "Osnovanie," pp. 285–286; Iu. Kh. Kopelevich, *Osnovanie*, pp. 32–38; P. P. Pekarskii, *Istoriia*, vol. 1, pp. xxi–xxiv; A. S. Lappo-Danilevskii, *Petr Velikii*, pp. 17–18; and V. I. Chuchmarev, "G. V. Leibnits."

7. P. P. Pekarskii, *Istoriia*, vol. 1, p. lix.

8. *Materialy dlia istorii Imperatorskoi Akademii*, vol. 1, pp. 5–6.

9. "Biographische Versuche," pp. 101–104; D. M. Lebedev, *Geografiia*, pp. 75–77; A. N. Pypin, "Russkaia nauka," part 1, p. 225; *Polnoe sobranie uchenykh puteshestvii*, vol. 1, pp. vi–vii.

10. For comments on, and a reproduction of, the map of the Caspian Sea, see G. Delisle l'Aîné, "Remarques," pp. 319–322.

11. K. S. Veselovskii, *Istoricheskoe obozrenie*, p. 4, and "Petr Velikii," p. 28.

12. *Spiritual Regulation*, p. 30.

13. A. N. Pypin, *Istoriia russkoi etnografii*, vol. 1, pp. 113–116.

14. For the text of the project, see *Materialy dlia istorii Imperatorskoi Akademii*, vol. 1, pp. 14–22; and K. V. Ostrovitianov et al., *Istoriia*, vol. 1, pp. 429–435.

15. P. P. Pekarskii, *Istoriia*, vol. 1, p. xlix.

16. For comments on the lowering of standards in selecting new academicians, see P. P. Pekarskii, *Istoriia*, vol. 1, p. lxi.

17. P. Ferliudin, *Istoricheskii obzor*, vol. 1, p. 5.

18. O. Spiess, *Leonhard Euler*, pp. 107–108.

19. For a concentrated and detailed discussion of Euler's work in the Academy, see S. N. Chernov, "Leonard Eiler." See also E. Ia. Kol'man, "Vklad."

20. For a brief discussion of the organization and achievements of this expedition, see L. G. Kamanin, "Velikaia Severnaia ekspeditsiia."

21. B. N. Menshutkin, *Russia's Lomonosov*, p. 16.

22. V. N. Tatishchev, *Izbrannye proizvedeniia*, p. 113.

23. D. A. Tolstoi, *Akademicheskii universitet*, p. 7

24. V. N. Tatishchev, "Razgovor," p. 112.

25. A. Satkevich, "Leonard Eiler," pp. 483–484.

26. Lomonosov's views on the Newtonian legacy are discussed in M. I. Radovskii, "N'iuton i Rossiia," pp. 99–102; and in P. S. Kudriavtsev, "Lomonosov," pp. 34–51.

27. S. I. Vavilov, "Mikhail Vasil'evich Lomonosov," p. 21. For an analysis of

Lomonosov's work in physics and chemistry, see B. N. Menshutkin, "Lomonosov kak fiziko-khimik" and *Russia's Lomonosov*, pp. 48–72; H. M. Leicester, Introduction; B. M. Kedrov, "Vazhneishie idei," pp. 25–30; B. G. Kuznetsov, *Tvorcheskii put'*, chaps. 10, 11, 14, 20; and V. Boss, *Newton and Russia*, chaps. 17–19.

28. As cited by H. M. Leicester, Introduction, p. 41.

29. Iu. Kh. Kopelevich, "Pervye otkliki," pp. 242–249.

30. For a careful, thorough, and well-documented Western analysis of Lomonosov's work in physics and chemistry, see H. M. Leicester, Introduction.

31. The best account of Lomonosov's activities in the Academy is presented in P. P. Pekarskii, *Istoriia*, vol. 2, pp. 259–892.

32. As cited in Iu. I. Solov'ev and N. N. Ushakova, *Otrazhenie*, p. 90.

33. Ibid., p. 89.

34. T. I. Rainov, "Russkoe estestvoznanie," p. 345.

35. *Materialy dlia istorii Imperatorskoi Akademii*, vol. 5, pp. 729, 743.

36. For more recent comments, see D. Anuchin, "Geografiia," pp. 108–109; and T. I. Rainov, "Teoriia i praktika," p. 19.

37. N. Bashkina et al., *United States and Russia*, p. 8.

38. C. Arness, *Naturalist*, p. 56.

39. For an excellent biography of Krasheninnikov, see A. I. Andreev, "Stepan Petrovich Krasheninnikov."

40. R. W. Home, Preface, p. ix.

41. N. Bashkina et al., *United States and Russia*, p. 3.

42. R. W. Home, Introduction, p. 47. A brief summary of Aepinus's scientific work is presented in Ia. G. Dorfman and M. I. Radovskii, "V. Franklin," pp. 302–312.

43. R. W. Home, Introduction, pp. 189–224.

44. Iu. Kh. Kopelevich, "Pervye otkliki," p. 241.

45. P. P. Pekarskii, *Istoriia*, vol. 2, p. xiv.

46. For more details, see M. I. Radovskii, *M. V. Lomonosov*, pp. 192–218.

47. Ibid., pp. 219–220.

48. A detailed history of the Geography Department is presented in V. F. Gnucheva, *Geograficheskii departament*. For a summary treatment, see I. P. Gerasimov, *Ocherki*, pp. 47–60.

49. V. F. Gnucheva, *Geograficheskii departament*, pp. 52–58; K. S. Veselovskii, *Istoricheskoe obozrenie*, pp. 13–14; Iu. Kh. Kopelevich, "Peterburgskaia Akademiia," p. 96.

50. K. S. Veselovskii, *Istoricheskoe obozrenie*, p. 7.

51. M. I. Radovskii, *M. V. Lomonosov*, pp. 222–224.

52. For details on the Russian translation of Buffon's popular work, see M. I. Sukhomlinov, *Istoriia*, vol. 2, pp. 210–230. Buffon's influence in Russia is treated in I. I. Kanaev, *Zhorzh Lui Leklerk de Biuffon*, chap. 11.

53. V. Boss, *Newton and Russia*, pp. 211–212.

54. M. J. A. N. de C. Condorcet, "Elogium," p. lxiv.

55. P. P. Pekarskii, "Ekaterina II," p. 66.

56. T. I. Rainov, "Russkoe estestvoznanie," pp. 341–342.

57. For basic biographical data on Pallas, see F. Keppen, "Pallas, Petr Simon"; and G. P. Dement'ev, "Petr Simon Pallas."

58. G. Cuvier, *Recueil des éloges*, vol. 2, p. 157.

59. For pertinent comments, see T. I. Rainov, "Russkoe estestvoznanie," p. 25.

60. For a discussion of Pallas as a precursor of Darwin, see B. E. Raikov, *Russkie biologi-evoliutsionisty*, vol. 1, pp. 42–103. See also E. Stresemann, "Leben und Werk," p. 255. Unlike Raikov, Stresemann would not go so far as to identify Pallas as "a pioneer of the transformation theory." For a detailed study, see F. Keppen, "Uchenye trudy."

61. "Biographische Versuche," p. 112. Wolff's contributions to the evolutionary theory are discussed in B. E. Raikov, *Russkie biologi-evoliutsionisty*, vol. 1, pp. 106–193. For a discussion of Wolff's activities in the St. Petersburg Academy, see A. E. Gaisinovich, *K. F. Vol'f*, chap. 11. See also N. F. Utkina, *Estestvennonauchnyi materializm*, pp. 187–188.

62. M. I. Sukhomlinov, *Istoriia*, vol. 4, pp. 2–5. For an endorsement and elaboration of this view, see A. N. Pypin, *Istoriia*, vol. 1, pp. 132–134. For a more modern endorsement, see G. E. Pavlova and A. S. Fedorov, *Mikhail Vasil'evich Lomonosov*, p. 267.

63. The influence of "Eulerism" on Russian scientific thought in the eighteenth century is analyzed in T. I. Rainov, "Russkoe estestvoznanie," pp. 363–369.

64. B. G. Kuznetsov, "Vvedenie," pp. 209–211. For information on the society's publication activities, see V. V. Oreshkin, *Vol'noe ekonomicheskoe obshchestvo*, p. 31.

65. M. I. Sukhomlinov, *Istoriia*, vol. 8, p. 366.

66. These changes are surveyed in K. V. Ostrovitianov et al., *Istoriia*, vol. 1, part 1, pp. 168–170. See also G. D. Komkov et al., *Akademiia*, 2d ed., vol. 1, pp. 138–154.

67. *PSZ*, series 1, vol. 17, no. 12, p. 750.

68. Details on Orlov's activities as director of the Academy are presented in V. Orlov-Davydov, "Biograficheskii ocherk," pp. 301–395.

69. K. Stählin, *Aus den Papieren*, p. 412; M. I. Sukhomlinov, "Piatidesiatiletnii i stoletnii iubilei," pp. 5–6.

70. B. E. Raikov, *Akademik Vasilii Zuev*, pp. 187–202.

71. K. S. Veselovskii, "Otnosheniia," pp. 237–238. According to the imperial decrees of September 16 and October 22, 1796, the censorship committees, stationed at the five major custom houses, consisted of three members, appointed by the Synod, the Senate, and the Academy of Sciences or Moscow University.

The Academy and Moscow University were obliged to appoint "prominent scholars" to censorship posts.

72. V. F. Gnucheva, *Geograficheskii departament*, pp. 92–99.

73. M. I. Sukhomlinov, *Rech'*, pp. 16–17.

74. For pertinent comments, see V. Boss, *Newton and Russia*, pp. 236–237.

75. A. N. Pypin, "Russkaia nauka," part 2, p. 549.

76. M. I. Sukhomlinov, *Issledovaniia*, vol. 2, p. 68.

77. For comments on the new charter, see M. I. Sukhomlinov, *Istoriia*, vol. 2, pp. 371–372.

78. For the text of the charter, see *PSZ*, vol. 27, pp. 786–800.

79. M. I. Sukhomlinov, *Istoriia*, vol. 2, pp. 361–367. See also V. P. Zubov, *Istoriografiia*, p. 163.

80. M. I. Sukhomlinov, *Istoriia*, vol. 2, pp. 363–364.

81. "Pis'ma K. F. Gaussa," pp. 220, 237.

82. M. I. Sukhomlinov, *Istoriia*, vol. 2, pp. 373–374.

83. *PSZ*, vol. 27, p. 786.

84. For more details, see K. S. Veselovskii, "Otnosheniia," pp. 5–9.

85. A. N. Pypin, *Obshchestvennoe dvizhenie*, p. 109.

86. A. Shtorkh and F. Adelung, *Sistematicheskoe obozrenie*, vol. 1, p. 340; vol. 2, p. 204.

87. M. I. Sukhomlinov, *Rech'*, p. 23. See also Iu. I. Solov'ev and N. N. Ushakova, *Otrazhenie*, pp. 14–15. In 1802 A. L. Schlözer, former member of the St. Petersburg Academy, published his personal memoirs, in which he made frequent reference to Lomonosov. Labeling him "the creator of Russian poetry," Schlözer considered Lomonosov a true genius but concluded that his tendency to work in too many fields had prevented him from making outstanding contributions (*Öffentliches und Privat-Leben*, pp. 217–222).

88. As cited in M. I. Sukhomlinov, *Materialy dlia istorii obrazovaniia*, vol. 2, p. 5.

89. L. A. Tarasevich, "Estestvoznanie," p. 294.

90. M. I. Sukhomlinov, *Materialy dlia istorii obrazovaniia*, vol. 1, p. 105; Z. A. Kamenskii, "I. Kant," pp. 52–63; A. N. Pypin, *Obshchestvennoe dvizhenie*, p. 107.

91. Z. A. Kamenskii, *Russkaia filosofiia*, p. 130.

92. M. I. Sukhomlinov, *Materialy dlia istorii obrazovaniia*, vol. 1, pp. 105–106.

93. M. I. Sukhomlinov, *Materialy dlia istorii obrazovaniia*, vol. 2, pp. 32–64. See also M. T. Nuzhin et al., *Kazanskii universitet*, pp. 10–12; and S. V. Rozhdestvenskii, *Istoricheskii obzor*, pp. 111–127. D. P. Runich's attack on "dangerous" professors at St. Petersburg Univeristy is described in M. I. Sukhomlinov, *Issledovaniia*, vol. 1, pp. 301–307.

94. D. Presniakov, "Ideologiia," p. 80.

95. N. N. Zagoskin, *Istoriia*, vol. 3, pp. 261–262.

96. S. V. Rozhdestvenskii, *Istoricheskii obzor*, pp. 334–335.

97. P. H. Fuss, *Coup d'oeil historique*, p. 7.

98. K. V. Ostrovitianov et al., *Istoriia*, vol. 2, p. 22.

99. G. K. Skriabin et al., *Ustavy*, pp. 114–115.

100. A. I. Gertsen, *Sobranie sochinenii*, vol. 3, pp. 100–116.

101. Ibid., p. 101; P. S. Shkurinov, *Pozitivizm*, pp. 70–73.

102. A. Herzen, *Selected Philosophical Works*, p. 77.

103. Ibid., p. 88.

104. "État du personnel en 1849," pp. i-x.

105. K. A. Timiriazev, "Probuzhdenie," p. 1.

106. K. S. Veselovskii, "Otchet": 1869, p. 2.

107. For details, see G. V. Bykov and Z. I. Sheptunova, "Nemetskii 'Zhurnal khimii.'"

108. For pertinent comments on D. I. Pisarev's views on cultural values encouraging the advancement of science, see E. Solov'ev, *D. I. Pisarev*, pp. 206–224.

109. A. Coquart, *Dmitri Pisarev*, pp. 336–341.

110. A. P. Shchapov, *Sotsial'no-pedagogicheskie usloviia*, pp. 143–145.

111. P. N. Berkov, "Lomonosovskii iubilei," p. 225.

112. Ia. Grot', "Dva slova," p. 150; E. V. Soboleva, *Bor'ba*, p. 144.

113. E. V. Soboleva, *Bor'ba*, pp. 150–151.

114. K. S. Veselovskii, *Istoricheskoe obozrenie*, pp. 39–40.

115. For a review of biographical literature on Lomonosov during the 1860s, see M. I. Radovskii, *M. V. Lomonosov*, pp. 238–243.

116. I. I. Sreznevskii, "Otchet o pervom prisuzhdeniiu," pp. 195–196.

117. "Lichnyi sostav," pp. 103–105.

118. For details on von Baer's activities outside the Academy, see B. E. Raikov, *Karl Ber*, pp. 335–370.

119. K. S. Veselovskii, "Otchet": 1865, pp. 47–48.

120. K. S. Veselovskii, "Otchet": 1862, p. 26.

121. J. F. Brandt, "Zoologicheskii i zootomicheskii muzei," p. 34.

122. As cited in S. Shtraikh, "I. I. Mechnikov," p. 111.

123. K. V. Ostrovitianov et al., *Istoriia*, vol. 2, p. 272; "O proekte ustava," pp. 149–150. For details on the preparatory work for the new charter, see "Materialy dlia novogo ustava," pp. 141–281.

124. "Izvlechenie iz zhurnalov," pp. 596–597.

125. "Zamechaniia," pp. 185–197; "O proekte ustava," pp. 142–165; E. V. Soboleva, *Bor'ba*, pp. 183–185.

126. A. E. Gaisinovich, "Biolog-shestidesiatnik," p. 391; A. Vucinich, "Russia," pp. 237–239; B. E. Raikov, "Iz istorii darvinizma," pp. 68–71; S. L. Sobol', "Iz istorii bor'by."

127. K. S. Veselovskii, "Otchet": 1869, pp. 2–3.

128. See particularly von Baer's "Zum Streit," pp. 1986–1988, and *Reden*, vol. 2, pp. 235–480.

129. C. Darwin, *Descent of Man*, pp. 175–176. See also L. Ia. Bliakher, "Ch. Darvin i brat'ia Kovalevskie," pp. 66–73.

130. K. A. Timiriazev, "Probuzhdenie," pp. 6–7.

131. Ibid,. p. 7. The famous neurophysiologist I. M. Sechenov wrote in 1883 that the frequent claims that Russian science had entered a phase of decline after the 1860s could not be justified by a careful study of relevant statistical data ("Nauchnaia deiatel'nost'," p. 330).

132. V. Modestov, "Russkaia nauka," p. 89. For details, see H. M. Leicester, "Mendeleev." See also A. F. Plate, "V. V. Markovnikov," p. 297.

133. A. M. Butlerov, *Sochineniia*, vol. 3, p. 128.

134. B. Meilakh, "Posleslovie," p. 196.

135. D. I. Mendeleev, "Kakaia zhe Akademiia nuzhna?" p. 184.

136. P. Miliukov, "Universitety," p. 796; V. V. Markovnikov, "Istoricheskii ocherk," p. 9. The philosopher Sergei Trubetskoi wrote in 1897 that the seventeen years of the new charter had rendered universities totally incapable of maintaining even a semblance of internal order without help from externally generated coercion. D. A. Tolstoi, he said, had transformed the Ministry of Public Education into a police institution (*Sobranie sochinenii*, vol. 1, pp. 86, 263–265).

137. A. Derman, 'Akademicheskii Intsident,' pp. 53–58.

138. For details, see G. A. Kniazev, "Maksim Gor'kii," pp. 23–24.

139. A. V. Predtechenskii and A. V. Kol'tsov, "Iz istorii," pp. 82–85; B. N. Menshutkin, *Zhizn'*, pp. 190–191; V. N. and I. I. Liubimenko, "Ivan Parfen'evich Borodin," pp. 26–27.

140. B. N. Menshutkin, *Zhizn'*, p. 190.

141. Ibid., p. 189. V. I. Vernadskii, *Stranitsy*, pp. 200–201; I. I. Mochalov, *Vladmir Ivanovich Vernadskii*, pp. 150–152. For similar pronouncements, see K. A. Timiriazev, *Nauka i demokratiia*, pp. 17–27; and N. Losskii, "Osvobozhdenie nauki," pp. 408–412.

142. V. I. Vernadskii, "Tri resheniia," pp. 172–173.

143. V. I. Vernadskii, *Pis'ma*, pp. 16–19.

144. V. I. Vernadskii, *Ocherki i rechi*, vol. 2, pp. 43–44.

145. James R. Millar, *Soviet Socialism*, p. 5.

146. For biographical data on Shaniavskii, see N. Speranskii, *Voznikovenie*, pp. 6–20. For information on Lesgaft, see S. Metal'nikov, "Petr Frantsevich Lesgaft," pp. 53–56.

147. N. Speranskii, *Voznikovenie*, pp. 20–21.

148. P. P. Lazarev, *Ocherki*, p. 66.

149. *Vremennik*, p. 45.

150. M. von Laue, *History of Physics*, p. 89.

151. M. I. Rostovtsev, "Russia's Contribution," p. 602.

152. P. P. Lazarev, *Ocherki*, p. 166.

153. Ibid., pp. 72–74.

154. M. S. Bastrakova, "Organizatsionnye tendentsii," pp. 162–163.

155. I. Pospelov, "Iz zhizni," p. 111.

156. "Moskovskii Nauchnyi institut," pp. 281–282.

157. See particularly V. I. Vernadskii's "O gosudarstvennoi seti."

158. V. I. Vernadskii, *Ocherki i rechi*, vol. 1, p. 146.

159. A. F. Ioffe, *Moia zhizn'*, p. 18; M. S. Sominskii, *Abram Fedorovich Ioffe*, pp. 190–192.

160. J. C. McClelland, *Autocrats*, pp. 45–46.

161. O. D. Khvol'son, *Znanie i vera*, pp. 3–15.

162. R. E. Regel', "Selektsiia," pp. 446–496; G. Mendel, *Opyty*.

163. A. V. Kol'tsov, "Proekty," pp. 294–295.

164. D. Anuchin, "Geografiia," p. 122.

165. V. I. Vernadskii, "Radievye instituty," p. 252.

166. *Otchet-I*: 1916, pp. 322–373.

167. B. A. Lindener, *Raboty*, p. 2.

168. V. I. Vernadskii, *Ocherki i rechi*, vol. 1, p. 157.

169. S. Nagibin, "Russkoe Botanicheskoe obshchestvo"; A. Riabinin, "O deiatel'nosti"; A. P., "Pervyi s'ezd"; and S. K., "Pervyi Vserossiiskii astronomicheskii s'ezd."

170. Nik. Kulagin, "Ob'edinenie," p. 1080.

171. For a discussion of the general features of the prerevolutionary heritage in Russian science on both institutional and disciplinary levels, see L. R. Graham, "Development of Science Policy," pp. 13–19.

TWO: THE SEARCH FOR NEW IDENTITY

1. S. Fitzpatrick, *Commissariat*, p. 73.

2. M. K. Korbut, *Kazanskii gosudarstvennyi universitet*, vol. 2, p. 303.

3. L. V. Ivanova, *Formirovanie*, p. 23.

4. V. P. Volgin, "Sovetskaia vlast'," p. 22.

5. P. S. Aleksandrov et al., *Istoriia Moskovskogo universiteta*, vol. 2, p. 81. For comments on the general role and successes of workers' faculties, see J. C. McClelland, "Bolshevik Approaches," pp. 825–826.

6. A. S. Butiagin and Iu. A. Saltanov, *Universitetskoe obrazovanie*, p. 52.

7. J. C. McClelland, "Bolshevik Approaches," p. 825.

8. V. I. Vernadskii, "O nauchnoi rabote," p. 5.

9. A. F. Ioffe and A. T. Grigor'ian, *Sbornik k stoletiiu Maksa Planka*, p. 77.

10. V. I. Vernadskii, "O gosudarstvennoi seti."

11. *Gosudarstvennye universitety*, p. 109.

12. For details on the nature and dynamics of sectoral institutes, see B. I. Kozlov, *Organizatsiia i razvitie*, pp. 8–9.

13. R. Olby, *Double Helix*, p. 17.

14. N. K. Kol'tsov, "Predislovie," p. 4.

15. A. F. Kononkov and A. N. Osinovskii, "Iz istorii," p. 315.

16. S. V. Kalesnik, *Geograficheskoe obshchestvo*, pp. 358–360.

17. V. V. Kozlov, *Ocherki*, p. 51.

18. S. A. Fediukin, *Velikii Oktiabr'*, pp. 197–207.

19. *Obshchestvennyi vrach*, 1918, nos. 5–6, pp. 50–51 (as cited by S. A. Fediukin, *Velikii Oktiabr'*, p. 189).

20. O. D. Khvol'son, "Chto dal Oktiabr'," p. 22.

21. *Priroda*, 1928, no. 2, p. 174.

22. T. P. Kravets, "VI Vsesoiuznyi s'ezd fizikov," p. 915.

23. *Otchet-S:* 1926, vol. 2, pp. iii–iv.

24. B. Zbarskii, "Na povorote," p. 67.

25. V. Vladimirskii, "III. Vserossiiskii s'ezd."

26. N. G. Fradkin, *Ocherki*, p. 9.

27. N. Ia. Marr, "Kraevedcheskaia rabota," p. 13.

28. B. A. Lindener, "Issledovatel'skaia rabota," p. 19.

29. T. V. Staniukovich, *Etnograficheskaia nauka*, p. 172.

30. N. V. Grave, "Kraevednye uchrezhdeniia," p. 127.

31. E. A. Preobrazhenskii, "Blizhaishie zadachi," p. 7. See also "Protokol Obshchego sobraniia," pp. 399–400.

32. J. Shapiro, *History of the Academy*, p. 16; I. S. Smirnov, *Lenin*, pp. 299–302.

33. M. N. Pokrovskii, "Primechaniia," p. 39; *Piat' let vlasti sovetov*, p. 509.

34. V. I. Nevskii, "Restavratsiia idealizma," pp. 119–122. For relevant biographical data related to this period of Vernadskii's life and work, see K. E. Bailes, "Science, Philosophy and Politics," pp. 281–285.

35. L. Trotsky, "Kul'tura," pp. 171–172, and *Sochineniia*, vol. 21, pp. 260–268; B. Knei-Paz, *Social and Political Thought*, pp. 481–482.

36. F. P. Maiorov, "O mirovozzrenii," p. 20.

37. N. I. Bukharin, *Ataka*, pp. 171–215.

38. G. E. Hutchison, "Biosphere," pp. 3, 11.

39. For representative interpretations, see A. A. Bogdanov, "Ob'ektivnoe ponimanie"; S. Iu. Semkovskii, "K sporu"; A. K. Timiriazev, "Teoriia otnositel'nosti," "Novaia neudachnaia popytka," and Review; and A. Gol'tsman, "Nastuplenie," and "Einshtein." Semkovskii emphasized the compatibility of the theories of relativity and dialectical materialism; Timiriazev rejected Einstein's ideas *in toto*; Bogdanov hailed the epistemological value of the theory of relativity and emphasized its applicability in the social sciences; Gol'tsman attacked the current efforts to anchor the theory of relativity in the philosophical legacy of Ernst Mach.

40. A. Serebrovskii, "Teoriia nasledstvennosti"; "Darvinizm i marksizm"; F. Duchinskii, "Darvinizm, lamarkizm"; A. Timiriazev, "Teoriia 'kvanta'"; Z. Tseitlin, "O 'misticheskoi' prirode"; and A. F. Samoilov, "Dialektika." The diversity of theoretical orientations in biology is discussed in A. E. Gaisinovich, "Origins," pp. 49–51.

41. "Ot Rossiiskoi assotsiatsii," pp. 255–256.

42. "Pervaia Vsesoiuznaia konferentsiia," p. 258.

43. V. I. Shevshchenko, "K voprosu ob uchete nauchnykh kadrov," p. 26.

44. A. M. Deborin, *Filosofiia i politika*, p. 195.

45. A. Gol'tsman, "Einshtein i materializm," p. 134.

46. In 1928 this society merged with the Society of Materialistic Friends of Hegelian Dialectics to form the Society of Militant Dialectical Materialists ("Pervoe Vsesoiuznoe soveshchanie obshchestv voinstvuiushchikh materialistov-dialektikov," p. 131).

47. The conflict between the mechanists and dialecticians forms the core of D. Joravsky's *Soviet Marxism*, the most comprehensive study in the field. See also L. N. Suvorov, "Rol' filosofskikh diskussii," pp. 38–45.

48. V. Egorshin, Review, p. 269.

49. P. V. Alekseev, "Diskussiia," p. 52.

50. A. M. Deborin, "Mekhanisty," p. 47.

51. For a summary of the philosophical views of the mechanists, see A. V. Shcheglov, "Filosofskie diskussii," p. 112. The philosophical positions of the dialecticians are surveyed in A. M. Deborin, "Mekhanisty," pp. 22–55.

52. For a philosophical analysis of the conflict between the mechanists and the dialecticians, see V. I. Ksenofontov, *Leninskie idei*, pp. 58–85.

53. D. Joravsky, *Soviet Marxism*, p. 131.

54. "Pervaia Vsesoiuznaia konferentsiia," pp. 290–291.

55. *Korotke zvidomlennia*, p. 4; A. Ia. Artems'kyi, *Shcho take Vseukraïns'ka Akademiia nauk*, p. 14; V. I. Vernadskii, "First Years," p. 21; and S. Timoshenko, *Vospominaniia*, pp. 154–155.

56. R. A. Lewis, "Government and Technological Sciences," p. 177.

57. A. K. Timiriazev, *Nauka v Sovetskoi Rossii*, p. 25.

58. G. D. Alekseeva, *Oktiabr'skaia revoliutsiia*, pp. 263–264.

59. N. I. Bukharin, *Etiudy*, pp. 171–215.

60. V. P. Volgin, "Sovetskaia vlast'," p. 21.

61. A. Gerasimov, "Rossiiskoe Mineralogicheskoe obshchestvo," p. 378. The unfavorable working conditions in I. P. Pavlov's laboratory at the Institute of Experimental Medicine are described in B. M. Zavadovskii, "Nauka v Sovetskoi Rossii," pp. 144–145. See also [M. Gorky], *Gor'kii i nauka*, pp. 100–101.

62. S. Mokshin, "Sovnarkomovskii paek," p. 10.

63. Among the recipients of TsEKUBU rations was Anna Mendeleeva, the

widow of the famous chemist D. I. Mendeleev (V. A. Volkov, "Novye doku-menty," pp. 139–140).

64. *La Commission centrale*, pp. 17–27.

65. "Iz istorii deiatel'nosti Doma Uchenykh," pp. 4–7; "Dom Uchenykh," pp. 36–38.

66. The fluidity of the situation is described in S. Fitzpatrick, *Commissariat*, pp. 68–73.

67. L. R. Graham, "Development of Science Policy," pp. 19–20.

68. *Pervaia Moskovskaia obshchegorodskaia konferentsiia*, p. 4.

69. A. V. Lunacharskii, *Vospominaniia*, p. 206.

70. K. A. Timiriazev, *Nauka i demokratiia*, p. 447.

71. V. L. Omel'ianskii, "Ocherednye zadachi," p. 27.

72. *Otchet-R:1920*, pp. 10–11; *Otchet-R:1921*, p. 5.

73. *Otchet-R:1922*, p. 11.

74. *Otchet-R:1920*, p. 12.

75. *Otchet-R:1919*, p. 22.

76. V. A. Steklov, *Galileo Galilei*, p. 99.

77. A. K. Timiriazev, *Nauka v Sovetskoi Rossii*, p. 25.

78. V. P. Volgin, *Akademiia nauk za chetyre goda*, p. 58.

79. A. S. Lappo-Danilevskii, *Metodologiia*, vol. 1, pp. 261–265.

80. A. S. Lappo-Danilevskii and A. Nol'de, "Ob Institute," pp. 1403–1404.

81. I. S. Smirnov, *Lenin*, p. 299.

82. M. N. Pokrovskii, "Obshchestvennye nauki," p. 24.

83. The academicians elected during this period are listed in G. K. Skriabin et al., *Akademiia nauk*, vol. 2, pp. 2–9.

84. V. I. Vernadskii, *Ocherki i rechi*.

85. V. I. Vernadskii, *Mysli*, pp. 1–17.

86. V. I. Vernadskii, *Ocherki i rechi*, vol. 2, pp. 22–24.

87. Ibid., pp. 20–24. See also V. I. Vernadskii, *Nachalo*, p. 57. For a more recent Marxist interpretation of Vernadskii's views on the history of science, see S. R. Mikulinskii, "Problemy analiza"; and V. P. Iakovlev, "Metodologicheskie voprosy."

88. V. I. Vernadskii, "Sovremennoe sostoianie," p. 646.

89. V. Egorshin, "Estestvoznanie," pp. 115–116.

90. V. I. Nevskii, "Restavratsiia," p. 122.

91. L. S. Berg, *Nauka*, chap. 7.

92. A. E. Fersman, *L'Académie*, p. 179. At a meeting of the Academy members, held on May 14, 1921, Vernadskii stated: "In Russia there is not a single orga-nization devoted to the study of the history of scientific and philosophical thought and to scientific creativity. In this respect we are placed in an unfavorable position in comparison with the West and America" ("Piatoe zasedanie," p. 10).

93. V. I. Vernadskii, "Raboty po istorii," p. 156.

94. V. A. Steklov, *Lomonosov.*

95. Among the more notable studies devoted to the life and work of the leading scientists are V. A. Steklov, *Lomonosov* and *Teoriia i praktika; Stoletie neevklidovoi geometrii Lobachevskogo;* A. A. Borisiak, *V. O. Kovalevskii;* V. I. Vernadskii, *Pamiati akademika K. M. fon-Bera;* L. A. Chugaev, *Dmitrii Ivanovich Mendeleev.*

96. A. V. Sidorenko et al., *50 let,* p. 25.

97. B. A. Lindener, *Raboty,* p. 88.

98. V. I. Vernadskii, "Ob ispol'zovanii," p. 84.

99. L. A. Chugaev, "O merakh," p. 167.

100. "Protokol sovmestnogo zasedaniia," p. 151.

101. [M. Gorky], *Gor'kii i nauka,* p. 20. See also V. Ipat'ev, "Rol' nauchnykh issledovanii," p. 12.

102. V. I. Vernadskii, "O gosudarstvennoi seti," p. 158.

103. The work of KEPS during the early years of Soviet rule is described in A. V. Kol'tsov, "Leninskii plan," pp. 68–74. See also B. A. Lindener, *Raboty;* I. S. Smirnov, *Lenin,* pp. 249-264.

104. "Akademiia nauk v pervyi god," p. 9.

105. I. S. Smirnov, *Lenin,* p. 251.

106. Ibid., pp. 254, 256.

107. N. M. Mitriakova, "Struktura," p. 205; S. I. Vavilov, *Materialy k istorii Akademii,* p. 21.

108. D. I. Shcherbakov, "Rol' V. I. Vernadskogo," p. 94.

109. *KPSS v rezoliutsiiakh,* vol. 2, pp. 52–53.

110. B. A. Lindener, *Raboty,* pp. 1–2; V. I. Vernadskii, "First Years," p. 11; L. R. Graham, "Development of Science Policy," p. 18; R. A. Lewis, "Government and the Technological Sciences," pp. 176–177, and *Science,* chap. 4; B. I. Kozlov, *Organizatsiia i razvitie,* pp. 28–29.

111. V. I. Lenin, *Polnoe sobranie sochinenii,* vol. 36, p. 228.

112. "Prilozhenie k protokolu VIII zasedaniia," p. 1391.

113. *Nikolai Semenovich Kurnakov,* pp. 28–35; G. B. Kauffman and A. Beck, "Nikolai Semenovich Kurnakov," p. 48.

114. *ON,* p. 130.

115. "Institut arkheologicheskoi tekhnologii," p. 35.

116. N. G. Fradkin, *Ocherki,* p. 19.

117. M. Ia. Lapirov-Skoblo, "Novye puti," p. 67.

118. A. P. Pavlov and V. I. Vernadskii, *Proekt Ustava.*

119. D. I. Shcherbakov, "Gor'kii," p. 273.

120. M. Gorky, "Rech' na publichnom zasedanii," p. 230; M. S. Bastrakova, *Stanovlenie,* p. 60; S. N. Valk et al., *Ocherki,* pp. 143–144.

121. M. Gorky, "Rech' na publichnom zasedanii," p. 229.

122. [M. Gorky], *Gor'kii i nauka,* p. 16. For pertinent details on Gorky's relations to the scientific community, see D. I. Shcherbakov, "Gor'kii."

123. [M. Gorky], *Gor'kii i nauka*, p. 20.

124. V. I. Lenin, *Polnoe sobranie sochinenii*, vol. 18, pp. 286–288.

125. A. E. Fersman, "Iz nauchnoi deiatel'nosti," p. 265, and "S'ezd," p. 138.

126. M. S. Bastrakova, *Stanovlenie*, pp. 106–107.

127. A. V. Kol'tsov, *Lenin*, p. 31.

128. V. I. Vernadskii, *Ocherki i rechi*, vol. 2, pp. 32–37 and vol. 1, p. 37.

129. S. F. Ol'denburg, "Voprosy organizatsii," pp. 8–14.

130. *ON*, p. 5. For pertinent details, see V. Ipat'ev, *Life of a Chemist*, pp. 359–362. See also K. B. Bailes, *Technology*, pp. 56–57; L. V. Zhigalova, "K istorii sozdaniia institutov," pp. 187–202; B. I. Kozlov, *Organizatsiia*, p. 31; N. P. Gorbunov, "Velikie plany," p. 278; E. A. Beliaev and N. S. Pyshkova, *Formirovanie*, pp. 112–123; V. N. Makeeva, "V. I. Lenin," pp. 48–51.

131. R. A. Lewis, *Science*, p. 38.

132. N. G. Fradkin, *Ocherki*, p. 7.

133. M. M. Novikov, *Ot Moskvy*, p. 314.

134. V. Ia. Frenkel', "Einshtein," p. 27.

135. E. A. Beliaev and N. S. Pyshkova, *Formirovanie*, p. 119.

136. I. S. Smirnov, *Lenin*, p. 290.

137. R. A. Lewis, *Science*, p. 45.

138. F. N. Petrov, "Ob organizatsii," p. 12; *ON*, pp. 39, 46–51.

139. L. G. Ivanova, *Formirovanie*, p. 326.

140. *ON*, p. 48.

141. Ibid., 68–70.

142. Ibid., pp. 70–75; R. A. Lewis, "Government and the Technological Sciences," p. 183.

143. A. K. Timiriazev, *Nauka v Sovetskoi Rossii*, p. 25.

144. *Piat' let vlasti sovetov*, p. 508.

145. I. S. Smirnov, *Lenin*, pp. 265–267.

146. A. V. Kol'tsov, *Lenin*, p. 31, and "Akademik V. P. Volgin," p. 106.

147. G. D. Komkov et al., *Akademiia nauk SSSR; Kratkii ocherk*, p. 277.

148. M. A. Diakonov, "Aleksandr Sergeevich Lappo-Danilevskii," p. 359.

149. V. I. Lenin, *Polnoe sobranie sochinenii*, vol. 36, pp. 228–231.

150. *ON*, p. 127.

151. B. N. Molas, "Iubileinye dni," p. 41.

152. G. D. Komkov et al., *Akademiia nauk SSSR—shtab*, p. 28.

153. A. V. Lunacharskii, *Vospominaniia*, p. 206. Lunacharskii used the festive atmosphere to emphasize the role of Marxism as a philosophy of science. Science, he said, needed Marxism for two reasons: its utilitarian orientation and its unifying—i.e. monistic—philosophy ("Rol' marksizma," pp. 19–21).

154. L. Trotsky, *Marxism and Science*, p. 20.

155. L. B. Kamenev, *Rabochaia revoliutsiia*, p. 5.

156. W. Bateson, "Science in Russia," pp. 682–683.

157. S. F. Ol'denburg, "Nauchno-kul'turnye itogi," p. 3, and "K dvukhsotle-tiiu," pp. 65–69.
158. P. P. Lazarev, "Chistaia nauka," p. 39.
159. V. L. Omel'ianskii, "Ocherednye zadachi," p. 27.
160. Otchet-S:1926, vol. 1, pp. ii–iii.
161. A. E. Fersman, Akademiia nauk, pp. 1–4; G. K. Skriabin et al., Ustavy, pp. 120–129.
162. A. A. Bogdanov, Nauka i rabochyi klass, p. 6, and Sotsializm nauki, pp. 92–94.
163. For an analysis of Bogdanov's philosophy of proletarian science, see J. C. McClelland, "Utopianism," pp. 411–412. For a more general view, see Z. A. Sochor, "Was Bogdanov Russia's Answer to Gramsci?" p. 76. For details on Bogdanov's relations to the Proletkul't movement, see S. Fitzpatrick, Commissariat, pp. 91–106.
164. A. A. Bogdanov, Vseobshchaia organizatsionnaia nauka. The first two volumes of this study were published originally in 1913–1917. The Soviet editions discussed much more extensively the multiple relations between organization theory and Marxist thought. For relevant comments on the reception of tektology in the Soviet Union, see G. Gorelik, "Bogdanov's Tektology," p. 334.
165. A. A. Bogdanov, "Uchenie ob analogiiakh," pp. 80, 95. Bogdanov also made an effort to present the history of human society as an evolution of the universal organizational principles of three major systems of social activity: technology, economy, and ideology ("Organizatsionnye printsipy").
166. K. B. Bailes, Technology, p. 49. See also J. C. McClelland, "Bolshevik Approaches," p. 830.
167. V. M. Selunskaia, "V. I. Lenin," p. 34.
168. V. I. Lenin, Polnoe sobranie sochinenii, vol. 38, p. 59.
169. KPSS v rezoliutsiiakh, vol. 2, p. 52.
170. E. A. Beliaev and N. S. Pyshkova, Formirovanie, p. 118.
171. ON, pp. 41, 272.
172. Ibid.
173. N. M. Mitriakova, "Struktura," p. 207.
174. Otchet-S:1926, p. iv.

THREE: THE FORGING OF A SOVIET INSTITUTION

1. The basic features of the new style of work are discussed in "Za reshitel'nuiu perestroiku," pp. 10–14. See also A. Iarilov, "Staraia i novaia Akademiia," p. 50; B. Keller, "Na vyezdnoi sessii," pp. 59–60; and A. N. Bakh, "Vsesoiuznaia Akademiia," pp. 21–22.
2. LP, May 12, 1928.
3. I. K. Luppol, "IV Plenum," p. 17.

4. Ibid., p. 19.

5. Ibid.

6. "Pervaia Vsesoiuznaia konferentsiia," *VKA*, no. 26, p. 249. See also Tur., "Pod kupolom."

7. I. K. Luppol, "K vyboram v Akademiiu," p. 5. For a general statement, see V. P. Volgin, "Akademiia nauk k 15 godovshchine," pp. 25-27.

8. For details, see L. R. Graham, *Soviet Academy of Sciences*, pp. 100-101. See also V. Ipat'ev, *Life of a Chemist*, p. 463-464.

9. N. I. Baronskii and V. P. Iakovlev, "Evoliutsiia," p. 27.

10. V. Ipat'ev, *Life of a Chemist*, p. 463. Zabolotnyi's scientific work is described in V. I. Bilai, *Zhizn'*, pp. 27-44.

11. V. A. Ul'ianovskaia, *Formirovanie*, p. 158; "Leningradskie uchenye."

12. *Izvestiia*, February 5, 1929; V. A. Ul'ianovskaia, *Formirovanie*, p. 161.

13. *VARNITSO: tseli i zadachi*.

14. B. Ul'ianskaia, "Vtoraia Vsesoiuznaia konferentsiia," p. 180.

15. A. N. Bakh, "Nauchno-issledovatel'skaia rabota," pp. 2-5, and "K voprosu o planirovanii," pp. 1-3.

16. *Izvestiia*, February 5, 1929; Ul'ianovskaia, *Formirovanie*, pp. 159-160.

17. Iu. P. Figatner, "Proverka aparata," pp. 73-75; "K obnaruzheniiu."

18. Academicians Likhachev, Liubavskii, Platonov, and Tarle were expelled officially by the Academy's General Assembly on February 1, 1931 ("Organizatsionno-administrativnaia khronika," p. 49).

19. L. R. Graham, *Soviet Academy of Sciences*, p. 152.

20. V. Sverdlov, "Piat' let," pp. 13-18.

21. "V tsentral'nom biuro VARNITSO," pp. 125-126. For more details, see P. Valeskal and V. Sverdlov, "Itogi," pp. 1-7.

22. A. V. Kol'tsov, "Akademik V. P. Volgin," pp. 106-107.

23. *Otchet-S:1933*, p. 338. Since only 18.7 percent of academicians were members or candidates of the party, this number must have included the chief nonacademic personnel and the leaders of local mass organizations as well. For additional details, see G. D. Komkov et al., *Akademiia*, vol. 2, 2d ed., p. 68.

24. "Rezoliutsiia X Plenuma IKKI"; A. Shabanov, "O teorii evoliutsii"; V. Bistrianskii, "Bukharin i Trotskii"; A. Martynov, "O neskromnoi pretenzii."

25. "Kritika teoreticheskikh osnov," pp. 232-250.

26. "Organizatsionno-administrativnaia khronika," p. 49.

27. G. D. Komkov et al., *Akademiia*, 2d ed., vol. 2, p. 68.

28. For details on the growth of the Technical Group, see R. A. Lewis, "Government and the Technological Sciences," pp. 195-197; N. M. Mitriakova, "Struktura," pp. 227-228. For general comments, see L. R. Graham, "Reorganization of U.S.S.R. Academy of Sciences," p. 137.

29. A. N. Bakh, "Vsesoiuznaia Akademiia," p. 22.

30. The growing role of the Academy in technical research at this time is discussed in pertinent detail in R. A. Lewis, *Science and Industrialization*, pp. 74–78.

31. V. P. Volgin, *Akademiia*, pp. 15–16.

32. B. Bykhovskii, "Bekon," pp. 98–99.

33. A. V. Lunacharskii, *Sobranie sochinenii*, vol. 6, p. 319.

34. V. Rodkevich, Review of F. Bacon, "O printsipakh," pp. 187–188.

35. V. F. Asmus, *Izbrannye filosofskie trudy*, vol. 2, p. 213. See also N. I. Bukharin, *Etiudy*, p. 41. While noting the emphasis on practicalism as a link between Bacon's philosophy and Marxism, B. Bykhovskii was concerned primarily with differences between Baconian and Marxist epistemology, logic, and psychology ("Bekon," pp. 98–99).

36. N. I. Bukharin, *Etiudy*, p. 279.

37. *OSN*, pp. 210–213.

38. V. Sologub, "Nauchnaia obshchestvennost'," p. 178.

39. E. M. Kreps, *Perepiska*, pp. 38–39. For a general statement on the role of central planning in the development of the scientific base of the national economy, see G. S. Tymanskii, "Lenin i nauka," pp. 13–17.

40. N. I. Bronskii and V. P. Iakovlev, "Evoliutsiia," p. 27.

41. I. M. Gubkin, "Osnovnye zadachi," pp. 9–10. See also V. S. Neapolitanskaia, *Perepiska: 1918–1939*, pp. 81–83.

42. V. P. Volgin, *Akademiia nauk*, pp. 15–16.

43. A. N. Bakh, "K voprosu o planirovanii," p. 1.

44. N. Smit, "Plannovoe vreditel'stvo," p. 29.

45. P. A. M. Dirac, *Principles*, p. 139.

46. S. I. Vavilov, "Neskol'ko predlozhenii," p. 40.

47. R. A. Lewis, *Science and Industrialization*, p. 79.

48. G. K. Skriabin et al., *Ustavy*, pp. 142, 195.

49. D. Nasledov, "Otchetnyi doklad," p. 22; S. Krum, "Fizika," p. 110.

50. I. Korel', "V Akademii nauk," pp. 265–266.

51. "General'nyi dogovor," pp. 237–238.

52. N. M. Mitriakova, "Struktura," pp. 230–231.

53. *VII Vsesoiuznomu s'ezdu*, pp. 33–34.

54. *Otchet-S:1934*, pp. 8–9.

55. For a summary description of the proceedings, see "Sessiia Akademii nauk SSSR," 1936.

56. A. F. Ioffe, "Otchet o rabote," pp. 858–859. See also "Rezoliutsiia Martovskoi sessii," pp. 839–840.

57. D. Nasledov, "Otchetnyi doklad," p. 22; S. Krum, "Fizika," pp. 108–111.

58. F. V., "Problemy," p. 30.

59. "Martovskaia sessiia," p. 838; S. Krum, "Fizika," pp. 108–111.

60. "Bor'ba za peredovuiu nauku," pp. 4–5.
61. S. I. Vavilov, *Materialy k istorii Akademii*, p. 191.
62. E. D. Esakov, *Sovetskaia nauka*, p. 159.
63. V. P. Volgin, *Akademiia nauk*, p. 7; G. K. Skriabin et al., *Ustavy*, p. 193; G. D. Komkov et al., *Akademiia—shtab*, p. 43.
64. *Otchet-S:*1934, p. 10.
65. *OSN*, pp. 169–192; V. P. Volgin, "Akademiia nauk na novom etape," pp. 22–24; G. S. Brega, *Sotrudnichestvo*, p. 48; V. A. Ul'ianovskaia, *Formirovanie*, pp. 174–175.
66. V. P. Volgin, *Akademiia nauk*, p. 13.
67. Ibid., p. 33; V. L. Komarov, *Izbrannye sochineniia*, pp. 481–483.
68. E. A. Beliaev and N. S. Pyshkova, *Formirovanie*, p. 176.
69. Sh., "V Akademii nauk," p. 108.
70. I. D. Bagbaia, "Vtoraia Zakavkazskaia konferentsiia," p. 272.
71. Sh., "V Akademii nauk," p. 108.
72. N. K. Karataev, "Podgotovka," pp. 74–75.
73. E. V., "Podgotovka," p. 383.
74. *OSN*, pp. 199–203.
75. E. V. "Podgotovka," p. 383; N. K. Karataev, "Podgotovka," p. 77.
76. E. V., "Podgotovka," p. 381. See also V. I. Shevshchenko, "K voprosu ob uchete nauchnykh kadrov," p. 26.
77. I. S. Taitslin, "Nauchnye kadry," p. 27. A survey of *all* persons—Communist or otherwise—engaged in scholarship showed a different picture: 31.5 percent of all scholars worked in the exact sciences, 29.6 percent worked in the humanities, 21 percent worked in medicine, and 17.8 percent worked in technical sciences. Three-fourths of persons engaged in scientific work or in humanistic research were situated in Moscow or Leningrad.
78. *Otchet-S:*1934, p. 566.
79. For pertinent information and general comments, see S. Fitzpatrick, *Education and Social Mobility*, chap. 11.
80. "Instruktsiia o poriadke," pp. 169–170.
81. A. F. Sukhanov, "Ustav," p. 85.
82. S. I. Vavilov, "Neskol'ko predlozhenii," p. 39; B. Shparo, "V Akademii," p. 39.
83. D. Nasledov, "Ochetnyi doklad," p. 21.
84. A. M. Terpigorev, "O vysshei shkole," p. 15.
85. V. F. Popov, "Podgotovka," p. 188.
86. N. K. Karataev, "Podgotovka," pp. 77–78.
87. N. M. Mitriakova, "Struktura," p. 224.
88. V. P. Volgin, "Nashi zadachi," p. 16.
89. G. M. Krzhizhanovskii, "Akademiia," p. 7, and "Programma," p. 1.
90. N. I. Bukharin, "Nauka v SSSR," p. 11.

91. E. Ia. Kol'man, "Politika, ekonomika i matematika," p. 53.
92. A. M. Deborin, "Lenin i krizis," p. 27.
93. V. F. Mitkevich, "Osnovnye vozzreniia," pp. 242–243.
94. "Vazhneishii istoricheskii dokument," p. 5.
95. "O zhurnale 'Pod znamenem marksizma,'" pp. 1–2.
96. "O raznoglasiiakh," pp. 78–79.
97. Ibid., p. 83.
98. "Materialy nauchnoi sessii Instituta filosofii Komakademii," p. 144.
99. N. I. Bronskii and V. P. Iakovlev, "Evoliutsiia," p. 27. This, however, did not prevent A. K. Timiriazev, the most stubborn defender of the mechanist orientation, from labeling Vernadskii a typical representative of "the idealistic-fideistic world view," thoroughly antagonistic to the cause of socialism ("Volna idealizma," pp. 119–122).
100. N. I. Bronskii and V. P. Iakovlev, "Evoliutsiia," p. 27.
101. S. A. Ianovskaia, "Ocherednye zadachi," p. 92.
102. A. A. Maksimov, "Bol'shaia Sovetskaia Entsiklopediia," pp. 340–342.
103. Za povorot na fronte estestvoznaniia, p. 80; P. V. Alekseev, Marksistsko-leninskaia filosofiia, pp. 139–140.
104. For an unqualified defense of Lamarckism, see E. S. Smirnov, "Novye dannye." Smirnov did not reject the basic theories in modern biology, but he was satisfied to give them a secondary position (p. 197).
105. Na bor'bu za materialisticheskuiu dialektiku v matematike. See particularly M. Vygodskii, "Problemy istorii matematiki," pp. 143–160.
106. L. N. Suvorov, "K otsenke leninskogo etapa," p. 119. In an effort to bring all theoretical and ideological aspects of dialectical and historical materialism into the Leninist fold, M. B. Mitin and I. Razumovskii published the first major Soviet textbook in Marxist philosophy (Dialekticheskii i istoricheskii materializm).
107. G. Iaffe, "Filosofskie osnovy," pp. 235–237. For a general discussion of this brand of "idealism," see L. Man'kovskii, "K voprosu o filosofskikh istokakh."
108. "O polozhenii na fronte estestvoznaniia," pp. 25–27.
109. Ibid., p. 30.
110. M. B. Mitin, Boevye voprosy, p. 49.
111. V. Egorshin, "Lenin i sovremennoe estestvoznanie," p. 10.
112. M. B. Mitin, "K voprosu o leninskom etape," pp. 24–25.
113. E. Ia. Kol'man, "Pis'mo," p. 165. A. A. Maksimov was only too ready to comment on Frenkel's reported challenge to Marxist philosophy. He wrote: "A group of old specialists continues actively to resist the reconstruction of science on the basis of Marxism-Leninism and the advent of Bolshevik science. At a recent conference on the structure of matter (December 1931), held in Leningrad, Frenkel' spoke strongly against the claim of Marxism-Leninism to occupy the leading position in physics and chemistry. Ready to recognize verbally the role

of Marxism in the social sciences, he categorically rejected the leading role of dialectical materialism in the natural sciences. Unfortunately, Frenkel' is not alone. There are other physicists who are our enemies and who fight against a reconstruction of science on the basis of Marxism-Leninism. In this struggle against Marxism-Leninism, Frenkel' and Co. have joined the international ideological reaction, particularly along the lines of Machism, a most thoroughly disguised orientation. P. Frank, the leader of the Austrian-German Machists, was fully aware of Professor Frenkel's unfriendly attitude toward materialism when he praised him in the March 1931 issue of the Italian journal *Scientia* and presented him as the only consistent Machist, unafraid to express Machist views in his textbook on electricity" ("Ob otnoshenii, p. 158).

114. I. E. Tamm, "O rabote filosofov-marksistov," pp. 220–221. Tamm's charges evoked an immediate rebuttal by one of the most active Soviet philosophers of science, who used the opportunity to emphasize the achievements of Marxist scholars in the philosophy of science and to underscore the strategic significance of the philosophical questions raised by the "crisis in physics" as interpreted by Lenin (V. Egorshin, "Kak I. E. Tamm kritikuet marksistov").

115. V. P. Volgin, "Akademiia v 1933 g.," p. 34.

116. V. F. Mitkevich, "Osnovnye vozzreniia," pp. 239–241.

117. A. F. Ioffe, "Nepreryvnoe i atomnoe stroenie," pp. 249–250.

118. S. I. Vavilov, "Novaia i staraia fizika," p. 218.

119. N. I. Bukharin, "Lenin," p. 93.

120. A. M. Deborin, "Lenin," pp. 174–246.

121. L. M. Rubanovskii, "Vtoroi zakon," p. 327.

122. A. F. Ioffe, "Razvitie atomisticheskikh vozzrenii," pp. 467–468.

123. A. F. Ioffe, "O polozhenii;" V. A. Fok, "K diskussii," p. 159, and "Chto vnesla teoriia kvantov," pp. 8–10; and Ia. I. Frenkel', "Teoriia otnositel'nosti," p. 38.

124. S. Iu. Semkovskii, "'Dialektika prirody' Engel'sa," pp. 12–14, and "O nekotorykh metodologicheskikh problemakh," p. 36.

125. L. E. Mandel'shtam, *Polnoe sobranie sochinenii*, p. 93.

126. V. I. Vernadskii, *Nachalo*, p. 57.

127. D. Novogrudskii, "Geokhimiia," p. 175.

128. Ibid., p. 203. At the same time, M. B. Mitin criticized Vernadskii's claim that science should be independent of "government and tribal constraints" as well as heavy practical burdens ("K voprosu o leninskom etape," p. 20).

129. A. E. Fersman, *Izbrannye trudy*, p. 407.

130. B. Hessen, "Social and Economic Roots," pp. 155–212. For relevant comments, see P. G. Werskey, *Visible College*, pp. 142–143.

131. B. M. Zavadovskii, "'Physical' and 'Biological,'" pp. 74–80.

132. E. Ia. Kol'man, "Dynamic and Statistical Regularities," pp. 90–94.

133. P. G. Werskey commented on the impact of Soviet papers on British scientists: "Zavadovsky's paper, for instance, greatly impressed [Joseph] Need-

ham, because its conclusions—so similar to his own—had apparently been derived from the axioms of dialectical materialism. The contribution of Hessen made an even deeper and wider impression. As [Lancelot] Hogben has since recalled, it 'reinforced my interest' in historical materialism 'as an intellectual tool for expository use.' Hyman Levy suddenly found most works on the history of science inadequate, because they did not 'give an account at the same time of man's social and economic background.' . . . But perhaps the most interesting convert was Joseph Needham" (Introduction, p. xii).

134. E. Ia. Kol'man, "Pis'mo," pp. 165–166. See also Kol'man's "Uzlovye problemy," p. 32.

135. E. Ia. Kol'man, "Present Crisis," p. 225.

136. E. Ia. Kol'man, "Pis'mo," p. 168.

137. B. M. Kedrov, "Istoriia khimii," p. 38.

138. V. L. Komarov, *Izbrannye sochineniia*, p. 575.

139. S. F. Cohen, *Bukharin*, p. 110.

140. G. Adamian, "Istoricheskii materializm," p. 708.

141. F. Konstantinov, "Sotsialisticheskoe obshchestvo," p. 47.

142. G. M. Enteen, *Soviet Scholar-Bureaucrat*, chaps. 9, 10, 11; M. Kammari, "Teoreticheskie korni," p. 3.

143. "Ot Instituta," p. 3.

144. Em. Iaroslavskii, "Antimarksistskie izvrashcheniia," p. 12.

145. See particularly "Ot Instituta"; Em. Iaroslavskii, "Antimarksistskie izvrashcheniia"; and A. Pankratova, "Razvitie."

146. For a broad comment, see G. Enteen, "Marxist Historians," p. 167.

147. N. P. Gorbunov, "Akademiia," p. 3. For details on Gorbunov's participation in the Pamir exploration, see A. V. Nedospasov, "K vysochaishei vershine strany," pp. 115–118. A sympathetic sketch of Gorburov's life and work is presented in E. P. Podvigina, "Akademik Gorbunov."

148. S. T. Beliakov, "Neopublikovannye pis'ma," p. 99.

149. A. N. Bakh, "Sovetskaia nauka," pp. 3–12.

150. G. A. Kniazev and A. V. Kol'tsov, *Kratkii ocherk*, p. 116.

151. "Resheniia Plenuma TsK VKP(b)," pp. 6, 10.

152. M. B. Adams, "Science, Ideology and Structure," p. 194.

153. "Usilit' revoliutsionnuiu bditel'nost'!" p. 4.

154. N. Bourbaki, *Éléments*, pp. 176–177.

155. "O vragakh v sovetskoi maske," *Pravda*; "Ob akademike Luzine," pp. 7–8; "Nauchnaia obshchestvennost' kleimit vragov"; "Protiv Luzina."

156. "Protiv Luzina," p. 124; "Noiabr' 1936 goda," pp. 6–7.

157. The names of other leading imprisoned physicists are given in R. A. Medvedev, *Let History Judge*, p. 226.

158. A. K. Timiriazev, "Eshche raz o volne idealizma," p. 152.

159. E. Ia. Kol'man, "Protiv lzhenauki," p. 47.

160. M. Bessarab, *Landau*, p. 51.

161. V. Ia. Frenkel', *Iakov Il'ich Frenkel'*, p. 213.

162. I. I. Prezent, "O lzhenauchnykh vozzreniiakh," p. 153.

163. L. C. Dunn, *Short History*, pp. 138–139.

164. I. V. Michurin, *Sochineniia*, vol. 1, pp. 213–214.

165. V. Kolbanovskii, "Obshchii obzor," pp. 127–140.

166. M. B. Mitin, "Za peredovuiu sovetskuiu geneticheskuiu nauku," p. 150. For a well-balanced summary of Vavilov's scientific work, see M. B. Adams, "Nikolai Ivanovich Vavilov," pp. 505–513.

167. I. I. Prezent, "O lzhenauchnykh vozzreniiakh," p. 147; "Bor'ba za peredovuiu nauku," p. 7.

168. V. L. Komarov, "Novoe popolnenie Akademii," p. 78.

169. "Obshchee sobranie Akademii (1936)," p. 80.

170. S. I. Vavilov, *Materialy k istorii Akademii*, p. 192.

171. A. Kolmogorov, "Matematika," pp. 359–360.

172. M. B. Mitin, "Za peredovuiu sovetskuiu geneticheskuiu nauku," p. 162. See also I. T. Frolov, *Genetika i dialektika*, p. 96.

173. V. A. Fok, "K diskussii," p. 159.

174. "80-letie I. P. Pavlova."

175. "Oblispolkom i Leningradsovet."

176. "O polozhenii na fronte estestvoznaniia," pp. 25–26.

177. "Resheniia Plenuma TsK VKP(b)," p. 17. It was only A. A. Maksimov, the heartiest and most inflexible of all Stalinist philosophers, who found a blemish on Shmidt's Marxist record. Shmidt's article on algebra in the *Great Soviet Encyclopedia* presented, according to Maksimov, a "thoroughly formalistic" rather than "materialistic" analysis (*"Bol'shaia Sovetskaia Entsiklopediia,"* p. 349).

178. T. I. Samuelian, *Search for Marxist Linguistics*, p. 242.

179. L. L. Thomas, "Some Notes," pp. 340–344; N. S. Derzhavin, "Iafeticheskaia teoriia," no. 2, p. 37.

180. M. N. Pokrovskii, "Obshchestvennye nauki," p. 26.

181. "Deiatel'nost' Kommunisticheskoi Akademii," pp. 265–267.

182. For a brief survey of the evolution of Vil'iams's thought, see A. Ia. Bush, "K 50-letnemu iubileiu." See also "Vasilii Robertovich Vil'iams," pp. 785–794.

183. I. V. Tiurin, *Achievements*, p. 9. Vil'iams was completely ignored by K. D. Glinka, the leading Soviet expert in soil science at the time, who contributed "Pochvovedenie v SSSR" to the volume on science prepared in celebration of the tenth anniversary of the October Revolution.

184. S. I. Vol'kovich et al., *Dmitrii Nikolaevich Prianishnikov*, pp. 29–30, 38–39, 115–116.

185. For an official list of the components of the Academy in 1937, see "Akademiia nauk v 1937 godu," pp. 345–347.

186. V. L. Komarov, "Novoe popolnenie," pp. 80–81.

187. "Obshchee sobranie Akademii (1938)," pp. 81–84. The eight departments

covered the following clusters of disciplines: biological sciences, physical and mathematical sciences, chemical sciences, geological and geographical sciences, economics and law, history and philosophy, literature and language, and technical sciences.

188. According to a historian of the institutional base of Soviet physiology: "As witnessed by his students and associates, I. P. Pavlov demanded that his staff give undivided attention to the problems and hypotheses advanced by himself. This demand was so rigid that his associates were forbidden to talk about other research topics. In full control of all research activities in his laboratories, he discouraged his associates from showing independent initiative and ability" (K. A. Lange, *Organizatsiia*, p. 201).

189. L. A. Orbeli, "Razvitie biologicheskikh nauk," p. 223.

190. *Vklad akademika A. F. Ioffe*, p. 7.

191. A. I. Oparin, "Institut biokhimii," pp. 199–200.

192. E. Segré, *From X-Rays to Quarks;* M. von Laue, *History of Physics*, pp. 105, 114; B. Hoffmann, *Strange Story*, p. 225.

193. R. Olby, *Double Helix*, p. 17.

194. V. F. Mitkevich, "Osnovnye vozzreniia," "O sovremennoi bor'be," "O pozitsii," and "Za Faradee-Maskvellovskuiu ustanovku"; A. K. Timiriazev, "Volna idealizma" and "Eshche raz o volne idealizma."

195. For an analysis of this orientation, see E. Ia. Kol'man, "Teoriia kvant" and "Teoriia otnositel'nosti"; A. A. Maksimov, "Lenin i krizis," p. 39; M. B. Mitin, *Boevye voprosy*, pp. 378–380.

196. V. A. Fok, "K diskussii po voprosam fiziki," pp. 158–159.

197. See particularly I. I. Shmal'gauzen, *Organizm* and *Puti*.

198. B. Tokin, "Teoreticheskaia biologiia," p. 51. Details on the evolution of Soviet biological theory between 1920 and 1946 are provided in M. B. Adams, "Sergei Chetverikov," pp. 244–272, "Severtsov," and "Founding of Population Genetics." See also T. Dobzhansky, "Birth of Genetic Theory"; and R. Goldschmidt, *Material Basis of Evolution*, pp. 336–340.

199. N. A. Maksimov was the first major biologist to hail Lysenko's "physiological" approach as a major contribution to science ("Razvitie," p. 487). M. M. Zavadovskii expressed the views of a typical contemporary biologist when he stated that "Darwinism and [Mendelian] genetics are not mutually exclusive but complementary theories" ("Darvin i ego nasledstvo," p. 53).

200. I. Agol, "Zadachi," p. 106.

201. E. Ia. Kol'man, "Boevye voprosy," pp. 67–68.

202. A. N. Bakh, "Sovetskaia nauka," pp. 10–11.

FOUR: THE TRIUMPH OF IDEOLOGY

1. In 1941, on the eve of World War II, the Academy had 76 research institutes, 11 laboratories, 6 observatories and 24 museums. Its personnel consisted of 118

full members, 5 honorary members, 182 corresponding members, and 4,700 scientific and technical associates and assistants (V. I. Salov, "Iz istorii," pp. 3–4).

2. [V. L. Komarov], "Rech' prezidenta Akademii nauk SSSR akademika V. L. Komarova na torzhestvennom zasedanii," p. 8.

3. A. F. Ioffe, "Fizika i voina," p. 92. During World War II, according to S. I. Vavilov, "Soviet theoretical physicists transferred their interests from intranuclear forces and quantum electrodynamics to ballistics, military acoustics, radio transmission, etc. Experimental physicists, postponing temporarily the most critical questions of cosmic radiation and spectroscopy, concerned themselves with detection of production defects, industrial spectral analysis, magnetic and acoustic mines, and radar" ("Ocherk razvitiia," p. 28). D. N. Prianishnikov, the leading expert in agrochemistry, was transferred to Samarkand, central Asia, where he stayed from October 1941 to the summer of 1943. Since the Samarkand laboratory was crowded and underequipped, he was compelled to shift his activities to "experimental work in agricultural fields," "practical work," "the planning of local economy," and "the popularization of knowledge." S. I. Vol'kovich et al., Prianishnikov, pp. 219–220.

4. V. I. Salov, "Iz istorii," p. 9.

5. Ibid., p. 13.

6. A. S. Fedorov, "Nauka i tekhnika," p. 11; V. L. Komarov, "K itogam Iubileinoi sessii," pp. 56–57.

7. S. V. Al'tshuler and G. P. Kulikov, "Organizatsiia," p. 25.

8. G. A. Kniazev, "Akademiia nauk za 30 let," p. 130.

9. Pravda, December 24, 1966; A. P. Aleksandrov, "Shest'desiat let," p. 24.

10. A. S. Fedorov, "Nauka i tekhnika," p. 12.

11. A. F. Ioffe, "Sovetskaia nauka v gody otechestvennoi voiny," p. 7.

12. S. I. Vavilov, Sobranie sochinenii, pp. 77–80.

13. E. Ia. Kol'man, Noveishie otkritiia, pp. 70–73.

14. G. V. Malkin, "Organizatsiia," pp. 261–262.

15. V. A. Krotov, "Organizatsiia," p. 280.

16. E. A. Beliaev and N. S. Pyshkova, Formirovanie, p. 197.

17. "Fakty i dokumenty," p. 49.

18. "Organizatsiia Kirgizskogo filiala," p. 122; K. I. Skriabin, Moia zhizn', pp. 339–342.

19. "Fakty i dokumenty," p. 49.

20. V. F. Kagan, Lobachevskii. See also [N. I. Lobachevskii], N. I. Lobachevskii: Sbornik stat'ei.

21. See particularly P. S. Aleksandrov, "Russkaia matematika"; A. N. Kolmogorov, "Rol' russkoi nauki"; L. S. Pontriagin, "Topologiia"; and B. N. Delone, "Razvitie teorii chisel."

22. M. I. Radovskii, "N'iuton i Rossiia," no. 2, pp. 130–132.

23. T. I. Rainov, "N'iuton"; T. P. Kravets, "N'iuton."

24. L. S. Berg, *Vsesoiuznoe Geograficheskoe obshchestvo*, p. 63.

25. "220-letie Akademii nauk," p. 36. For details on the celebration, as seen by a Western observer, see E. Ashby, *Scientist in Russia*, chap. 6.

26. "220-letie Akademii nauk," p. 51.

27. G. D. Komkov et al., *Akademiia*, 2d ed., vol. 2, p. 256.

28. M. Korsunsky, *Atomic Nucleus*, p. 206.

29. [V. L. Komarov], "Rech' prezidenta," pp. 9–10.

30. N. P. Dubinin, *Vechnoe dvizhenie*, p. 265.

31. M. M. Zavadovskii, "Osnovnye etapy," pp. 54–55. See also I. S. Travin, "Sovremennye tsentry," pp. 245–250.

32. M. A. Markov, "O prirode fizicheskogo znaniia," pp. 140–176.

33. V. S. Neapolitanskaia, *Perepiska (1940–1944)*, pp. 121–122.

34. A. S. Borovik-Romanov, Foreword, p. xxiv.

35. A. A. Zhdanov, *Vystuplenie*.

36. "Diskussiia o prirode fizicheskogo znaniia," no. 3, p. 222.

37. Ibid., p. 235.

38. Loren Graham has advanced cogent arguments in favor of the thesis that Stalin, rather than Zhdanov, masterminded the massive attacks on science, philosophy, and the arts generally subsumed under the category of "zhdanovshchina" (*Science and Philosophy*, appendix 2).

39. "Obrashchenie k tovarishchu Stalinu," p. 20. The phenomenon of Lysenkoism has been subjected to detailed and critical scrutiny in D. Joravsky, *Lysenko Affair;* and Z. Medvedev, *Rise and Fall of Lysenko*. See also T. Dobzhansky, "Soviet Biology"; L. R. Graham, *Science and Philosophy*, chap. 6; D. Lecourt, *Proletarian Science;* F. Belardelli, "The 'Lysenko Affair'"; and R. Lewontin and R. Levins, "Problem of Lysenkoism." The most thorough Soviet studies are I. T. Frolov, *Genetika i dialektika;* and N. P. Dubinin, *Vechnoe dvizhenie*.

40. L. G. Leibson, *Leon Abgarovich Orbeli*, pp. 333–334. On the theoretical front, Orbeli was accused of placing too strong an emphasis on the autonomy of the nervous system in the higher mental activity, of underestimating the role of the cortex in the functioning of internal organs, of claiming that only sympathetic nerves exercise an influence on the organs of trophic activity, and of misinterpreting Pavlov's views on the interaction of subjective and objective factors in the higher nervous activity.

41. "Postanovlenie Prezidiuma," p. 23.

42. F. Dvoriankin, "Pobeda," p. 39. See also "Za rastsvet."

43. V. M. Molotov, *31-ia godovshchina*, p. 20.

44. "Za rastsvet," p. 9.

45. N. A. Maksimov, Review of *Trudy Instituta*, p. 106.

46. "Za protsvetanie," p. 15.

47. O. B. Lepeshinskaia, *Proiskhozhdenie kletok*, p. 23; T. D. Lysenko, "Raboty," pp. 191–193.

48. N. N. Zhukov-Verezhnikov, I. N. Maiskii, and L. A. Kalinichenko, "O nekletochnykh formakh," pp. 320–322.

49. L. N. Pliushch, "Ob osnovakh," p. 185.

50. B. G. Kuznetsov, *Patriotizm*, p. 230.

51. O. B. Lepeshinskaia, *Proiskhozhdenie kletok*, pp. 5–14.

52. O. B. Lepeshinskaia, "Tvorcheskoe znachenie."

53. A. I. Oparin, "Problema proiskhozhdeniia zhizni," p. 285.

54. K. A. Lange, *Razvitie*, p. 265.

55. *Nauchnaia sessiia*, p. 166.

56. A. G. Ivanov-Smolenskii, "Po pavlovskomu puti," p. 87.

57. S. Petrushevskii, "Voinstvuiushchii materializm," p. 24.

58. [S. I. Vavilov], "Vstupitel'noe slovo," p. 7.

59. V. P. Iagunkova, "Ob osnovnykh printsipakh," pp. 112, 115.

60. *Nauchnaia sessiia*, p. 713. Pavlov's theory, trimmed of accumulated digressions, is presented in G. F. Aleksandrov, "Uchenie I. P. Pavlova."

61. "Vil'iams, Vasilii Robertovich," p. 84. A systematic survey of Vil'iams's "biological orientation in soil science" is presented in V. P. Bushinskii, "Uchenie V. R. Vil'iamsa."

62. D. M. Troshin, "Znachenie truda," pp. 212–213.

63. I. A. Prokof'eva, "Konferentsiia." For a succinct but pertinent description of the state of Soviet cosmology at this time, see S. Müller-Markus, *Einstein*, vol. 2, pp. 351–355.

64. B. M. Kedrov, "O putiakh perekhoda," p. 94.

65. B. Iu. Levin, *Origin of the Earth*, pp. 32–33.

66. O. Iu. Shmidt, "Problema," p. 122.

67. S. G. Suvorov, "Leninskaia teoriia," p. 48.

68. V. L'vov, "Trubadury" and "Protiv idealizma."

69. V. A. Fok, "Kritika vzgliadov Bora," p. 13. The influence of the Copenhagen school on Soviet physicists is scrutinized in L. R. Graham, "Quantum Mechanics" and *Science and Philosophy*, chap. 3. See also M. Jammer, *Philosophy of Quantum Mechanics*, pp. 248–251.

70. S. G. Suvorov and R. Ia. Shteinman, "Za posledovatel'no-materialisticheskuiu traktovku," pp. 428–439. The first systematic survey of the process of harmonizing the theory of modern physics with the Leninist "theory of reflection" was presented in S. G. Suvorov, "Leninskaia teoriia."

71. A. A. Maksimov et al., *Filosofskie voprosy sovremennoi fiziki*.

72. The attack on the idealistic leanings of selected Soviet physicists was brought into focus in M. E. Omel'ianovskii, "Bor'ba materializma," pp. 152–166.

73. M. B. Mitin, "Rol' i znachenie," p. 28.

74. M. E. Omel'ianovskii, "Bor'ba materializma," p. 156. For a point-by-point refutation of Einstein's philosophical views, see M. M. Karpov, "O filosofskikh vzgliadakh." Karpov claimed that Einstein's work on the unified field theory could not gain ground because it was based on faulty (idealistic) premises, which led Einstein to exaggerate the role of mathematics in physical theory (p. 136).

75. As cited in V. A. Fok, "Protiv nevezhestvennoi kritiki," p. 72.

76. I. V. Kuznetsov, "Sovetskaia fizika," p. 62. See also B. M. Kedrov, "Neudachnaia kniga," pp. 374–377.

77. The basic criticisms of "physical idealism" are surveyed in M. E. Omel'ianovskii, "Bor'ba materializma," pp. 150–167.

78. E. Ia. Kol'man, "Kuda vedet fizikov sub'ektivizm," p. 189.

79. V. M. Tataevskii and M. I. Shakhparonov, "Ob odnoi makhistskoi teorii"; O. A. Reutov, "O knige G. V. Chelintseva." See also G. V. Chelintsev, "O novoi pozitsii"; and O. A. Reutov, "K voprosu o formalizme."

80. "Ot redaktsii," pp. 194–196.

81. D. N. Kursanov et al., "Present State of Structural Theory," p. 213; "Soveshchanie po teorii khimicheskogo stroeniia," pp. 113–123; L. Pauling, "Fifty Years," pp. 291–293.

82. "Soveshchanie po teorii khimicheskogo stroeniia," p. 123. For more information, see L. R. Graham, *Science and Philosophy*, pp. 317–322.

83. I. L. Knuniants, "O nekotorykh teoreticheskikh problemakh," pp. 18–19.

84. I. I. Potekhin, "Zadachi," p. 25.

85. *Bol'shevik*, the party's theoretical journal, greeted Voznesenskii's book as the most profound analysis of "the superiority of the socialist economy" and of postwar developments in Soviet economic life ("Novye uspekhi sovetskoi nauki," p. 6).

86. V. V. Kolotov, *Nikolai Alekseevich Voznesenskii*, p. 346. For more details on Voznesenskii's downfall, see W. Hahn, *Postwar Soviet Politics*, pp. 129–135.

87. F. Chernov, "Burzhuaznyi kosmopolitizm," pp. 34–37.

88. B. N. Gimmel'farb, "Mirovoe znachenie," p. 80.

89. *Materialy po istorii otechestvennoi khimii*, p. 4.

90. I. I. Potekhin, "Kosmopolitizm," pp. 36–37.

91. B. M. Kedrov, "O znachenii," p. 22.

92. I. I. Potekhin, "Kosmopolitizm," pp. 31–32.

93. S. P. Tolstov, "Krizis," p. 12.

94. D. A. Ol'derogge and I. I. Potekhin, "Funktsional'naia shkola," pp. 54–55.

95. "Za patrioticheskuiu sovetskuiu nauku," p. 7.

96. Ibid., p. 8.

97. W. S. Vucinich, "Structure of Soviet Orientology," p. 70.

98. I. V. Kuznetsov, *Kharakternye cherty*, p. 5.

99. Published in the same year, the papers read at the conference produced a volume of over 800 printed pages.

100. S. I. Vavilov, "Vstupitel'noe slovo," p. 13.

101. Ibid. At this time philosophers and physicists were reminded that a true understanding of Einstein's "positive" contributions to physics could be achieved only by recognizing the role of N. I. Lobachevskii in validating the notion of "relativity," and that of P. N. Lebedev in paving the way for the law of the equivalence of mass and energy (G. A. Kursanov, "K kriticheskoi otsenke," p. 174).

102. S. I. Vavilov et al., *Voprosy istorii*, p. 882.

103. *O polozhenii v biologicheskoi nauke*, p. 513.

104. S. I. Vavilov et al., *Voprosy istorii*, p. 101.

105. A. A. Maksimov, "O znachenii truda I. V. Stalina," p. 9.

106. A. A. Zvorykin, "O sovetskom prioritete," p. 23. See also N. A. Figurovskii, "Nekotorye zadachi," p. 12.

107. A graphic example of the recounting of Russian priorities in science and technology is given in the following excerpt from a paper presented by V. V. Danilevskii at a January 1949 session of the Academy devoted to the history of Russian science: "Led by Lomonosov, Russian inventors made significant contributions to the theory and practice of metallurgy as early as the eighteenth century. They built the largest and most advanced blast furnace in the world, which made Russia the chief supplier of metal on the world market. The industrial revolution in the eighteenth century, a turning point in world history, was made possible by Russian metal. The contribution of P. P. Anosov and D. K. Chernov gave Russia a priority in transforming the production of glass from a craft to a science. At the present time, the entire world uses the following Russian discoveries made in the nineteenth century: P. R. Bagration's gold-mining technique, N. N. Beketov's and D. A. Peniakov's improvements in the production of aluminum and A. A. Musin-Pushkin's and P. G. Sobolevskii's techniques of extracting and processing platinum. To Russia belongs the priority in the construction of ship motors and electrically powered ships. Russia is the original home of electrical light, telegraph, electrical welding, electrical transmission and radio. By combining theory and practice, Russian inventors have contributed to the development of aviation more than inventors of any other country. Mendeleev invented the stratospheric balloon in 1875, A. F. Mozhaiskii constructed the first airplane in 1881 and G. E. Kotel'nikov made the first packed parachute in 1911 . . . and G. N. Nesterov opened the era of high-altitude flights in 1917" ("Sessiia Akademii nauk SSSR posviashchennaia istorii otechestvennoi nauki," p. 92). Danilevskii's aim was not merely to establish an honorable place for Russia in the evolution of modern science and technology, but to give Russia a

position of undisputed superiority and to make Russian contributions pure products of national genius, uninfluenced by outside developments.

108. B. M. Kedrov, "Kriticheskie zametki," pp. 57–58.

109. D. D. Ivanenko, "K itogam diskussii," p. 68.

110. Iu. G. Perel', "O knige B. E. Raikova," p. 196.

111. V. L. Komarov, "O zadachakh," p. 31. Typical works published during this phase were V. F. Kagan, *Lobachevskii*, and B. G. Kuznetsov, *Lomonosov*, both characterized by a cultivated respect for historical detail.

112. M. M. Zavadovskii, "Osnovnye etapy," p. 55.

113. I. S. Travin, "Sovremennye tsentry," pp. 245–250.

114. L. S. Berg, *Vsesoiuznoe Geograficheskoe obshchestvo*, pp. 109–118.

115. G. Vasetskii, "O knige A. A. Maksimova," p. 94. "It is necessary to emphasize," wrote I. V. Kuznetsov in 1947, "that the materialism of the stalwarts of Russian science was one-sided and narrow. Influenced by the bourgeois world view and by the prejudices of the surrounding society, they could extend their materialism to the study of nature, but not to social inquiry." (I. V. Kuznetsov, *Kharakternye cherty*, p. 18.)

116. E. Chernakov, "K istorii," p. 57.

117. See, for example, I. A. Poliakov, "Otechestvennye biologi," pp. 4–8.

118. S. Petrushevskii, "Trudy velikogo fiziologa-materialista," p. 78. Exaggerated but closer to the truth is Bertrand Russell's identification of Pavlov's work on conditioned reflexes with "the behavioristic school in psychology," grounded in positivism (*Wisdom*, p. 299).

119. B. M. Kedrov, "Nekotorye voprosy," p. 658.

120. For a typical example, see I. V. Kuznetsov, "Ser'eznye oshibki," p. 71.

121. S. A. Ianovskaia, *Peredovye idei*, pp. 3–4.

122. M. B. Vil'nitskii, "Za posledovatel'no-materialisticheskuiu traktovku," p. 185.

123. A. A. Maksimov, "Obsuzhdenie," p. 386.

124. I. V. Kuznetsov, "Ser'eznye oshibki," pp. 77–80.

125. "Za patrioticheskuiu sovetskuiu nauku," p. 9.

126. D. M. Troshin, "Marksizm-Leninizm," p. 226.

127. Materialist, "Komu sluzhit kibernetika," pp. 216–217.

128. E. Ia. Kol'man, "Kuda vedet fizikov sub'ektivizm," p. 172. See also D. M. Troshin, "V plenu," p. 44.

129. V. N. Stoletov, "'Materializm,'" p. 237.

130. A. I. Oparin, "Problema proiskhozhdeniia," 1951, p. 283.

131. Ibid., pp. 276–279.

132. Ibid., p. 285.

133. Oparin published two articles under the title "Problema proiskhozhdeniia zhizni v sovremennom estestvoznanii." In the first article, published in 1951,

he lauded both Lysenko's war against the "reactionary idealistic views" of A. Weismann and T. H. Morgan and Lepeshinskaia's "dialectical" approach to the study of noncellular origins of cells ("Problema proiskhozhdeniia," 1953).

134. A. I. Oparin, *Zhizn' kak forma dvizheniia*, p. 17. See also J. Farley, *Spontaneous Generation Controversy*, p. 182.

135. A. M. Kuzin, "O nekotorykh zadachakh," p. 42.

136. Lysenkoist theory of the formation of new species is presented in T. D. Lysenko, "Novoe v nauke," pp. 14–15.

137. V. I. Dvoriashin, "Uchenye na velikikh stroikakh," pp. 44–49.

138. A. N. Nesmeianov, "Velikaia sila," p. 14.

139. Ibid., p. 3.

140. A. F. Ioffe, "Razvitie tochnykh nauk," p. 151.

141. S. I. Vavilov, *Tridtsat' let*, p. 39.

142. E. A. Chudakov, "K perestroike," p. 7.

143. S. I. Vavilov, *Tridtsat' let*, p. 39.

144. A. N. Nesmeianov, "Velikaia sila," p. 6.

145. V. K. Nikol'skii and N. F. Iakovlev, "Osnovnye polozheniia," p. 285.

146. I. V. Stalin, "Ekonomicheskie problemy," pp. 3–7.

147. Ibid., p. 34.

148. For a detailed analysis of Stalin's essay, see P. F. Iudin, "Trud I. V. Stalina." See also G. V. Platonov, "O nekotorykh voprosakh filosofii."

149. A. A. Maksimov, "O znachenii truda I. V. Stalina," p. 8.

150. Within a year after the publication of Stalin's critique of Marr, the Institute of Philosophy sponsored 2,000 lectures and published twenty-nine articles on the theme of Stalinist linguistics. The institute held fifteen conferences on the subject. The Institute of Linguistics revamped its entire annual research plan, shifting toward more formal—that is, grammatical-syntactic—aspects of language ("Sessii i zasedaniia," pp. 36–37, 75). In 1952 the Institute of Philosophy published *Voprosy dialekticheskogo i istoricheskogo materializma v trude I. V. Stalina 'Marksizm i voprosy iazykoznaniia,'* which covered a wide assortment of topics from the "creative character of Marxism-Leninism" to "the objective character of Stalin's linguistic pronouncements."

151. "K semidesiatiletiiu," p. 8.

152. M. B. Mitin, "Rol' i znachenie," p. 40.

153. "Za razvitie peredovoi sovetskoi fiziki," p. vi. See also "Uluchshit' rabotu," pp. 5–15.

154. A. N. Nesmeianov, "Velikaia sila," p. 6.

155. Fok was particularly opposed to Einstein's initial claim that the general theory of relativity was a generalized version of the special theory of relativity (V. A. Fok, "Problema dvizheniia mass," p. 32).

156. G. I. Naan, "K voprosu o printsipe otnositel'nosti," p. 57. For historical comments, see K. Kh. Delokarov, "Teoriia otnositel'nosti i sovetskaia filosofskaia nauka," pp. 558–559.

157. I. P. Bazarov, "Za dialektiko-materialisticheskoe ponimanie," p. 175. In 1953 A. D. Aleksandrov publicly attacked A. A. Maksimov's "idealistic" distortions of both Einstein's theories and dialectical materialism ("Po povodu nekotorykh vzgliadov," pp. 239–245). The attack precipitated a rapid decline in Maksimov's philosophical fortunes.

158. A. D. Aleksandrov, "O nekotorykh obshchikh voprosakh," pp. 13–14.

159. A. D. Aleksandrov, "Ob idealizme," no. 8, p. 6.

160. "Matematika," p. 126.

161. D. N. Kursanov et al., "Present State of Structural Theory," pp. 11–12.

162. The German title of this work is *Die Wandlungen in den Grundlagen der Naturwissenschaft*.

163. "Soveshchanie po teorii khimicheskogo stroeniia," p. 123.

164. L. R. Graham, *Between Science and Values*, p. 96.

165. "Razvertyvat' kritiku."

166. N. P. Dubinin, *Vechnoe dvizhenie*, p. 276.

167. "Rasshirennoe zasedanie," pp. 92–96.

168. D. Joravsky, *Lysenko Affair*, p. 189.

169. N. D. Ivanov, "O novom uchenii," p. 841. For a Lysenkoist answer to Ivanov's charges, see G. V. Platonov, "Nekotorye filosofskie voprosy," pp. 118–120.

170. N. V. Turbin, "Darvinizm," p. 808.

171. Ibid., p. 817.

172. See, for example, G. F. Aleksandrov, "Uchenie I. P. Pavlova," pp. 135–159.

173. N. D. Sokolov, "Soveshchanie," p. 290.

174. D. Joravsky, *Lysenko Affair*, pp. 394–395.

175. V. N. Sukachev, "O vnutrividovykh i mezhvidovykh vzaimootnosheniiakh."

176. N. V. Dylis, "K 100-letiiu," p. 73.

177. See, for example, O. A. Reutov, "K voprosu o formalizme"; and G. V. Chelintsev, "O novoi pozitsii."

178. B. M. Kedrov, "Kriticheskie zametki."

179. M. Karasev and V. Nozdrev, "O knige M. E. Omel'ianovskogo," p. 342.

180. D. D. Ivanenko, "K itogam diskussii."

181. A. A. Maksimov, "Obsuzhdenie," p. 387.

182. N. P. Dubinin, *Vechnoe dvizhenie*, p. 278.

183. A. Ia. Khinchin, "Poniatie entropii."

FIVE: THE THAW AND THE SCIENTIFIC AND TECHNOLOGICAL REVOLUTION

1. "Nauchnye sessii posviashchennye 50-letiiu KPSS," p. 28.

2. B. M. Kedrov, "D. I. Mendeleev ob internatsional'nom kharaktere nauki,"

p. 63. See also the author's "O zakonakh," p. 101, and "D. I. Mendeleev o roli uchenykh," pp. 21–23.

3. D. Danin, "Kvanty pamiati," pp. 135–136. See also M. B. Adams, "Biology after Stalin," p. 57.

4. I. E. Tamm, "Rol'," p. 13.

5. N. N. Semenov, *Nauka i obshchestvo*, p. 222.

6. *Spravochnik partiinogo rabotnika*, pp. 459–465.

7. "Godichnoe sobranie Akademii nauk SSSR (1–2 fevralia 1965 g.)," p. 97.

8. N. N. Semenov, "Nauka ne terpit sub'ektivizma," p. 42.

9. "Doklad Komissii," p. 31.

10. Ibid., p. 127. See also L. Ia. Bliakher, *Problema nasledovaniia*, pp. 7–8, 206–207.

11. B. M. Kedrov, "Vzaimodeistvie nauk," pp. 562–563, and "Puti poznaniia istiny," pp. 232–233.

12. V. V. Skripchinskii, "O razlichii," pp. 108–115. See also N. P. Dubinin, *Vechnoe dvizhenie*, p. 201; and S. I. Alikhanian, *Teoreticheskie osnovy*, pp. 48–60.

13. P. Terent'ev, "Zhizn'."

14. A. V. Topchiev, "Sovetskaia nauka v pervom godu," p. 9. See also M. B. Adams, "Biology after Stalin," p. 55.

15. N. P. Dubinin, *Teoreticheskie osnovy*, pp. 172–173, and "Genetika," p. 271. The biologists lost no time in recasting the place of Lysenko in modern science. In 1966 the Academy of Sciences published a collection of articles on the development of evolutionary theories in biology. The volume made only one reference to Lysenko: "The views of T. D. Lysenko received a particularly strong emphasis during the last two decades. According to Lysenko, changes in species occur as a result of direct adjustment of organisms to the environment and full inheritance of acquired characteristics. These views can be treated only as a return to the naive Lamarckism of a mechanist variety" (V. I. and Iu. I. Polianskii, *Istoriia*, p. 287).

16. V. S. Nemchinov, *Izbrannye proizvedeniia*, p. 374.

17. D. G. Protskaia, "K voprosu o perezhitkakh," p. 128.

18. P. S. Romashkin, "O nekotorykh problemakh," p. 323. The reasons behind the reemergence and modernization of sociology are discussed in R. Ahlberg, "Einleitung," pp. 19–20. According to Ahlberg, pressing social and economic needs rather than enlightenment on the part of the architects of the national science policy were behind the rebirth of sociology in the Soviet Union.

19. For general comments on the early Stalinist phase of Soviet social psychology, see D. Joravsky, "Construction of the Stalinist Psyche," pp. 120–126.

20. "O filosofskikh voprosakh psikhologii," pp. 185–189.

21. L. G. Leibson, *Leon Abgarovich Orbeli*, p. 343.

22. A. N. Leont'ev and A. R. Luriia, "Psikhologicheskie vozzreniia," p. 4. For a detailed description of the new interpretation of Vygotskii's theories, see A. A. Smirnov, *Razvitie*, pp. 168–178; and A. R. Luriia, "Teoriia razvitiia."

23. S. L. Rubinshtein, *Bytie;* A. N. Leont'ev, *Problemy.*
24. P. N. Fedoseev et al., *Filosofskie voprosy fiziologii,* pp. 757–762.
25. A. R. Luriia, "Psikhologiia v sisteme estestvoznaniia," p. 221.
26. A. R. Luriia, "Mozg i psikhika," pp. 113–114.
27. L. R. Graham, *Science and Philosophy,* pp. 427–428.
28. N. A. Bernshtein, "Novye linii," p. 322.
29. L. Pauling, "Fifty Years," p. 293.
30. A. N. Nesmeianov, "Ob osnovnykh napravleniiakh," p. 20.
31. N. Wiener, *Human Use of Human Beings,* p. 193.
32. [L. F. Il'ichev], "Doklad," p. 24.
33. L. R. Kerschner, "Western Translations."
34. E. Ia. Kol'man, "Chto takoe kibernetika," p. 159.
35. V. A. Engel'gardt, "Novyi etap," p. 25.
36. N. N. Semenov, "Budushchee cheloveka," p. 102.
37. For comments from Western sources, see W. Kneale and M. Kneale, *Development of Logic,* pp. 522, 705, 726; E. W. Beth, *Foundations of Mathematics,* pp. 363, 510; J. von Heijenoort, *From Frege to Gödel,* pp. 355–365, 414–437.
38. E. Ia. Kol'man, "Znachenie," p. 16.
39. S. A. Ianovskaia, "Osnovaniia," p. 45.
40. A. A. Markov, "Matematicheskaia logika," *FE,* pp. 340–342.
41. See, for example, N. I. Stiazhkin, "Elementy," pp. 21–31.
42. E. Ia. Kol'man, "Znachenie," p. 18.
43. A. A. Markov, "Matematicheskaia logika," *BSE,* p. 480.
44. B. M. Kedrov, "Revoliutsiia v estestvoznanii," p. 75.
45. M. V. Keldysh, "Estestvennye nauki," p. 39.
46. B. M. Kedrov, "Puti razvitiia nauki," p. 5; N. N. Semenov, *Nauka i obshchestvo,* pp. 259, 452.
47. V. I. Shinkaruk, "Integrativnaia funktsiia," p. 41.
48. B. M. Kedrov, S. Mikulinskii, and I. Frolov, "Filosofsko-sotsiologicheskie problemy," p. 45.
49. A. A. Zvorykin, "Nauka," p. 53.
50. E. M. Babasov, "Nauchno-tekhnicheskaia revoliutsiia," p. 23.
51. S. R. Mikulinskii, "From the Revolution," p. 77.
52. D. L. Jassby, "Princeton Fusion Experiment," p. 370.
53. A. Kramish, *Atomic Energy,* pp. 183, 200–202.
54. S. Trapeznikov, "Leninizm," pp. 7–11.
55. For a brief survey of Kapitsa's life and work, see A. Parry, *Russian Scientist,* pp. 115–138.
56. P. L. Kapitsa, *Eksperiment,* p. 235.
57. Ibid., pp. 183–184.
58. Z. A. Medvedev, *Medvedev Papers,* p. 168.
59. For major types of informal groupings and activities, see I. I. Leiman, *Nauka,* pp. 161–162.

60. P. L. Kapitsa, *Experiment*, p. 172.

61. L. A. Artsimovich, "Fizik," p. 203.

62. I. V. Kuznetsov and A. S. Fedorov, "60 let," p. 5.

63. "Nauchnye issledovaniia po fiziologii," pp. 48–49.

64. P. L. Kapitsa, "Ob organizatsii," p. 100.

65. E. M. Kreps, *Perepiska*, pp. 38–39.

66. A. A. Zvorykin and E. I. Rabinovich, "Organizatsionnaia struktura," p. 99.

67. R. W. Davies and R. Amann, "Science Policy," p. 25.

68. [L. A. Artsimovich], "Nekotorye voprosy," p. 5.

69. A. N. Kosygin, "Za tesnuiu sviaz'," p. 95.

70. L. A. Artsimovich, "Fizik," p. 203.

71. Ibid.

72. M. A. Leontovich and V. L. Ginzburg, "Nekotorye voprosy," p. 132.

73. G. D. Komkov et al., *Akademiia*, 2d ed., vol. 2, p. 291; V. I. Duzhenkov, "Nauchnye tsentry," p. 241. In 1973, to give an exact number, the Siberian Department had 5,959 "scientific workers" (*Narodnoe khoziaistvo SSSR v 1973 g.*, p. 183).

74. P. M. Cocks, *Science Policy*, p. 49. See also L. R. Graham, "Place of the Academy in Soviet Science," pp. 54–55.

75. G. Marchuk, "Krai bol'shoi nauki," p. 34; A. Trofimuk, "Strategiia," p. 41.

76. M. A. Lavrent'ev and A. Trofimuk, "Nauka v Sibiri," pp. 67–68.

77. D. Gvishiani, S. R. Mikulinskii, and S. A. Kugel, *Scientific Intelligentsia*, p. 168.

78. A curricular breakdown of mathematics teaching at Novosibirsk is presented in I. Z. Shtokalo et al., *Istoriia matematicheskogo obrazovaniia*, pp. 167–168, 181.

79. I. N. Vekua, "Vysshaia shkola," p. 20.

80. "Razvitie nauchnykh uchrezhdenii," p. 3.

81. M. V. Keldysh, "Mnogonatsional'nyi Soiuz," p. 32.

82. V. I. Duzhenkov, *Problemy*, p. 197.

83. Ibid., p. 177.

84. M. D. Millionshchikov, "Shirokie gorizonty."

85. Y. Sheinin, *Science Policy*, p. 145.

86. V. I. Duzhenkov, "Nauchnye tsentry," p. 249. See also Iu. Zhdanov, "Ob'edinenie."

87. V. I. Duzhenkov, *Problemy*, p. 119.

88. "Nauchnyi tsenter."

89. G. R. Ivanitskii et al., "Nauchnye uchrezhdeniia," pp. 54–55.

90. [L. A. Artsimovich], "Nekotorye voprosy," p. 13.

91. G. Marchuk, "Sibirskomu otdeleniiu," p. 637.

92. G. D. Komkov et al., *Akademiia*, 2d ed., vol. 2, p. 385.

93. The new trend in the geographical expansion of scientific institutions is discussed in B. A. Ruble, "Expansion of Soviet Science," p. 534.

94. V. I. Duzhenkov, *Problemy*, p. 144. See also G. M. Dobrov et al., *Potentsial nauki*, p. 45; E. A. Beliaev, "Sovershenstvovanie," p. 265.

95. "O sostoianii podgotovki doktorov nauk," p. 64.

96. E. A. Beliaev, "Voprosy territorial'nogo razmeshcheniia nauki," p. 341.

97. I. P. Gureev, V. V. Zverev, and A. I. Kolchin, "Nauchnye kadry," p. 125.

98. D. Gvishiani, S. R. Mikulinskii, and S. A. Kugel, *Scientific Intelligentsia*, p. 174.

99. A. K. Romanov, L. A. Androsova, and A. F. Felinger, *Nauchnye kadry*, p. 36.

100. D. Gvishiani, S. R. Mikulinskii, and S. A. Kugel, *Scientific Intelligentsia*, p. 125.

101. V. P. Eliutin, "Konkurs"; A. Vucinich, "Peasants," pp. 313–314. See also A. N. Kolmogorov, "Matematika na poroge vuza," p. 246.

102. Vl. Kantorovich, "Sotsiologiia i literatura," p. 127. See also N. A. Aitov, "Sotsial'nye aspekty," p. 193; M. P. Rutkevich, "Problemy izmeneniia," p. 52, and "Sotsial'nye istochniki," p. 19.

103. I. P. Gureev, V. V. Zverev, and A. I. Kolchin, "Nauchnye kadry," p. 124; Y. Sheinin, *Science Policy*, p. 159; D. Gvishiani, "Sotsial'naia rol' nauki," p. 57.

104. V. G. Afanas'ev, *Nauchno-tekhnicheskaia revoliutsiia*, p. 276.

105. S. R. Mikulinskii, "Nekotorye problemy," p. 143.

106. G. M. Dobrov et al., *Potentsial nauki*, pp. 31–32. See also G. M. Dobrov, *Nauka o nauke*, pp. 138–141. At the same time, the ratio of doctors of science to candidates of science underwent a similar change. In 1950 there were five candidates of science for each doctor of science; in 1965 there were nine candidates of science for each doctor of science (I. I. Leiman, *Nauka*, p. 149).

107. M. M. Karpov et al., *Sotsiologiia nauki*, p. 92.

108. V. Krasil'nikov, "Vykhod—v planirovanii," p. 13. See also V. A. Rassudovskii and V. P. Rassokhin, *Pravo*, p. 93.

109. G. D. Komkov et al., *Akademiia nauk SSSR: Kratkii istoricheskii ocherk*, p. 458.

110. For a perceptive and careful assessment of the comparative development of Soviet technology, see R. Amann, "Some Approaches," pp. 23–24. The problem of Soviet dependence on Western technology has received extensive treatment in A. C. Sutton, *Western Technology*.

111. A. Parry, *Peter Kapitsa*, pp. 218–219. V. G. Afanas'ev has noted that the productivity of Soviet scientists is "significantly lower" than that of American scientists (*Nauchno-tekhnicheskaia revoliutsiia*, p. 277).

112. S. R. Mikulinskii, "Nekotorye problemy," p. 143.

113. G. M. Dobrov et al., *Potentsial nauki*, p. 48.

114. V. G. Afanas'ev, *Nauchno-tekhnicheskaia revoliutsiia*, pp. 289–290.

115. "Obsuzhdenie doklada M. V. Keldysha," pp. 41–42.

116. S. A. Kugel' et al., *Nauchnye kadry,* p. 105.

117. V. A. Rassudovskii and V. P. Rassokhin, *Pravo,* p. 94.

118. A. N. Nesmeianov, "Ob osnovnykh napravleniiakh," p. 38.

119. N. N. Semenov, "Nauka segodnia i zavtra," p. 3. See also Semenov's speech at the December 1959 Plenum of the Central Committee of the Communist party (*Plenum Tsentral'nogo Komiteta KPSS,* pp. 629–630).

120. Cited in S. R. Mikulinskii, *Nauka segodnia,* pp. 263–264.

121. A. Korol, *Soviet Research,* p. 28.

122. For general comments on the role of the Academy in technological progress, see R. Amann, M. J. Berry, and R. W. Davies, "Science and Industry," pp. 430–431.

123. G. D. Komkov et al., *Akademiia,* 2d ed., vol. 2, p. 278; V. P. Rassokhin, "Problemy razvitiia," p. 66.

124. M. V. Keldysh, "Voprosy organizatsii," p. 18.

125. K. Meyer, *Das wissenschaftliche Leben,* p. 11.

126. R. Amann, "Soviet Research," p. 233. The functions of the committee are outlined in N. de Witt, "Reorganization," pp. 1988–1989.

127. *Izvestiia,* April 12, 1961, p. 1.

128. H. Wienert, "Organization and Planning," p. 205.

129. The reforms and their chief proponents and opponents are discussed in L. R. Graham, "Reorganization," pp. 137–159.

130. *KPSS v rezoliutsiiakh,* vol. 8, pp. 396–402. For further analysis and pertinent details, see A. Korol, *Soviet Research,* pp. 40–44.

131. V. A. Rassudovskii, *Gosudarstvennaia organizatsiia,* pp. 25–30.

132. E. A. Beliaev and N. S. Pyshkova, *Formirovanie,* p. 137.

133. A. A. Zvorykin and E. I. Rabinovich, "Organizatsionnaia struktura," p. 90.

134. G. M. Dobrov, "Gosudarstvennaia politika," p. 164; P. M. Cocks, *Science Policy,* pp. 37–41. For detailed information on the jurisdiction and function of the committee, see G. A. Dorokhova, "Obshchegosudarstvennaia organizatsiia," pp. 157–170.

135. V. A. Rassudovskii and M. S. Bastrakova, "Akademiia nauk," p. 18.

136. A. Korol, *Soviet Research,* p. 41; I. M. Muksinov and I. L. Davitnidze, "Akademiia," p. 212; V. A. Rassudovskii, *Gosudarstvennaia organizatsiia,* p. 38.

137. Y. Sheinin, *Science Policy,* p. 211.

138. V. G. Shorin and A. A. Popova, "Organizatsiia," p. 209. For a discussion of the general problems of scientific research in institutions of higher education, see "Nauchnaia rabota v vuzakh," pp. 3–16.

139. V. P. Eliutin, "Nauka i vysshee obrazovanie," p. 171.

140. "Obsuzhdenie doklada M. V. Keldysha," pp. 42–43.

141. V. A. Rassudovskii and V. P. Rassokhin, *Pravo,* p. 93.

142. V. G. Shorin and A. A. Popova, "Organizatsiia"; V. A. Rassudovskii and V. P. Rassokhin, *Pravo*, pp. 99–100.

143. V. G. Afanas'ev, *Nauchno-tekhnicheskaia revoliutsiia*, pp. 298–299.

144. V. P. Eliutin, "Osnovnye puti," p. 32.

145. "Nekotorye voprosy razvitiia sovremennoi fiziki," pp. 14, 25, 42.

146. B. Parrott, "Organizational Environment," p. 79.

147. Z. A. Medvedev, *Soviet Science*, p. 131.

148. "Obshchee sobranie Akademii nauk SSSR (29–30 iiunia 1962 g.)," p. 7.

149. For the full text of the 1963 charter, see G. K. Skriabin et al., *Ustavy*, pp. 166–183.

150. Ibid., pp. 176–177.

151. A. Korol, *Soviet Research*, pp. 43–44.

152. "Patrioticheskii dolg," pp. 4–6.

153. According to the August 1967 decision of the Presidium of the Academy, the councils were assigned the following tasks: analysis of the present state of science in the USSR and abroad; determination of basic orientations and lines of research in selected areas of primary importance; recommendations for application of specific scientific results in the national economy; maintenance of relations with other scientific councils for the purpose of coordinating research on related problems; and improvement of planning procedures in science and coordination of research carried on by various institutions regardless of their official affiliation (K. A. Lange, *Organizatsiia*, p. 173; see also N. I. Golubtsova, "Organizatsiia," pp. 189–191).

154. K. A. Lange, *Organizatsiia*, p. 172.

155. Ibid., p. 174.

156. "Molodezh v nauke"; I. I. Leiman, *Nauka*, p. 162.

157. V. A. Rassudovskii, *Gosudarstvennaia organizatsiia*, p. 132.

158. P. M. Cocks, "Policy Process," p. 178.

159. For a cogent and critical analysis of the Soviet setting of the matrix system, see P. M. Cocks, "Retooling the Directed Society," p. 71.

SIX: SCIENCE AND CHANGING VALUES

1. S. R. Mikulinskii, "Nekotorye problemy," pp. 148–149. For a systematic and critical survey of *naukovedenie*, see L. L. Lubrano, *Soviet Sociology of Science*, pp. 3–30. See also K. Meyer, "Wissenschaftspolitik," pp. 184–187. The expression "science of science" was introduced by M. Ossowska and S. Ossowski, "Science of Science." For details, see R. Scharff, *Wissenschaftsorganisation*, p. 59; and G. M. Dobrov, *Nauka o nauke*, pp. 13–14.

2. G. M. Dobrov, "Osnovy nauchnoi politiki," p. 6.

3. S. R. Mikulinskii and N. I. Rodnyi, "Nauka kak predmet," p. 26.

4. D. I. Gordeev and E. A. Kurazhkovskaia, "Problemy naukovedeniia," p.

12. According to Mikulinskii, as cited by L. L. Lubrano, "the purpose of *nau-kovedenie* is to create an interdisciplinary theory of science, one that would integrate the social aspects of science with its political, economic, historical and philosophical dimensions, while at the same time supporting the Marxist-Leninist framework of analysis" ("Scientific Collectives," pp. 131–132). See also K. E. Bailes, "Technical Specialists," p. 119.

5. S. R. Mikulinskii, "Nekotorye problemy," p. 145.

6. For a systematic and critical survey of the development of the science of science, see Y. M. Rabkin, "'Naukovedenie.'" See also E. M. Mirskii, "Nauko-vedenie," pp. 86–98; and G. M. Dobrov et al., *Potentsial nauki,* chap. 1.

7. S. R. Mikulinskii, "O naukovedenii," pp. 34–36.

8. For a thorough survey of Soviet interests and accomplishments in scientometrics, see V. V. Nalimov and Z. M. Mul'chenko, *Naukometriia.*

9. D. Gvishiani, "Nauchno-tekhnicheskaia revoliutsiia," p. 52.

10. S. R. Mikulinskii and N. I. Rodnyi, "Istoriia nauki i naukovedenie," p. 58.

11. L. L. Lubrano, *Soviet Sociology of Science,* p. 93. Lubrano's book provides a good survey and assessment of theoretical and methodological developments in the Soviet sociology of science and related disciplines.

12. A. D. Sakharov, *Progress,* pp. 84–85.

13. N. A. Figurovskii, "Zadachi issledovaniia istorii nauki," p. 31.

14. S. Ia. Plotkin, "Organizatsiia," p. 8.

15. E. Ia. Kol'man, "K sporam o teorii otnositel'nosti," p. 189.

16. A. L. Takhtadzhian, "Tektologiia," pp. 205–208, 274–275. M. I. Setrov treated Bogdanov's theory as a beginning of the modern search for the general principles of organization and recommended a republication of his *Essays on the Universal Organization Theory* ("Ob obshchikh elementakh," p. 60). This marked the first effort to rescue Bogdanov's theoretical legacy from total rejection imposed by Leninist-Stalinist philosophers.

17. V. A. Ambartsumian, "Kosmogoniia," pp. 349–350; B. Iu. Levin, "Raz-vitie," pp. 306–313; B. M. Kedrov, *Lenin i nauchnye revoliutsii,* pp. 253–255.

18. N. V. Tsitsin, "I. V. Michurin," pp. 97–111.

19. B. M. Kedrov, "Ob otnoshenii marksizma," pp. 149–166.

20. I. T. Frolov, "Modern Science," p. 124.

21. J. Monod, *Chance and Necessity,* p. 40.

22. B. M. Kedrov, *O marksistskoi istorii estestvoznaniia,* p. 18. See also Kedrov's "Predmet i zadachi," pp. 8–9.

23. A. D. Aleksandrov, "Raz uzh zagovorili," p. 210. See also I. I. Leiman, *Nauka,* p. 57.

24. L. A. Markova, "Ob istorii estestvoznaniia," p. 133; N. I. Rodnyi, *Ocherki,* pp. 206–210.

25. L. R. Graham, "Development of Science Policy," p. 33.

26. M. V. Keldysh, "Estestvennye nauki," p. 29.

27. I. S. Kon, *Sotsiologiia lichnosti*, pp. 89, 92; Y. Sheinin, *Science Policy*, p. 40.
28. M. Petrov and A. Potemkin, "Nauka," p. 248.
29. "K itogam diskussii," p. 135.
30. M. V. Keldysh, "Estestvennye nauki," pp. 45–46. The papers commemorating Einstein's death in 1955 demonstrated the profound respect of the leading Soviet physicists for the scientific legacy of the great man. These papers marked the beginning of a rapid disappearance of the breed of philosophers who made their careers by interpreting dialectical materialism as an antirelativist philosophy (see particularly A. F. Ioffe, "Pamiati Al'berta Einshteina"; E. V. Shpol'skii, "Al'bert Einshtein"). In 1956 two publications made a particularly strong contribution to Einstein's popularity: Leopold Infeld's warm reminiscences about his Princeton days with Einstein ("Moi vospominaniia") and Einstein's autobiography (A. Einstein, "Tvorcheskaia avtobiografiia"), both in Russian translation.
31. A. D. Aleksandrov, "Dialektika i nauka," pp. 11–16. In 1959 *Voprosy filosofii* published another article by A. D. Aleksandrov, which summarized the main arguments of the new dialectical-materialistic interpretation of the theory of relativity. According to Aleksandrov, the theory of relativity represents (1) a confirmation and advancement of the views of dialectical materialism on "space and time as forms of the existence of matter"; (2) a corroboration and elaboration of the dialectical-materialistic theory of the inseparability of matter and motion; (3) a confirmation of the dialectical unity of the absolute and relative views of nature; and (4) an illustration of the material unity of nature ("Filosofskoe soderzhanie," p. 84).
32. For papers presented at this conference and for related discussion material, see P. N. Fedoseev et al., *Filosofskie problemy sovremennogo estestvoznaniia*.
33. V. A. Ambartsumian, *Filosofskie voprosy*, pp. 35–36.
34. For comments on and summaries of the proceedings of the conference, see "Vsesoiuznoe soveshchanie po filosofskim voprosam estestvoznaniia"; and E. N. Chesnakov, "All-Union Conference."
35. B. M. Kedrov, "Revoliutsiia v estestvoznanii i filosofii," p. 73.
36. A. D. Aleksandrov, "Dialektika i nauka," p. 16. For a Western comment on Aleksandrov's philosophical effort, see S. Müller-Markus, "Diamat and Einstein," pp. 76–77.
37. [L. F. Il'ichev], "Doklad," p. 41.
38. M. V. Keldysh, "Estestvennye nauki," p. 29.
39. [L. F. Il'ichev], "Doklad," p. 20.
40. *Ibid.*, p. 41.
41. M. E. Omel'ianovskii, "Dialekticheskii materializm i sovremennaia fizika," p. 21.
42. The changing attitude toward Bohr is discussed in S. Müller-Markus, "Niels Bohr," pp. 87–91.
43. M. E. Born, *Physics and Politics*, chap. 2.

44. M. E. Omel'ianovskii, "Dialekticheskii materializm i problema real'nosti," pp. 12–37.

45. M. E. Omel'ianovskii, "Lenin," pp. 130, 146–147.

46. P. S. Dyshlevyi, V. I. Lenin, pp. 143–147; V. L. Ginzburg, "Eksperimental'naia proverka," p. 134.

47. For a succinct statement of Fok's reinterpretation of the general theory of relativity, see V. A. Fok, "Poniatiia odnorodnosti," pp. 131–135. See also M. G. Veselov, "Nauchnaia deiatel'nost'," p. 23.

48. P. S. Dyshlevyi and A. Z. Petrov, "Filosofskie voprosy," p. 96.

49. Iu. V. Sachkov, "Obsuzhdenie," pp. 74–75. See also Ya. P. Terletsky, "On the Problem of Spatial Structure," pp. 149–150.

50. R. V. Ryvkina, "O strukture," p. 12.

51. D. M. Ugrinovich, "O strukture," p. 12.

52. E. S. Markarian, Ocherki, pp. 76–81, "Chelovecheskoe obshchestvo," pp. 63–66, and Voprosy sistemnogo issledovaniia, pp. 44–60.

53. L. F. Il'ichev, "Metodologicheskie problemy," p. 137.

54. E. D. Modrzhinskaia and Ts. A. Stepanian, Future Society, p. 101.

55. G. Vagnov, "Nauka," p. 12.

56. "Vystupleniia," p. 20.

57. "Godichnoe sobranie Akademii nauk SSSR (1–2 fevralia 1965 g.)," p. 97.

58. For his attack on Frenkel' and Landau in 1957, see S. T. Meliukhin, "Neischerpaemost'," pp. 194–195.

59. "Obshchee sobranie Akademii nauk SSSR (19–20 oktiabria 1962 g.)," p. 48.

60. A. A. Zinov'ev, "Logical and Philosophical Implications," pp. 91–92.

61. W. F. Boeselager, Soviet Critique, p. 49.

62. V. A. Smirnov and P. V. Tavanets, "O vzaimootnoshenii," p. 32.

63. Soviet work in constructive mathematics is discussed in A. Mostowski, Thirty Years, p. 105.

64. B. G. Kuznetsov, "O kvantovo-reliativistskoi logike"; I. P. Stakhanov, "Logika 'vozmozhnogo'"; A. A. Zinov'ev, "O logike mikrofiziki." A solid contribution to many-valued logic, Zinov'ev's article presents arguments against the need for a special logic for microphysics.

65. B. Russell, Wisdom, p. 289.

66. A. A. Zinov'ev, "O printsipakh determinizma," p. 60.

67. A. A. Liubishchev, "K klassifikatsii," p. 208.

68. M. N. Alekseev et al., "Predmet," pp. 288, 299–301.

69. A. F. Okulov, Sovetskaia filosofskaia nauka, pp. 70–71.

70. D. D. Comey, "Current Trends," p. 105.

71. M. N. Alekseev et al., "Predmet dialekticheskoi logiki," p. 287.

72. A. Ia. Sharov, "Prezidium," pp. 134–139.

73. L. A. Vladislavskii and V. I. Kuraev, "Obsuzhdenie," pp. 116–118.

74. A. L. Subbotin, "Po sledam 'Novogo Organona,'" p. 106.
75. Particularly notable and refreshingly original was the work of E. S. Markarian. See his *Ocherki teorii kul'tury* and *Voprosy sistemnogo issledovaniia*.
76. M. B. Mitin, "Problema," p. 17.
77. P. V. Kopnin, "O prirode," pp. 127–128; M. B. Mitin, "Leninskii etap," pp. 10–12.
78. M. Berthelot, *Science et morale*, p. 2; K. A. Timiriazev, *Sochineniia*, p. 306.
79. V. I. Vernadskii, *Ocherki i rechi*, vol. 2, p. 49.
80. V. P. Tugarinov, *O tsennostiakh*, p. 113.
81. O. G. Drobnitskii, "Priroda," pp. 36–37.
82. E. G. Fedorenko, *Osnovy*, p. 94.
83. For comments on Brecht's *Galileo*, see V. I. Tolstykh, "Galileo," pp. 332–357; and A. V. Gulyga, "Krizis," pp. 358–385.
84. E. Feinberg, "Obyknovennoe," p. 228.
85. A. N. Nesmeianov, "Vozmozhnosti nauki," p. 11.
86. A. D. Aleksandrov, "Nravstvennoe znachenie nauki."
87. P. S. Aleksandrov, "Neobkhodimo vdokhnovenie," p. 12.
88. H. Poincaré, *Mathematics and Science*, p. 113.
89. V. I. Tolstykh, "Nauka," pp. 79–80.
90. Iu. Shreider, "Nauka," p. 225.
91. Ibid., p. 221.
92. A. Einstein, *Essays*, p. 67.
93. For more information on the relations of science to morality, see L. L. Lubrano, *Soviet Sociology of Science*, pp. 89–91.
94. For an interesting view on the "thaw" in Soviet aesthetics, see I. Frizer, "Theory of Beauty."
95. A. Moszkowski, *Einstein*, p. 185.
96. B. G. Kuznetsov, "Einshtein i Dostoevskii," and *Etiudy ob Einshteine*, pp. 119–134. See also D. Danin, "Vozmozhnye resheniia," pp. 92–96.
97. B. S. Meilakh, *Na rubezhe*, p. 160.
98. Ibid., pp. 218–219.
99. O. E. Mandel'shtam, "Zametki," p. 443.
100. B. Runin, *Vechnyi poisk*, p. 169.
101. Ibid., p. 68.
102. A. S. Arsen'ev, "Nauka i chelovek," p. 155.
103. B. Runin, *Lichnost' i tvorchestvo*, p. 117.
104. A. Herzen, *Selected Philosophical Works*, pp. 82–83.
105. P. S. Aleksandrov, "O prizvanii uchenogo," p. 123.

CONCLUSION

1. The conflict between the physicists and philosophers has been analyzed in A. Vucinich, "Soviet Physicists."

2. F. V. Konstantinov, "Sovremennye problemy," p. 41.

3. I. T. Frolov, *Genetika i dialektika*.

4. When it was published in 1965, N. N. Semenov's "Nauka ne terpit sub'ektivizma" provided the sharpest criticism of Lysenkoism yet to be launched by a Soviet scholar, calling it a pseudoscience and a flagrant attack on the ethos of science. When the article was republished in 1973—as a chapter in *Nauka i obshchestvo*, a collection of Semenov's papers—all references to Lysenko and Lysenkoism were eliminated (*Nauka i obshchestvo*, pp. 207–209).

5. A few years later Frolov asserted that although socialism provided constitutional guarantees for the freedom of research and for the humanistic aspirations of scientists, it did not do away with all contradictions connected with the realization of these goals. After rejecting attempts to interpret Lysenkoism as an inevitable product of socialism, he noted: "Such phenomena [as Lysenkoism] do not follow from the nature of socialism. More than that, they contradict it, and if they have taken place at a definite state of historical development, it should not be forgotten that we ourselves have criticized them and cast them overboard" ("Modern Science and Humanism," p. 124). The essence of Frolov's argument seems to be that since Lysenkoism was the result of—and could have taken place only at—a "definite stage" in the evolution of socialism, no institutional safeguards are necessary to prevent the rise of a future Lysenkoism.

6. "Chestvovanie akademika M. B. Mitina," p. 113.

BIBLIOGRAPHY

Adaman, G. "Istoricheskii materializm." *BSE* 29 (1935): 676–762.

Adams, Mark B. "Biology After Stalin." *Survey* 23 (1977): 53–58.

———. "The Founding of Population Genetics: Contributions of the Chetverikov School 1924–1934." *Journal of the History of Biology* 1, no. 1 (1968): 23–39.

———. "Nikolai Ivanovich Vavilov." In *Dictionary of Scientific Biography*, supp. vol. 16: 505–513.

———. "Science, Ideology and Structure: The Kol'tsov Institute, 1900–1970." In *The Social Context*, 173–204. See Lubrano and Solomon 1980.

———. "Sergei Chetverikov, the Kol'tsov Institute, and the Evolutionary Synthesis." In *The Evolutionary Synthesis*, 242–278. See Mayr and Provine 1980.

———. "Severtsov and Schmalhausen: Russian Morphology and the Evolutionary Synthesis." In *The Evolutionary Synthesis*, 193–225. See Mayr and Provine 1980.

Aepinus, F. U. T. *Essay on the Theory of Electricity and Magnetism*. Trans. by P. J. Connor. Princeton, N. J.: Princeton University Press, 1979.

Afanas'ev, V. G. *Nauchno-tekhnicheskaia revoliutsiia, upravlenie, obrazovanie*. Moscow, 1972.

———. *The Scientific and Technological Revolution—Its Impact on Management and Education*. Moscow, 1975.

Agol, I. "Zadachi marksistov-lenintsev v biologii." *PZM*, 1930, no. 5: 95–111.

Ahlberg, René. "Einleitung: Die Entwicklung der sowjetischen Soziologie." In *Soziologie*, 9–49. See Ahlberg 1969.

———. ed. *Soziologie in der Sowjetunion: Ausgewählte sowjetische Abhandlungen zu Problemen der sozialistischen Gesellschaft*. Freiburg, 1969.

Aitov, N. A. "Sotsial'nye aspekty polucheniia obrazovaniia v SSSR." *Sotsial'nye issledovaniia* 2 (1968): 187–196.

"Akademiia nauk v pervyi god Velikogo Oktiabria." *VIE-T*, 1977, no. 58: 8–19.

"Akademiia nauk SSSR v 1937 godu." *VAN*, 1937, nos. 11–12: 345–347.

Aleksandrov, A. D. "Dialektika i nauka." *VAN*, 1957, no. 6: 3–17.

————. "Filosofskoe soderzhanie i znachenie teorii otnositel'nosti." *VF*, 1959, no. 1, pp. 67–84.

————. "Nravstvennoe znachenie nauki." *LG*, March 29, 1967.

————. "Ob idealizme v matematike." *Priroda*, 1951, no. 7: 3–11; no. 8: 3–9.

————. "O nekotorykh obshchikh voprosakh nauchnoi raboty i prepodavaniia matematiki." *VLU*, 1950, no. 1: 3–20.

————. "Po povodu o nekotorykh vzgliadov na teoriiu otnositel'nosti." *VF*, 1953, no. 5: 225–245.

————. "Raz uzh zagovorili o nauke." *NM*, 1970, no. 10: 204–220.

Aleksandrov, A. P. "Shest'desiat let sovetskoi nauki." In *Oktiabr' i nauka*, 5–36. *See* A. P. Aleksandrov et al. 1977.

Aleksandrov, A. P., et al., eds. *Oktiabr' i nauka*. Moscow, 1977.

Aleksandrov, G. F. "I. V. Stalin o roli nadstroiki v razvitii obshchestva." *VF*, 1952, no. 1: 3–22.

————. "Uchenie I. P. Pavlova—velikii vklad v nauku." In *Filosofskie voprosy*, 135–159. *See* Novinskii and Platonov 1951.

Aleksandrov, G. F., et al., eds. *Voprosy dialekticheskogo i istoricheskogo materializma v trude I. V. Stalina 'Marksizm i voprosy iazykoznaniia,'* Moscow, 1951.

Aleksandrov, P. S. "Neobkhodimo vdokhnovenie." *LG*, January 25, 1967: 12.

————. "O prizvanii uchenogo." In *Nauka i chelovechestvo: 1971–1972*, 120–127. Moscow, 1972,

————. "Russkaia matematika v XIX i XX vv. i ee vlianie na mirovuiu nauku." *UZMU*, 1946, no. 92: 3–34.

Aleksandrov, P. S., et al., eds. *Istoriia Moskovskogo universiteta*. Vol. 2. Moscow, 1955.

Alekseev, M. N., et al. "Predmet dialekticheskoi logiki." In *Dialektika i logika*, 286–301. *See* Konstantinov et al. 1966.

Alekseev, P. V. "Diskussiia s mekhanistami po probleme vzaimosviazi filosofii i estestvoznaniia."*VF*, 1966, no. 4: 44–54.

————. *Marksistsko-leninskaia filosofiia i meditsina v SSSR*. Moscow, 1970.

Alekseeva, G. D. *Oktiabr'skaia revoliutsiia i istoricheskaia nauka, 1917–1923 gg.* Moscow, 1968.

Alikhanian, S. I. *Teoreticheskie osnovy ucheniia Michurina o peredelke rasteniia.* Moscow, 1966.

Al'tshuler, S. V., and G. P. Kulikov. "Organizatsiia nauchnykh issledovanii v gody voiny: beseda s professorom S. V. Kaftanovym." *VIE-T* 2 (1975): 24–30.

Amann, Roland. "Some Approaches to the Comparative Assessment of Soviet Technology: Its Level and Rate of Development." In: *The Technological Level*, chap. 1. *See* Amann et al. 1977.

————. "The Soviet Research and Development System: The Pressures of Academic Tradition and Rapid Industrialization." *Minerva* 8, no. 2 (1970): 217–241.

Amann, Roland, M. J. Berry, and R. W. Davies. "Science and Industry in the USSR." In *Science Policy*, 381–490. See Zaleski et al. 1969.

Amann, Roland, Julian Cooper, and R. W. Davies, eds. *The Technological Level of Soviet Industry*. New Haven, Conn.: Yale University Press, 1977.

Ambartsumian, V. A. *Filosofskie voprosy nauki o Vselennoi*. Erevan, 1973.

————. "Kosmogoniia." In *Astronomiia*, 347–364. See Mikhailov 1960.

Ambartsumian, V. A., et al., eds. *Razvitie astronomii v SSSR*. Moscow, 1967.

Ananichev, K. V., et al., eds. *Osnovnye printsipy i obshchie problemy upravleniia naukoi*. Moscow, 1973.

Andreev, A. I. "Osnovanie Akademii nauk v Peterburge." In: *Petr Velikii*, 284–333. See A. I. Andreev 1947.

————. "Stepan Petrovich Krasheninnikov." *LRN*:geo, 345–355.

————, ed. *Petr Velikii: Sbornik statei*. Vol. 1. Moscow-Leningrad, 1947.

Andreev, A. I., and L. B. Modzalevskii, eds. *Lomonosov: Sbornik statei*. Vol. 1, Moscow-Leningrad, 1940.

Andreev, I. D., et al., eds. *Voprosy dialekticheskogo i istoricheskogo materializma v trude I. V. Stalina 'Marksizm i voprosy iazykoznaniia.'* Vol. 2. Moscow, 1952.

Anuchin, D. "Geografiia XVIII-go veka i Lomonosov." In *Prazdnovanie*, 94–122.

Anweiler, Oskar, and Karl-Heinz Ruffmann, eds. *Kulturpolitik der Sowjetunion*. Stuttgart, 1973.

Aris'ian, L. "Protiv odnoi opportunisticheskoi revizii istoricheskogo materializma." *PZM*, 1931, nos. 4–5: 75–98.

Arkhiptsev, F. T., et al., eds. *Leninskii etap razvitiia filosofii marksizma*. Moscow, 1972.

Arness, Carol, ed. *A Naturalist in Russia*. Minneapolis: University of Minnesota Press, 1967.

Arsen'ev, A. S. "Nauka i chelovek (Filosofskii aspekt)." In *Nauka*, 114–158. See Tolstykh 1971.

Artems'kyi, A. Ia. *Shcho take Vseukrains'ka Akademiia nauk*. Kiev, 1931.

Artsimovich, L. A. "Fizik nashego vremeni. (Zametki o nauke i ee mesto v obshchestve)." *NM*, 1967, no. 1: 190–203.

[————.] "Nekotorye voprosy razvitiia sovremennoi fiziki." *VAN*, 1965, no. 2: 3–14.

Ashby, Eric. *Scientist in Russia*. New York: Penguin, 1947.

Asmus, V. F. *Izbrannye filosofskie trudy*. Vol. 2. Moscow, 1971.

Babasov, E. M. "Nauchno-tekhnicheskaia revoliutsiia i kommunizm." In *Dialektika material'noi zhizni*, 22–35. See Konstantinov et al. 1966.

Baer, Karl von. "Zum Streit über der Darwinismus." *Augsburger allgemeine Zeitung*, 1873: 1986–1988.

————. *Reden und kleinere Aufsätze.* Vol. 2. St. Petersburg, 1876.

Bagbaia, I. D. "Vtoraia Zakavkazskaia konferentsiia po istorii i teorii nauki." *Istoriia i metodologiia estestvennykh nauk (fizika)* 8 (1970): 271–277.

Bailes, Kendall E. "Science, Philosophy and Politics in Soviet History: The Case of Vladimir Vernadskii." *Russian Review* 40, no. 3 (1981): 278–299.

————. "The Technical Specialists: Social Composition." In *The Social Context,* 137–172. *See* Lubrano and Solomon 1980.

————. *Technology and Society Under Lenin and Stalin.* Princeton, N.J.: Princeton University Press, 1978.

Bakh, A. N. "K voprosu o planirovanii raboty nauchno-issledovatel'skikh uchrezhdenii Soiuza." *VAR,* 1929, no. 3: 1–3.

————. "Nauchno-issledovatel'skaia rabota i promyshlennost'." *VAR,* 1929, nos. 6–7: 2–5.

————. "Sovetskaia nauka." *SN,* 1938, no. 1: 3–12.

————. "Vsesoiuznaia Akademiia nauk i sotsialisticheskoe stroitel'stvo." *FNiT,* 1932, nos. 11–12: 20–24.

Bashkina, Nina N., et al., eds. *The United States and Russia. The Beginning of Relations, 1765–1815.* Washington, D.C.: Government Printing Office, 1981.

Bastrakova, M. S. "Organizatsionnye tendentsii russkoi nauki v nachale XX v." In *Organizatsiia,* 150–186. *See* Beliaev et al. 1968.

————. *Stanovlenie sovetskoi sistemy organizatsii nauki (1917–1922).* Moscow, 1973.

Bateson, W. "Science in Russia." *Nature* 116, no. 2923 (1925): 681–683.

Bazarov, I. P. "Za dialektiko-materialisticheskoe ponimanie i razvitie teorii otnositel'nosti." *VF,* 1952, no. 4: 175–185.

Belardelli, Filippo. "The 'Lysenko Affair' in the Framework of the Relations Between Marxism and the Natural Sciences." *Scientia* 112 (1977): 33–50.

Beliaev, E. A. "Sovershenstvovanie territorial'nogo razmeshcheniia nauchnykh uchrezhdenii SSSR." In *Osnovnye printsipy,* 257–273. *See* Ananichev et al. 1973.

————. "Voprosy territorial'nogo razmeshcheniia nauki." In *Sotsiologicheskie problemy nauki,* 316–347. *See* Kelle and Mikulinskii, 1974.

Beliaev, E. A., and N. S. Pyshkova. *Formirovanie i razvitie seti nauchnykh uchrezhdenii SSSR.* Moscow, 1979.

Beliaev, E. A., et al., eds. *Organizatsiia nauchnoi deiatel'nosti.* Moscow, 1968.

Beliakov, S. T. "Neopublikovannye pis'ma N. P. Gorbunova V. I. Leninu." *VIET,* 1965, no. 18: 99–103.

Berg, L. S. *Nauka: ee smysl, soderzhanie i klassifikatsiia.* Petrograd, 1922.

————. *Vsesoiuznoe Geograficheskoe obshchestvo.* Moscow-Leningrad, 1946.

Berkov, P. N. "Lomonosovskii iubilei 1865 g." In *Lomonosov,* 216–247. *See* A. I. Andreev and Modzalevskii 1946.

Bernshtein, N. A. "Novye linii razvitiia v fiziologii i ikh sootnoshenii s kibernetikoi." In *Filosofskie voprosy fiziologii,* 299–322. *See* Fedoseev et al. 1963.

Berthelot, M. *Science et morale.* Paris, 1897.

Bessarab, Maria. *Landau: Stranitsy zhizni.* Moscow, 1971.

Beth, Evert W. *The Foundations of Mathematics.* New York: Harper, 1966.

Bialika, B. A., et al., eds. *Gor'kii i nauka.* Moscow, 1964.

Bibler, V. S ., et al., eds. *Ocherki istorii i teorii razvitiia nauki.* Moscow, 1969.

Bilai, V. I. *Zhizn' otdannaia liudam.* Kiev, 1966.

"Biographische Versuche." *Recueil des actes de la séance publique de L'Académie impériale des sciences de St. Pétersbourg tenue le 29 décembre 1831,* 101–117.

Bistrianskii, V. "Bukharin i Trotskii v roli apologetov kulaka." *LP,* November 13, 1929.

Bliakher, L. Ia. "Ch. Darvin i brat'ia Kovalevskie." *VIE-E,* 1959, no. 8: 66–73.

————. *Problema nasledovaniia priobretennykh priznakov: istoriia apriornykh i empiricheskikh popytok ee resheniia.* Moscow, 1971.

Blokh, M. A., et al. *Tvorchestvo.* Vol. 1. Petrograd, 1923.

Boeselager, W. F. *The Soviet Critique of Neopositivism.* Dordrecht, Holland: Reidel, 1975.

Bogdanov, A. A. *Nauka i rabochii klass.* Moscow, 1922.

————. "Ob'ektivnoe ponimanie printsipa otnositel'nosti." *VKA,* 1924, no. 8: 332–347.

————. "Organizatsionnye printsipy sotsial'noi tekhniki i ekonomiki." *VKA,* 1923, no. 4: 272–284.

————. *Sotsializm nauki.* Moscow, 1918.

————. "Uchenie ob analogiiakh." *VKA,* 1923, no. 2: 78–97.

————. *Vseobshchaia organizatsionnaia nauka (tektologiia).* Vol. 1 (3d ed.); vol. 2 (3d ed.); vol. 3 (2d ed.). Leningrad-Moscow, 1925–1929.

Bogoslovskii, M. M. *Petr I.* Vol. 2. Moscow, 1941.

Bondarenko, P. P., et al., eds. *Protiv mekhanisticheskogo materializma i men'shevistvuiushchego idealizma v biologii.* Moscow-Leningrad, 1931.

"Bor'ba za peredovuiu nauku." *VAN,* 1938, no. 6: 3–9.

Borisiak, A. A. *V. O. Kovalevskii.* Leningrad, 1928.

Borisov, V. N., et al., eds. *Priroda soznaniia i zakonomernosti ego razvitiia.* Novosibirsk, 1968.

Born, Max, *Physics and Politics.* London: Oliver and Bond, 1962.

Borovik-Romanov, A. S. Foreword. In *Experiment,* xxi–xxvi. *See* Kapitsa 1980.

Boss, Valentin. *Newton and Russia: The Early Influence, 1698–1796.* Cambridge, Mass.: Harvard University Press, 1972.

Bourbaki, Nicolas. *Eléments d'histoire des mathématiques.* Paris, 1960.

Brandt, J. F. "Zoologicheskii i zootomicheskii muzei." *ZAN* 7 (1865): 1–35.

Brega, G. S. *Sotrudnichestvo sovetskikh respublik v oblasti nauki v gody pervoi piatiletki.* Kiev, 1978.

Bronskii, N. I., and V. P. Iakovlev. "Evoliutsiia sotsial'nykh i filosofskikh vozzrenii V. I. Vernadskogo." In *V. I. Vernadskii,* 7–43. *See* Bronskii, Reznikov, and Iakovlev 1963.

Bronskii, N. I., A. P. Reznikov, and V. P. Iakovlev. *V. I. Vernadskii, k stoletiiu so dnia rozhdeniia.* Rostov-on-the-Don, 1963.

Bukharin, N. I. *Ataka: Sbornik teoreticheskikh statei,* 2d ed. Moscow, n.d.

———. *Etiudy.* Moscow-Leningrad, 1932.

———. "Lenin i ego filosofskoe uchenie." In *Pamiati V. I. Lenina,* 71–93. *See* Bukharin and Deborin 1934.

———. "Nauka v SSSR." In *Nauka i tekhnika.* Vol. 1, 3–16. *See* Ioffe et al. 1928.

Bukharin, N. I., et al. *Science at the Cross Roads,* 2d ed. London, 1971.

Bukharin, N. I., and A. M. Deborin, eds. *Pamiati Karla Marksa: Sbornik statei k piatidesiatiletiiu so dnia smerti (1883–1933).* Leningrad, 1933.

———, eds. *Pamiati V. I. Lenina: Sbornik statei k desiatiletiiu so dnia smerti, 1924–1934.* Moscow-Leningrad, 1934.

Bush, V. Ia. "K 50-letnemu iubileiu deiatel'nosti akademika V. R. Vil'iamsa." *VAN,* 1935, no. 2: 24–29.

Bushinskii, V. P. "Uchenie V. R. Vil'iamsa—boevoe oruzhie sovetskoi agronomii." *VAN,* 1950, no. 4: 46–58.

Butiagin, A. S., and Iu. A. Saltanov. *Universitetskoe obrazovanie v SSSR.* Moscow, 1957.

Butlerov, A. M. *Sochineniia.* Vol. 3. Moscow, 1958.

Bykhovskii, B. "Bekon i ego mesto v istorii filosofii." *PZM,* 1931, no. 6: 73–108.

Bykov, G. V., and Z. I. Sheptunova. "Nemetskii 'Zhurnal khimii' i russkie khimiki." *TIIE-T* 30 (1960): 97–110.

Chelintsev, G. V. "O novoi pozitsii khimikov makhistov." *VF,* 1950, no. 2: 170–180.

Chernakov, E. "K istorii russkoi biologii." *Bol,* 1948, no. 12: 55–58.

Chernov, F. "Burzhuaznyi kosmopolitizm i ego reaktsionnaia rol'." *Bol,* 1949, no. 5: 30–41.

Chernov, S. N. "Leonard Eiler i Akademiia nauk." In *Leonard Eiler,* 163–238. *See* Deborin 1935.

Chesnakov, E. N. "All-Union Conference on Philosophical Problems of Natural Science." In *Filosofskie problemy sovremennogo estestvoznaniia,* pp. 650–661. *See* Fedoseev 1959.

"Chestvovanie akademika M. B. Mitina." *VAN,* 1961, no. 9: 113–114.

Chuchmarev, V. I. "G. V. Leibnits i russkaia kul'tura nachala 18 stoletiia." *VIMK,* 1957, no. 4: 120–132.

Chudakov, E. A. "K perestroike akademicheskoi nauchnoi raboty v oblasti tekhnicheskikh nauk." *VAN,* 1948, no. 12: 6–9.

Chugaev, L. A. *Dmitrii Ivanovich Mendeleev.* Petrograd, 1924.

———. "O merakh k sodeistviiu issledovaniia po chistoi i prikladnoi khimii v Rossii." *Otchety o deiatel'nosti Komissii,* 1917, no. 8: 166–172.

Cocks, Paul M. "The Policy Process and Bureaucratic Politics." In *Dynamics,* 156–178. *See* Cocks, Daniels, and Heer 1976.

———. "Retooling the Directed Society: Administrative Modernization and

Developed Socialism." In *Political Development*, 53–92. *See* Triska and Cocks 1977.

———. *Science Policy in the Soviet Union*. Washington, D.C.: NSF-Division of International Programs, 1980.

Cocks, Paul M., R. V. Daniels, and N. W. Heer, eds. *The Dynamics of Soviet Politics*. Cambridge, Mass.: Harvard University Press, 1976.

Cohen, Stephen F. *Bukharin and the Bolshevik Revolution: A Political Biography, 1888–1938*. New York: Vintage, 1973.

Comey, D. D. "Current Trends in Soviet Logic." *Inquiry* 9 (1966): 94–108.

Condorcet, M. J. A. N. de Caritat, "Elogium of Euler." In *Letters*, xxxiii–lxii. *See* Euler 1802.

Coquart, A. *Dmitri Pisarev (1840–1866) et l'idéologie du nihilisme russe*. Paris, 1946.

Cuvier, G. *Recueil des éloges historiques lus dans les séances publiques de l'Institut Royal de France*. Vol. 2. Paris, 1819.

Danin, D. "Kvanty pamiati." *NiZh*, 1977, no. 10: 132–139.

———. "Vozmozhnye resheniia. (Iz dnevnika literatora)." *VL*, 1964, no. 8: 88–111.

Darwin, Charles. *The Descent of Man*. New York: Collier, 1900.

"Darvinizm i marksizm." *VKA*, 1926, no. 14: 226–274.

Davies, R. W., and R. Amann. "Science Policy in the U.S.S.R." *Scientific American* 220, no. 6 (1969): 19–29.

Deborin, A. M. *Dialektika i estestvoznanie*. Moscow-Leningrad, 1929.

———. *Filosofiia i politika*. Moscow, 1961.

———. "Lenin i krizis noveishei fiziki." *Otchet-S* 2 (1929): appendix, 1–29.

———. "Lenin i sovremennost'." In *Pamiati V. I. Lenina*, 147–246. *See* Bukharin and Deborin 1934.

———. "Mekhanisty v bor'be s dialektikoi." *VKA* 19 (1927): 21–61.

———, ed. *Leonard Eiler, 1707–1783: Sbornik statei i materialov k 150-letiiu so dnia smerti*. Moscow-Leningrad, 1935.

"Deiatel'nost' Kommunisticheskoi Akademii." *VKA* 30 (1928): 247–267.

Delisle l'Aîné, G. "Remarques sur la carte de la Caspienne, envoyée à l'Académie par Sa Majesté Czarienne." In *Histoire de l'Académie Royale des Sciences*, 319–332. Amsterdam, 1725.

Delokarov, K. Kh., ed. *Einshtein i filosofskie problemy fiziki xx veka*. Moscow, 1979.

———. "Puti stanovleniia soiuza filosofii i fiziki v SSSR (20-30-e gody)." *FN*, 1971, no. 1: 111–120.

———. "Teoriia otnositel'nosti i sovetskaia filosofskaia nauka." In *Einshtein*, edited by K. Kh. Delokarov, 520–566.

Delone, B. N. "Razvitie teorii chisel v Rossii." *UZMU*, 1946, no. 92: 77–96.

Dement'ev, G. P. "Petr Simon Pallas." In *LRN:bio*, 34–44.

Depenchuk, N. P., ed. *Nauchno-tekhnicheskaia revoliutsiia i sovremennoe estestvoznanie*. Kiev, 1978.

Derman, A. *'Akademicheskii intsident' (Istoriia ukhoda iz Akademii nauk V. G. Koro-*

lenko i A. P. Chekhova v sviazi s 'raz'iasneniem' M. Gor'kogo). Simferopol', 1923.

Derzhavin, N. S. "Iafeticheskaia teoriia akademika N. Ia. Marra." *IVS*, 1930, no. 1: 3–39; no. 2: 3–37.

de Witt, Nicholas. "Reorganization of Science and Research in the U.S.S.R." *Science* 133, no. 3469 (1961): 1981–1991.

Diakonov, M. A. "Aleksandr Sergeevich Lappo-Danilevskii." *IRAN*, ser. 6, 13 (1919): 135–166.

Dirac, P. A. M. *The Principles of Quantum Mechanics*, 4th ed. Oxford: Clarendon, 1958.

"Direktiva vsem organam VARNITSO po povodu perevyborov prepodavatel'skogo sostava vuzov." *VAR*, 1929, no. 4: 11.

"Diskussiia o prirode fizicheskogo znaniia: Obsuzhdenie stat'i M. A. Markova." *VF*, 1948, no. 1: 203–232; no. 3: 222–235.

Dobrov, G. M. "Gosudarstvennaia politika tekhnicheskogo obespecheniia issledovanii v sisteme tselei i sredstv upravleniia naukoi." In *Osnovnye printsipy*, 163–179. *See* Ananichev et al. 1973.

————. *Nauka o nauke*, 2d ed. Kiev, 1970.

————. "Osnovy nauchnoi politiki." *Naukovedenie*, 1973, no. 8: 3–19.

Dobrov, G. M., et al. *Potentsial nauki*. Kiev, 1969.

Dobzhansky, T. "The Birth of the Genetic Theory of Evolution in the Soviet Union in the 1920's." In *The Evolutionary Synthesis*, 229–242. *See* Mayr and Provine 1980.

————. "Soviet Biology and the Powers That Were." *Science* 164 (1969): 1507–1509.

"Doklad Komissii Akademii nauk SSSR o rezul'tatakh proverky deiatel'nosti eksperimental'noi nauchno-issledovatel'skoi bazy 'Gorki Leninskie' i ee podsobnogo nauchno-proizvodstvennogo khoziaistva." *VAN*, 1965, no. 11: 4–128.

"Dom Uchenykh: rezume i osnovnye polozheniia dokladov prochitannykh v Klube uchenykh." *NiR*, 1921, no. 5: 36–39.

Dorfman, Ia. G., and M. I. Radovskii. "V. Franklin i russkie elektriki xviii v." *TIIE-T* 19 (1957): 290–312.

Dorokhova, G. A. "Obshchegosudarstvennaia organizatsiia upravleniia naukoi v SSSR i soiuznykh respublikakh." In *Organizatsionno-pravovye voprosy*, 126-207. *See* Piskotin et al. 1973.

Drobizhev, V. Z., et al., eds. *V. I. Lenin vo glave velikogo stroitel'stva*. Moscow, 1960.

Drobnitskii, O. G. "Priroda moral'nogo soznaiia." *VF*, 1968, no. 2: 26–37.

Dubinin, N. P. "Genetika." In *Oktiabr' i nauchnyi progress*. Vol. 2, edited by M. V. Keldysh et al., 257–280.

————. *Teoreticheskie osnovy i metody rabot I. V. Michurina*. Moscow, 1966.

————. *Vechnoe dvizhenie*. Moscow, 1973.

Duchinskii, F. "Darvinizm, lamarkizm i neodarvinizm." *PZM*, 1926, nos. 7–8: 95–122.

Dunn, L. C. *A Short History of Genetics.* New York: McGraw-Hill, 1965.

Duzhenkov, V. I. "Nauchnye tsentry v SSSR i ikh rol' v razvitii nauki." In *Osnovnye printsipy*, 236–256. See Ananichev et al. 1973.

_____*Problemy organizatsii nauki (regional'nye aspekty).* Moscow, 1978.

Dvoriankin, F. "Pobeda michurinskoi biologicheskoi nauki." *Bol*, 1948, no. 16: 30–39.

Dvoriashin, V. I. "Uchenye na velikikh stroikakh kommunizma." *VAN*, 1952, no. 1: 44–49.

"220-letie Akademii nauk SSSR." *VAN*, 1945, nos. 7–8: 35–105.

Dylis, N. V. "K 100-letiiu so dnia rozhdeniia V. N. Sukacheva." *VIE-T*, 1980, no. 3: 71–78.

Dyshlevyi, P. S. *V. I. Lenin i filosofskie problemy reliativistskoi fiziki.* Kiev, 1969.

Dyshlevyi, P. S., and A. Z. Petrov. "Filosofskie voprosy teorii otnositel'nosti." *VF*, 1968, no. 1: 87–97.

Egorshin, V. "Estestvoznanie i klassovaia bor'ba." *PZM*, 1926, no. 6: 108–136.

_____. "Kak I. E. Tamm kritikuet marksistov." *PZM*, 1932, no. 2: 236–260.

_____. "Lenin i sovremennoe estestvoznanie. (K 25-letiiu 'Materializma i empiriokrititsizma')." *FNiT*, 1933, no. 12: 7–17.

_____. Review of *Dialektika i priroda.* *VKA* 27 (1928): 269–282.

Einstein, Albert. *Essays in Science.* New York: Philosophical Library, 1936.

[_____.] Einshtein, A. "Tvorcheskaia avtobiografiia." *UFN*, 59, no. 1: 71–105.

Eliutin, V. P. "Konkurs, student, professiia." *Pravda*, July, 19, 1969.

_____. "Nauka i vysshee obrazovanie." In *Upravlenie.* Vol. 1, 157–181. See Gvishiani et al. 1970.

_____. "Osnovnye puti razvitiia sovetskoi vysshei shkoly." *VIE-T*, 1977, no. 2: 30–34.

Enteen, George M. "Marxist Historians during the Cultural Revolution: A Case Study of Professional In-fighting." In *Cultural Revolution*, pp. 154–168. See Fitzpatrick 1978.

_____. *The Soviet Scholar-Bureaucrat: M. N. Pokrovskii and the Society of Marxist Historians.* University Park: Pennsylvania State University Press, 1978.

Engel'gardt. V. A. "Novyi etap nastupleniia sovetskoi nauki." *VAN*, 1961, no. 8: 24–27.

Esakov, E. D. *Sovetskaia nauka v gody pervoi piatiletki.* Moscow, 1971.

"Etat du personnel de l'Académie impériale des sciences de St.-Péterbourg en 1849." In *Recueil des actes de la séance publique de l'Académie impériale des sciences de Saint-Péterbourg*, i–xvi. St. Petersburg, 1849.

[Euler, Leonhard.] *Letters of Euler on Different Subjects in Physics and Philosophy, Addressed to a German Princess*, 2d ed. Translated from French by H. Hunter. Vol. 1. London: Murray & Highley, 1802.

"Fakty i dokumenty." *VIE-T*, 1975, no. 2: 43–50.

Farley, John. *The Spontaneous Generation Controversy from Descartes to Oparin.* Baltimore: Johns Hopkins University Press, 1977.

Fediukin, S. A. *Velikii Oktiabr' i intelligentsia.* Moscow, 1972.

Fedorenko, E. G. *Osnovy marksistsko-leninskoi etiki.* Kiev, 1965.

Fedorov, A. S. "Nauka i tekhnika v gody Otechestvennoi voiny." *VIE-T,* 1957, no. 2: 3–12.

Fedoseev, P. N. et al., eds. *Filosofskie problemy sovremennogo estestvoznaniia.* Moscow, 1959.

Fedoseev, P. N., et al., eds. *Filosofskie voprosy fiziologii vysshei nervnoi deiatel'nosti i psikhologii.* Moscow, 1963.

Feinberg, E. "Obyknovennoe i neobychnoe." *NM,* 1965, no. 8: 207–229.

Ferliudin, P. *Istoricheskii obzor mer po vysshemu obrazovaniiu v Rossii.* Vol. 1. Saratov, 1894.

Fersman, A. E. *Izbrannye trudy.* Vol. 5. Moscow, 1959.

——. "Iz nauchnoi deiatel'nosti na Ukraine." *Priroda,* 1919, nos. 4–6: pp. 265–272.

——. "S'ezd po izucheniiu proizvoditel'nykh sil i narodnogo khoziaistva Ukrainy." *Priroda,* 1925, nos. 1–3: 138–140.

——, ed. *Akademiia nauk SSSR za desiat' let, 1917–1927.* Leningrad, 1927.

——, ed. *L'Académie des sciences de l'U.S.S.R.* Leningrad, 1927.

Figatner, Iu. P. "Proverka aparata Akademii nauk." *VAR,* 1930, no. 2: 73–76.

Figurovskii, N. A. "Nekotorye zadachi sovetskikh istorikov khimii." In *Materialy po istorii otechestvennoi khimii,* pp. 9–21.

——. "Zadachi issledovaniia po istorii nauki." *VAN,* 1959, no. 11: 28–32.

——, ed. *Istoriia estestvoznaniia v SSSR.* Vol. 3. Moscow, 1952. (Preliminary publication for discussion.)

Figurovskii, N. A., et al., eds. *Istoriia estestvoznaniia v Rossii.* Vols. 1 (parts 1 and 2) and 2. Moscow, 1957–1962.

Fitzpatrick, Sheila. *The Commissariat of Enlightenment.* Cambridge: Cambridge University Press, 1970.

——. *Education and Social Mobility in the Soviet Union, 1921–1934.* Cambridge: Cambridge University Press, 1979.

——, ed. *Cultural Revolution in Russia, 1928–1931.* Bloomington: Indiana University Press, 1978.

Fok, V. A. "Chto vnesla teoriia kvantov v osnovnye predstavleniia fiziki." *SRiN,* 1935, no. 6: 3–10.

——. "K diskussii po voprosam fiziki." *PZM,* 1938, no. 1: 149–159.

——. "Kritika vzgliadov Bora na kvantovuiu mekhaniku." *UFN,* 45, no. 1 (1951): 3–14.

——. "Poniatiia odnorodnosti, kovariantnosti i otnositel'nosti v teorii prostranstva i vremeni." *VF,* 1955, no. 4: 131–135.

——. "Problema dvizheniia mass i teoriia tiagoteniia Einshteina." In *Sbornik,* 31–43. *See* Lukirskii et al. 1950.

_____. "Protiv nevezhestvennoi kritiki sovremennykh fizicheskikh teorii." *VF*, 1953, no. 1: 168–174.

Fradkin, N. G. *Ocherki po istorii fiziko-geograficheskikh issledovanii territorii SSSR (1917–1927)*. Moscow, 1961.

Frenkel', Ia. I. "Teoriia otnositel'nosti Einshteina." *FNiT*, 1935, nos. 5–6: 38–48.

Frenkel', V. Ia. "Einshtein i sovetskie fiziki." *VIE-T*, 1975, no. 3: 25–30.

_____. *Iakov Il'ich Frenkel'*. Moscow, 1966.

Frizer, I. "The Theory of Objective Beauty in Soviet Education." In *Philosophy in the Soviet Union*, edited by E. Laszlo, 149–160. New York: Praeger, 1967.

Frolov, I. T. *Genetika i dialektika*. Moscow, 1969.

_____. "Modern Science and Humanism." In *Science*, 109–124. *See* Velikhov, Gvishiani, and Mikulinskii 1980.

_____, ed. *Filosofiia i sovremennost'*. Moscow, 1973.

Fuss, P. H. *Coup d'oeil historique sur le dernier quart-de-siècle de l'existence de l'Académie impériale des sciences de Saint-Pétersbourg*. St. Petersburg, 1843.

Gaisinovich, A. E. "Biolog-shestidesiatnik N. V. Nozhin i ego vremia v razvitii embriologii i darvinizma." *ZhOB* 13, no. 5 (1952): 377–392.

_____. "The Origins of Soviet Genetics and the Struggle with Lamarckism, 1922–1929." *Journal of the History of Biology* 13, no. 1 (1980): 1–51.

_____. *K. F. Vol'f i uchenie o razvitii organizmov*. Moscow, 1961.

"General'nyi dogovor Akademii nauk Soiuza SSSR s Narkomatom tiazheloi promyshlennosti na 1932 g." *SRiN*, 1932, no. 4: 237–238.

Gerasimov, A. "Rossiiskoe Mineralogicheskoe obshchestvo v 1918–1919 gg." *Priroda*, 1919, nos. 7–9: 378–379.

Gerasimov, I. P., ed. *Ocherki istorii geograficheskoi nauki v SSSR*. Moscow, 1976.

Gertsen, A. I. *Sobranie sochinenii v tridtsati tomakh*. Vol. 3. Moscow, 1954.

Gessen, B. M. *See* Hessen, Boris.

Gimmel'farb, B. N. "Mirovoe znachenie sovetskoi astronomii." *Priroda*, 1948, no. 7: 76–80.

Ginetsinskii, A. G. "Osnovnye napravleniia v fiziologii tsentral'noi nervnoi sistemy." In *Uspekhi*, 27–34. *See* Orbeli 1945.

Ginzburg, V. L. "Eksperimental'naia proverka obshchei teorii otnositel'nosti." In *Einshtein*, pp. 117–135. *See* Grigor'ian et al. 1962.

_____. "Novye fizicheskie zakony i astronomiia." *VF*, 1972, no. 11: 14–19.

Glick, Thomas F., ed. *The Comparative Reception of Darwinism*. Austin: University of Texas Press, 1974.

Glinka, K. D. "Pochvovedenie v SSSR za poslednee desiatiletie." In *Nauka i tekhnika*. Vol. 2, 273–307. *See* Ioffe et al. 1928.

Gnucheva, V. F. *Geograficheskii departament Akademii nauk XVIII veka*. Moscow-Leningrad, 1946.

"Godichnoe sobranie Akademii nauk SSSR (1–2 fevralia 1965 g.)." *VAN*, 1965, no. 3: 3–119.

Goldschmidt, Richard. *The Material Basis of Evolution*. New Haven, Conn.: Yale

University Press, 1982.

Gol'tsman, A. "Einshtein i materializm." In *Teoriia otnositel'nosti*, 118–134.

———. "Nastuplenie na materializm." *PZM*, 1923, no. 1: 83–101.

Golubtsova, N. I. "Organizatsiia nauchnoi raboty v vuzakh." In *Osnovnye print-sipy*, pp. 180–198. See Ananichev et al. 1973.

Gorbunov, N. P. "Akademiia nauk na perelome." *VAN*, 1936, no. 6: 2–24.

———. "Velikie plany razvitiia nauki i tekhniki." In *V. I. Lenin*, 172–186. See Drobizhev et al. 1960.

Gordeev, D. I., and E. A. Kurazhkovskaia. "Problemy naukovedeniia." *VMU*, 1968, no. 1: 11–19.

Gorelik, George. "Bogdanov's Tektology: Its Basic Concepts and Relevance to Modern Generalizing Sciences." *Human Systems Management* 1 (1980): 327–337.

Gorky, Maksim. "Rech' na publichnom zasedanii Svobodnoi assotsiatsii dlia razvitiia i rasprostraneniia polozhitel'nykh nauk." *Letopis'*, 1917, nos. 5–6: 223–231.

[———.] *Gor'kii i nauka: Stat'i, rechi, pis'ma, vospominaniia*. Moscow, 1946.

Gosudarstvennye universitety. Moscow, 1934.

Graham, Loren R. *Between Science and Values*. New York: Columbia University Press, 1981.

———. "The Development of Science Policy in the Soviet Union." In *Science Policies of Industrial Nations*, edited by T. D. Long and C. Wright, 12–58. New York: Praeger, 1975.

———. "The Formation of Soviet Research Institutes: A Combination of Revolutionary Innovation and International Borrowing." *Social Studies in Science* 5 (1975): 303–329.

———. "The Place of the Academy of Sciences in the Overall Organization of Soviet Science." In *Soviet Science*, 44–62. See Thomas and Kruse-Vaucienne 1977.

———. "Quantum Mechanics and Dialectical Materialism." *SR* 25 (1966): 381–340.

———. "Reorganization of the U.S.S.R. Academy of Sciences." In *Soviet Policy-Making: Studies of Communism in Transition*, edited by P. H. Juviler and H. W. Morton, 133–161. New York: Praeger, 1967.

———. *Science and Philosophy in the Soviet Union*. New York: Knopf, 1972.

———. *The Soviet Academy of Sciences and the Communist Party, 1927–1932*. Princeton, N.J.: Princeton University Press, 1967.

Grave, N. V. "Kraevednye uchrezhdeniia SSSR." *Priroda*, 1926, nos. 1–2: 127–128.

Grekov, B. et al., eds. *Protiv istoricheskoi kontseptsii M. N. Pokrovskogo: Sbornik statei*. 2 vols. Moscow-Leningrad, 1939–1940.

Grigor'ian, A. T., ed. *Einshtein i razvitie fiziko-matematicheskoi mysli*. Moscow, 1962.

Grot', Ia. "Dva slova ob Akademii nauk." *Russkii vestnik* 32, no. 4, sect. 2 (1861): 149–152.

Gubkin, I. M. "Osnovnye zadachi i organizatsionnye formy SOPS." *VAN,* 1931, no. 3: 7–16.

Guerrier, W. *Leibniz in seinen Beziehungen zu Russland und Peter dem Grossen.* St. Petersburg, Leipzig, 1873.

[_____.] Ger'e, V. I. *Otnosheniia Leibnitsa k Rossii i Petru Velikomu.* St. Petersburg, 1871.

Gulyga, A. V. "Krizis—nravstvennyi ili sotsial'nyi?" In *Nauka,* 358–385. *See* Tolstykh 1971.

Gureev, I. P., V. V. Zverev, and A. I. Kolchin. "Nauchnye kadry SSSR." In *Osnovnye printsipy,* 119–134. *See* Ananichev et al. 1973.

Gvishiani, D. "Nauchno-tekhnicheskaia revoliutsiia i problemy nauki." In *Nauchno-tekhnicheskaia revoliutsiia i sotsial'nyi progress,* 49–66.

_____. "Sotsial'naia rol' nauki i politika gosudarstva v oblasti nauki." In *Upravlenie,* Vol. 1, 21–77. *See* Gvishiani et al. 1970.

Gvishiani, D., et al., eds. *Upravlenie, planirovanie i organizatsiia nauchnykh i tekhnicheskikh issledovanii.* 5 vols. Moscow, 1970.

Gvishiani, D., S. R. Mikulinskii, and S. A. Kugel, eds. *The Scientific Intelligentsia. (Structure and Dynamics of Personnel).* Translated by Jane Sayers. Moscow, 1976.

Hahn, Werner. *Postwar Soviet Politics: The Fall of Zhdanov and the Defeat of Moderation, 1946–53.* Ithaca, N.Y.: Cornell University Press, 1982.

Heijenoort, Jean von, ed. *From Frege to Gödel.* Cambridge, Mass.: Harvard University Press, 1967.

Herzen, Alexander. *Selected Philosophical Works.* Translated by L. Navrozov. Moscow, 1956.

_____. *See* Gertsen, A. I.

Hessen, Boris. "The Social and Economic Roots of Newton's *Principia.*" In *Science at the Cross Roads,* 149–212. *See* Bukharin et al. 1971.

Hoffman, Banesh. *The Strange Story of the Quantum,* 2d ed. New York: Dover, 1959.

Holloway, David. "Entering the Nuclear Arms Race: The Soviet Decision to Build the Atomic Bomb, 1939–45." *Social Studies of Science* 11 (1981): 159–197.

Holton, Gerald, ed. *The Twentieth-Century Science.* New York: Norton, 1970.

Home, R. W. Introduction to *Essay,* 3–224. *See* Aepinus 1979.

_____. Preface to *Essay,* ix–xii. *See* Aepinus 1979.

Hutchison, G. Evelyn. "The Biosphere." In *The Biosphere,* 3–11. San Francisco: Freeman, 1970.

Huygens, Christian. *The Celestial Worlds Discovered.* London: F. Cass, 1968.

Iachevskii, A. A., ed. *Iubileinyi sbornik posviashchennyi I. P. Borodinu.* Leningrad, 1927.

Iaffe, G. "Filosofskie osnovy neodarvinizma." *PZM*, 1932, nos. 7–8: 196–237.

Iagunkova, V. A. "Ob osnovnykh printsipakh reflektornoi teorii I. P. Pavlova." *VF*, 1953, no. 3: 109–119.

Iakovlev, V. P. "Metodologicheskie voprosy istorii estestvoznaniia v osveshchenii V. I. Vernadskogo." In *V. I. Vernadskii*, 86–103. *See* Bronskii et al. 1963.

Ianovskaia, S. A. "Ocherednye zadachi matematikov-marksistov." *PZM*, 1930, no. 5: 88–94.

———. "Osnovaniia matematiki i matematicheskaia logika." In A. G. Kurosh, *Matematika*, 11–52. *See* Kurash et al. 1948.

———. *Peredovye idei N. I. Lobachevskogo—orudie bor'by protiv idealizma v matematike*. Moscow-Leningrad, 1950.

Iarilov, A. "Staraia i novaia Akademiia nauk." *FNiT*, 1931, nos. 7–8: 50–59.

Iaroslavskii, Em. "Antimarksistskie izvrashcheniia i vul'garizatorstvo tak nazyvaemoi 'shkoly' Pokrovskogo." In *Protiv istoricheskoi kontseptsii*. Vol. 2, 5–24. *See* Grekov et al. 1940.

Il'ichev, L. F. "Metodologicheskie problemy estestvoznaniia i obshchestvennykh nauk." In *Metodologicheskie problemy*, 7–190.

———. "Doklad." *VAN*, 1963, no. 11: 4–46.

Il'in, A. Ia., et al. *Filosofiia i teoriia evoliutsii*. Moscow, 1974.

Infeld, Leopold. "Moi vospominaniia ob Einshteine." *UFN* 59, no. 1 (1956): 135–184.

"Institut arkheologicheskoi tekhnologii pri Akademii istorii material'noi kul'tury." *NiR*, 1921, no. 4: 35–36.

"Instruktsiia o poriadke primeneniia postanovleniia SNK SSSR ot 20/III, 1937 g." *SN*, 1938, no. 3: 169–171.

Ioffe, A. F. "Fizika v rekonstruktivnyi period." *VAN*, 1931, no. 1: 25–38.

———. "Fizika i voina." *PZM*, 1942, nos. 5–6: 90–99.

———. *Moia zhizn' i rabota*. Moscow, 1933.

———. "Nepreryvnoe i atomnoe stroenie materii." In *Pamiati Karla Marksa*, 245–254. *See* Bukharin and Deborin 1933.

———. "O polozhenii na filosofskom fronte sovetskoi fiziki." *PZM*, 1937, nos. 11–12: 131–143.

———. "Otchet o rabote Fiziko-tekhnicheskogo instituta." *UFN*, 16, no. 7 (1936): 847–871.

———. "Pamiati Al'berta Einshteina." *UFN* 57, no. 2 (1955): 187–192.

———. "Razvitie atomisticheskikh vozzrenii." In *Pamiati V. I. Lenina*, 449–468. *See* Bukharin and Deborin 1934.

———. "Razvitie tochnykh nauk v SSSR za 25 let." *PZM*, 1942, nos. 11–12: 150–163.

———. "Sovetskaia nauka v gody otechestvennoi voiny." *VAN*, 1943, nos. 7–8: 6–8.

Ioffe, A. F., ed. *Fiziko-matematicheskie nauki*. Moscow-Leningrad, 1945.

Ioffe, A. F., and A. T. Grigor'ian, eds. *Sbornik k stoletiiu so dnia rozhdeniia Maksa Planka*. Moscow, 1958.

Ioffe, A. F., et al., eds. *Nauka i tekhnika SSSR: 1917–1927.* 3 vols. Leningrad, 1928.

[Ipat'ev, V.] Ipatieff, V. *The Life of a Chemist*. Translated by Haensel and Lusher. Stanford, Calif.: Stanford University Press, 1946.

————. "Rol' nauchnykh issledovanii v razvitii promyshlennoi zhizni strany." *NiR*, 1921, no. 3: 10–16.

Istoriia Rossii v XIX veke. Brothers A. and I. Granat and Co. Series, vol. 7. St. Petersburg, n.d.

Iubileinaia sessiia Akademii nauk SSSR posviashchennaia 25 letiiu Velikoi Oktiabr'skoi Sotsialisticheskoi Revoliutsii. Moscow-Leningrad, 1943.

Iudin, P. F. "Trud I. V. Stalina 'Ekonomicheskie problemy sotsializma v SSSR'— osnova dal'neishego razvitiia obshchestvennykh nauk." *VAN*, 1953, no. 3: 52–74.

Ivanenko, D. D. "K itogam diskussii po knige B. M. Kedrova 'Engel's i estest-voznanie.'" *Bol*, 1948, no. 8: 66–71.

Ivanitskii, G. R., et al. "Nauchnye uchrezhdeniia—srednei shkole: O pushinskom eksperimente." *VAN*, 1979, no. 11: 54–56.

Ivanov, N. D . "O novom uchenii T. D. Lysenko o vide." *BZh* 37, no. 6 (1952): 819–842.

Ivanova, L. V. *Formirovanie sovetskoi nauchoi intelligentsii, 1917–1927*. Moscow, 1980.

Ivanov-Smolenskii, A. G. "Po pavlovskomu puti." *VAN*, 1952, no. 11: 82–90.

"Iz istorii i deiatel'nosti Doma Uchenykh." *NiR*, 1921, no. 2: 3–10.

"Izvlechenie iz zhurnalov sobranii Soveta Imperatorskogo S.-Peterburgskogo universiteta. Zasedanie 22-go fevralia 1865 goda." *ZhMNP* 126, sec. 2 (1865): 559–605.

Jammer, Max. *The Philosophy of Quantum Mechanics*. New York: John Wiley, 1974.

Jassby, D. L. "Princeton Fusion Experiment." *Science* 202 (1978): 370.

Joravsky, David. "The Construction of the Stalinist Psyche." In *Cultural Revolution*, 105–128. *See* Fitzpatrick 1978.

————. *The Lysenko Affair*. Cambridge, Mass.: Harvard University Press, 1970.

————. *Soviet Marxism and Natural Science: 1917–1932*. New York: Columbia University Press, 1961.

————. "The Vavilov Brothers." *SR* 24, no. 3 (1965): 381–394.

Juviler, Peter H., and H. W. Morton, eds. *Soviet Policy-Making: Studies of Communism in Transition*. New York: Praeger, 1967.

Kagan, V. F. *Lobachevskii*. Moscow, 1944.

Kaganov, V. M., et al., eds. *Nekotorye filosofskie voprosy estestvoznaniia*. Moscow, 1957.

Kalesnik, S. V., ed. *Geograficheskoe obshchestvo za 125 let*. Leningrad, 1970.

Kamanin, L. G. "Velikaia Severnaia ekspeditsiia (Vtoraia Kamchatskaia ekspeditsiia Beringa)." In *Ocherki*, 30–46. *See* Gerasimov 1976.

Kammari, M. "Teoreticheskie korni oshibochnykh istoricheskikh vzgliadov M. N. Pokrovskogo." *PZM*, 1936, no. 4: 1–18.

Kamenev, L. B. *Rabochaia revoliutsiia i nauka*. Moscow, 1925.

Kamenskii, Z. A. "I. Kant v russkoi filosofii nachala 19 veka." *VIMK*, 1960, no. 1: 49–64.

———. *Russkaia filosofiia nachala XIX veka i Shelling*. Moscow, 1980.

Kanaev, I. I. *Zhorzh Lui Leklerk de Biuffon, 1707–1788*. Moscow-Leningrad, 1966.

Kantorovich, Vl. "Sotsiologiia i literatura." *NM*, 1967, no. 12: 148–173.

Kapitsa, P. L. *Eksperiment, teoriia, praktika*. Moscow, 1974.

———. *Experiment, Theory, Practice: Articles and Addresses*. Translated from 2d Russian ed. Dordrecht, Holland: Reidel, 1980.

———. "Lomonosov i mirovaia nauka." *UFN*, 1965, no. 1: 155–168.

———. "Ob organizatsii nauchnoi raboty." *PZM*, nos. 7–8: 90–101.

———. "Problems of Soviet Scientific Policy." *Minerva* 4, no. 3 (1966): 391–397.

Karasev, M., and V. Nozdrev. "O knige M. E. Omel'ianovskogo 'V. I. Lenin i fizika XX veka.'" *VF*, 1949, no. 1: 338–342.

Karataev, N. K. "Podgotovka molodykh sovetskikh uchenykh." *VAN*, no. 1: 73–82.

Karpov, M. M. "O filosofskikh vzgliadakh A. Einshteina." *VF*, 1951, no. 1: 130–141.

Karpov, M. M., et al., eds. *Sotsiologiia nauki*. Rostov-on-the-Don, 1968.

Kauffman, G. B., and A. Beck. "Nikolai Semenovich Kurnakov." *Journal of Chemical Education* 39, no. 1 (1962): 44–49.

Kedrov, B. M. "D. I. Mendeleev ob internatsional'nom kharaktere nauki." *VIMK*, 1957, no. 1: 63–78.

———. "D. I. Mendeleev o roli uchenykh raznykh stran v razvitii nauki." *VIMK*, 1958, no. 2: 3–23.

———. "Istoriia khimii i marksizm." *SN*, 1940, no. 11: 27–47.

———. "Kriticheskie zametki na filosofskie temy." *VF*, 1948, no. 1: 53–71.

———. *Lenin i nauchnye revoliutsii. Estestvoznanie. Fizika*. Moscow, 1980.

———. "Nekotorye voprosy razrabotki marksistskoi istorii estestvoznaniia." In *Voprosy istorii*, 639–662. *See* Vavilov et al. 1949.

———. "Neudachnaia kniga po istorii fiziki." *VF*, 1950, no. 1: 365–378.

———. "Ob otnoshenii marksizma k darvinizmu v sviazi s problemoi vidoobrazovaniia." *VF*, 1955, no. 6: 149–166.

———. *O marksistskoi istorii estestvoznaniia*. Moscow, 1968.

———. "O putiakh perekhoda ot starogo kachestva k novomu." In *Voprosy dialekticheskogo i istoricheskogo materializma*, 70–146. *See* G. F. Aleksandrov et al. 1951.

———. "O zakonakh razvitiia nauki." In *Upravlenie*. Vol. 1, 78–101. *See* Gvishiani et al. 1970.

———. "O znachenii leninskikh 'Filosofskikh tetradei' dlia estestvoznaniia." *VAN*, 1947, no. 1: 5–22.

———. "Predmet i zadachi istorii estestvoznaniia i tekhniki. (O primenenii marksistskogo metoda k razrabotke istorii nauki)." *VIE-T*, 1967, no. 21: 5–19.

———. "Puti poznaniia istiny: razdum'ia o sud'bakh estestvoznaniia." *NM*, 1965, no. 1: 213–235.

———. "Puti razvitiia nauki i tekhniki." *VIE-T*, 1971, no. 3: 3–7.

———. "Revoliutsiia v estestvoznanii i filosofii." *VAN*, 1970, no. 11: 69–75.

———. "Vazhneishie idei M. V. Lomonosova v oblasti fiziki i khimii." *LRN:mat*, 25–30.

———. "Vzaimodeistvie nauk i nekotorye filosofskie voprosy biologii." In *Aktual'nye voprosy sovremennoi genetiki*, edited by S. I. Alikhanian, 540–563. Moscow, 1966.

———, ed. *Nauchno-tekhnicheskaia revoliutsiia i sotsializm*. Moscow, 1973.

———, ed. *Protivorechiia v razvitii estestvoznaniia*. Moscow, 1965.

Kedrov, B. M., S. R. Mikulinskii, and I. T. Frolov. "Filosofsko-sotsiologicheskie problemy nauchno-tekhnicheskoi revoliutsii." In *Nauchno-tekhnicheskaia revoliutsiia*, 39–48.

Keldysh, M. V. "Estestvennye nauki i ikh znachenie dlia razvitiia mirovozzreniia i tekhnicheskogo progressa." *Kom*, 1966, no. 17: 29–47.

———. "Mnogonatsional'nyi Soiuz Sovetskikh Sotsialisticheskikh Respublik i razvitie nauki." In *Nauka Soiuza SSR*, 7–45. *See* Keldysh et al. 1972.

———. "Voprosy organizatsii nauchnoi raboty." *VAN*, 1961, no. 7: 18–39.

Keldysh, M. V., et al., eds. *Nauka Soiuza SSR*. Moscow, 1972.

———. *Oktiabr' i nauchnyi progress*. 2 vols. Moscow, 1967.

Kelle, V. Zh., and S. R. Mikulinskii, eds. *Sotsiologicheskie problemy nauki*. Moscow, 1974.

Keppen, F. "Uchenye trudy P. S. Pallasa." *ZhMNP* 298, sec. 2 (1895): 386–437.

———. "Pallas, Petr Simon." *RBS*, 13 (1904): 153–162.

Kerschner, L. R. "Western Translations of Soviet Publications on Cybernetics." *Studies in Soviet Thought* 4, no. 2 (1964): 162–177.

Khinchin, A. Ia. "Poniatie entropii v teorii veroiatnostei." *UMN* 8, no. 3 (1953).

Khrushchov, G. K. "K itogakh konferentsii po probleme razvitiia kletochnykh i nekletochnykh form zhivogo veshchestva." *VAN*, 1952, no. 9: 92–95.

Khvol'son, O. D. *Znanie i vera v fizike*. Petrograd, 1916.

"K itogam diskussii po teorii otnositel'nosti." *VF*, 1955, no. 1: 134–138.

"K itogam Vsesoiuznogo soveshchaniia OVMD." *PZM*, 1931, no. 3: 1–11.

Kneale, W., and M. Kneale. *The Development of Logic*. Oxford: Clarendon, 1962.

Knei-Paz, Baruch. *The Social and Political Thought of Leon Trotsky*. Oxford: Clarendon, 1978.

Kniazev. G. A. "Akademiia nauk za 30 let." *VAN*, 1947, no. 11: 117–133.

_____. "Maksim Gor'kii i tsarskoe pravitel'stvo." *VAN*, 1932, no. 2: 23–44.

Kniazev, G. A., and A. V. Kol'tsov. *Kratkii ocherk istorii Akademii nauk SSSR*, 3d ed. Moscow-Leningrad, 1964.

Knuniants, I. L. "O nekotorykh teoreticheskikh problemakh sovremennoi organicheskoi khimii." *VAN*, 1953, no. 4: 15–29.

"K obnaruzheniiu politicheskikh dokumentov v Akademii nauk." *LP*, November 14, 1929.

Kolbanovskii, V. "Obshchii obzor." *PZM*, 1939, no. 11: 127–140.

Kol'man, Ernest Ia. "Boevye voprosy estestvoznaniia i tekhniki v rekonstruktivnyi period." *PZM*, 1931, no. 3: 56–78.

_____. "Chto takoe kibernetika." *VF*, 1955, no. 4: 148–159.

[_____.] Colman, E. "Dynamic and Statistical Regularities in Physics and Biology." In *Science*, 83–94. See Bukharin et al. 1971.

_____. "K sporam o teorii otnositel'nosti." *VF*, 1954, no. 5: 178–189.

_____. "Kuda vedet fizikov sub'ektivizm." *VF*, 1953, no. 6: 171–189.

_____. *Noveishie otkrytiia sovremennoi fiziki v svete dialekticheskogo materializma*. Moscow, 1943.

_____. "Pis'mo tov. Stalina i zadachi fronta estestvoznaniia i meditsiny." *PZM*, 1931, nos. 9–10: 163–172.

_____. "Politika, ekonomika i matematika." In *Na bor'bu za materialisticheskuiu dialektiku v matematike*, 53–68.

[_____.] Colman, E. "The Present Crisis in the Mathematical Sciences." In *Science*, 215–229. See Bukharin et al 1971.

_____. "Protiv lzhenauki." *SN*, 1938, no. 1: 39–50.

_____. "Teoriia kvant i dialekticheskii materializm." *PZM*, 1939, no. 10: 129–145.

_____. "Teoriia otnositel'nosti i dialekticheskii materializm." *PZM*, 1939, no. 6: 106–120.

_____. "Triumf marksizma—nauki proletariata." *MiE*, 1933: 62–79.

_____. "Uzlovye problemy sovremennoi atomnoi fiziki." *FNiT*, 1936, no. 2: 24–34.

_____. "Vklad Eilera v razvitie matematiki v Rossii." *VIE-T*, 1957, no. 4: 15–25.

_____. "Za marksistsko-leninskuiu nauku." *FNiT*, 1932, no. 1: 18–25.

_____. "Znachenie simvolicheskoi logiki." In *Logicheskie issledovaniia*, 3–19. See Kol'man et al. 1959.

Kol'man, Ernest Ia., et al., eds. *Logicheskie issledovaniia*. Moscow, 1959.

Kolmogorov, A. N. "Matematika." In *BSE*. Vol. 38, 359–401. Moscow, 1938.

_____. "Matematika na poroge vuza." In *Nauka segodnia*, 242–248. See Mikulinskii 1969.

_____. "Rol' russkoi nauki v razvitii teorii veroiatnostei." *UZMU*, no. 92: 53–64.

Kolotov, V. V. *Nikolai Alekseevich Voznesenskii.* Moscow, 1974.

Kol'tsov, A. V. "Akademik V. P. Volgin—organizator sovetskoi nauki." *VAN,* 1979, no. 8: 105–111.

_____. "K 250-letiiu Akademii nauk SSSR." *VIE-T,* 1974, no. 1: 3–21.

_____. *Lenin i stanovlenie Akademii nauk kak tsentra sovetskoi nauki.* Leningrad, 1969.

_____. "Leninskii plan nauchno-tekhnicheskikh rabot i Akademiia nauk." *VAN,* 1968, no. 4: 67–74.

_____. "Proekty organizatsii lomonosovskogo instituta v Akademii nauk v nachale XX v." In *Lomonosov: Sbornik statei i materialov.* Vol. 6, 294–300. Moscow-Leningrad, 1965.

Kol'tsov, N. K. "Predislovie." *Izvestiia Instituta eksperimental'noi biologii,* 1921, no. 1: 3–6.

Komarov, V. L. *Izbrannye sochineniia.* Vol. 11. Moscow-Leningrad, 1948.

_____. "O zadachakh Instituta Istorii estestvoznaniia Akademii nauk SSSR." *VAN,* 1945, nos. 1–2: 30–33.

_____. "K itogam Iubileinoi sessii Akademii nauk SSSR." *VAN,* 1945, nos. 7–8: 56–59.

_____. "Novoe popolnenie Akademii nauk SSSR." *SN,* 1939, no. 3: 74–81.

[_____] "Rech' prezidenta Akademii nauk SSSR akademika V. L. Komarova na torzhestvennom zasedanii iubileinoi sessii Akademii nauk." *VAN,* 1945, nos. 5–6: 3–10.

Komkov, G. D., et al. *Akademiia nauk SSSR; Kratkii istoricheskii ocherk.* Moscow, 1967.

Komkov, G. D., et al. *Akademiia nauk SSSR: Kratkii istoricheskii ocherk,* 2d ed. 2 vols. Moscow, 1977.

Komkov, G. D., et al. *Akademiia nauk SSSR—shtab sovetskoi nauki.* Moscow, 1968.

Kon, I. S. *Sotsiologiia lichnosti.* Moscow, 1967.

Kononkov, A. F. and A. N. Osinovskii. "Iz istorii nauchnykh sviazei sovetskikh i zarubezhnykh fizikov v pervye gody sovetskoi vlasti." *IMEN,* 1974, no. 15: 113–116.

Konstantinov, F. V. "Sotsialisticheskoe obshchestvo i istoricheskii materializm." *PZM,* 1936, no. 12: 45–66.

_____. "Sovremennye problemy marksistsko-leninskoi filosofii i zadachi filosofskoi obshchestvennosti." In *Filosofiia i sovremennost',* 37–54. *See* Frolov 1973.

Konstantinov, F. V., et al., eds. *Dialektika i logika nauchnogo poznaniia.* Moscow, 1966.

Konstantinov, F. V., et al., eds. *Dialektika material'noi i dukhovnoi zhizni obshchestva v period stroitel'stva kommunizma.* Moscow, 1966.

Kopelevich, Iu. Kh. *Osnovanie Peterburgskoi Akademii nauk.* Leningrad, 1977.

———. "Pervye otkliki zarubezhnoi pechaty na raboty Lomonosova." In *Lomonosov: Sbornik*. Vol. 5, 241–250.

———. "Peterburgskaia Akademiia nauk—novyi tip gosudarstvennogo uchrezhdeniia." *Priroda*, 1978, no. 4: 91–97.

Kopnin, P. V. "O prirode i osobennostiakh filosofskogo znaniia." *VF*, 1969, no. 4: 123–133.

Korbut, M. K. *Kazanskii Gosudarstvennyi Universitet*. Vol. 2. Kazan, 1930.

Korel', I. "V Akademii nauk SSSR." *SRiN*, 1931, nos. 2–3: 265–266.

Korol, Alexander. *Soviet Research and Development: Its Organization, Personnel and Funds*. Cambridge, Mass.: M.I.T. Press, 1965.

Korotke zvidomlennia diial'nist' Ukraïns'koï Akademii nauk v Kyïvi za 1919–1924 rr. Kiev, 1925.

Korovin, E. "'Uchenye' vrediteli i zadachi VARNITSO." *VAR*, 1930, nos. 9–10: 22–27.

Korsunsky, M. *The Atomic Nucleus*. Translated by G. Yanovsky. New York: Dover, 1963.

Korzheva, E. M. "Kategoriia kollektivnogo soznaniia i ee rol' v kontseptsii Emilia Diurkgeima." *VMU*, 1968, no. 4: 91–100.

Kosygin, A. N. "Za tesnuiu sviaz' nauki s zhizn'iu." *VAN*, 1961, no. 7: 90–106.

Kozlov, B. I., ed. *Organizatsiia i razvitie otraslevykh nauchno-issledovatel'skikh institutov Leningrada, 1917–1977*. Leningrad, 1979.

Kozlov, V. V. *Ocherki istorii khimicheskikh obshchestv SSSR*. Moscow, 1958.

KPSS v rezoliutsiiakh i resheniiakh s'ezdov, konferentsii i plenumov TsK. 9 vols. Moscow, 1970–1972.

Kramish, Arnold. *Atomic Energy in the Soviet Union*. Stanford, Calif.: Stanford University Press, 1959.

Krasil'nikov, V. "Vykhod—v planirovanii." *LG*, 1974, no. 20: 13.

Kravets, T. P. "N'iuton i izuchenie ego trudov v Rossii." In *Isaak N'iuton*, edited by S. I. Vavilov, 312–328.

———. "VI Vsesoiuznyi s'ezd fizikov." *Priroda*, 1928, no. 10: 913–920.

Kreps, E. M., ed. *Perepiska I. P. Pavlova*. Leningrad, 1970.

"Kritika teoreticheskikh osnov bukharinskoi kontseptsii istoricheskogo materializma." *VKA*, 1929, nos. 35–36: 227–296.

Krotov, V. A. "Organizatsiia i deiatel'nost' Vostochno-Sibirskogo filiala Akademii nauk SSSR." In *Akademiia nauk*, 276–285. *See* Okladnikov et al. 1977.

Krum, S. "Fizika na sessii Akademii nauk." *SRiN*, 1936, no. 5: 105–116.

Krzhizhanovskii, G. M. "Akademiia nauk v 1936 godu." *FNiT*, 1936, no. 4: 7–18.

———. "Programma rabot Akademii nauk SSSR na 1936 g." *VAN*, 1936, nos. 4–5: 1–17.

K., S. "Pervyi Vserossiiskii astronomicheski s'ezd." *Priroda*, 1917, nos. 5–6: 703–706.

"K semidesiatiletiiu so dnia rozhdeniia velikogo revoliutsionera i genial'nogo uchenogo Iosifa Vissarionovicha Stalina." *VAN*, 1949, no. 12: 8–12.

Ksenofontov, V. I. *Leninskie idei v sovetskoi filosofskoi nauke 20-kh godov.* Leningrad, 1975.

Kudriavtsev, P. S. "Lomonosov i N'iuton." *TIIE-T* 5 (1955): 32–51.

Kugel', S. A., et al., eds. *Nauchnye kadry Leningrada: Struktura kadrov i sotsial'nye problemy organizatsii truda.* Leningrad, 1973.

Kuhn, Thomas S. *The Structure of Scientific Revolutions.* Chicago: University of Chicago Press, 1970.

[————.] Kun, T. *Struktura nauchnykh revoliutsii.* Translated from the English by I. Z. Naletova. Moscow, 1975.

Kulagin, Nik. "Ob'edinenie (assotsiatsiia) russkikh estestvoispytatelei i vrachei." *Priroda*, 1916, no. 9: 1080–1082.

Kurosh, A. G., et al. *Matematika v SSSR za tridtsat' let.* Moscow-Leningrad, 1948.

Kursanov, D. N., et al. "The Present State of the Chemical Structural Theory." *Journal of Chemical Education* (1952): 2–13.

Kursanov, G. A. "K kriticheskoi otsenke teorii otnositel'nosti." *VF*, 1952, no. 1: 169–174.

Kuzin, A. M. "O nekotorykh zadachakh i perspektivakh razvitiia sovetskoi biologii." *VAN*, 1952, no. 3: 36–43.

Kuznetsov, B. G. "Einshtein i Dostoevskii." *NiZh*, 1965, no. 6: 38–45.

————. *Etiudy ob Einshteine.* Moscow, 1965.

————. *Lomonosov, Lobachevskii, Mendeleev.* Moscow-Leningrad, 1945.

————. "O kvantovo-reliativistskoi logike." *VF*, 1970, no. 2: 118–122.

————. *Patriotizm russkikh estestvoispytatelei i ikh vklad v nauku*, 2d ed. Moscow, 1951.

————. *Tvorcheskii put' M. V. Lomonosova.* Moscow-Leningrad, 1957.

————. "Vvedenie." In *Istoriia estestvoznaniia v Rossii*, edited by N. A. Figurovskii, 185–214. Vol. 1, part 1.

Kuznetsov, I. V. *Kharakternye cherty russkogo estestvoznaniia.* Moscow, 1948.

————. "Ser'eznye oshibki v osveshchenii istorii fiziki." *Bol*, 1950, no. 6: 70–80.

————. "Sovetskaia fizika i dialekticheskii materializm." In *Filosofskie voprosy*, 31–86. See A. A. Maksimov 1952.

————, ed. *Liudi russkoi nauki*, 2 vols. Moscow-Leningrad, 1948.

Kuznetsov, I. V., et al. *Philosophical Problems of Elementary-Particle Physics.* Translated from the Russian by G. Yanovskii. Moscow, 1968.

Kuznetsov, I. V., and M. E. Omel'ianovskii, eds. *Filosofskie voprosy sovremennoi fiziki.* Moscow, 1958.

Kuznetsov, I. V., and A. S. Fedorov. "60 let sovetskoi nauki." *VIE-T*, 1977, no. 1: 3–7.

————. *Filosofskie voprosy sovremennoi fiziki.* Moscow, 1959.

Kuzovkov, D. V. "Nauchno-issledovatel'skie instituty v sisteme nauchnykh uchrezhdenii." *NR*, 1925, no. 2: 40–55.

La Commission centrale pour l'amélioration des conditions de viè des hommes de science. Moscow, 1927.

Lange, K. A. *Organizatsiia upravleniia nauchnymi issledovaniami.* Leningrad, 1971.

————. *Razvitie i organizatsiia fiziologicheskoi nauki v SSSR.* Leningrad, 1978.

Lapirov-Skoblo, M. Ia. "Novye puti nauki i tekhniki SSSR." In *Nauka.* Vol. 1, 55–71. *See* Ioffe et al. 1928.

Lappo-Danilevskii, A. S. *Metodologiia istorii,* 3d ed. Vol. 1. Petrograd, 1923.

————. *Petr' Velikii, osnovatel' Imperatorskoi Akademii nauk v S.-Peterburge.* St. Petersburg, 1914.

Lappo-Danilevskii, A. S., and A. Nol'de. "Ob Institute sotsial'nykh nauk." *IRAN* 12, series 6 (1918): 1402–1403.

Laue, Max von. *History of Physics.* Translated by R. Oesper. New York: Academic Press, 1950.

Lavrent'ev, M. A., and A. A. Trofimuk. "Nauka v Sibiri." In *Nauka,* 67–68. *See* Millionshchikov et al. 1969.

Lazarev, P. P. "Chistaia nauka i prakticheskaia zhizn'." *NR,* 1925, no. 2: 31–39.

————. *Ocherki istorii russkoi nauki.* Moscow-Leningrad, 1950.

Lebedev, D. M. *Geografiia Rossii petrovskogo vremeni.* Moscow-Leningrad, 1950.

Lecourt, D. *Proletarian Science: The Case of Lysenko.* Translated by B. Brewster. London: NLB, 1977.

Leibson, L. G. *Leon Abgarovich Orbeli.* Leningrad, 1973.

Leicester, Henry M. Introduction to *Mikhail Vasil'evich Lomonosov's Corpuscular Theory,* 3–48. *See* [Lomonosov] 1970.

————. "Mendeleev and the Russian Academy of Sciences." *Journal of Chemical Education* 25 (1948): 439–442.

Leiman, I. I. *Nauka kak sotsial'nyi institut.* Leningrad, 1971.

Lenin, V. I. *Polnoe sobranie sochinenii,* 5th ed. Moscow.

"Leningradskie uchenye obsuzhdaiut blok nekotorykh novykh akademikov s reaktsionerami." *LP,* February 7, 1929.

Leont'ev, A. N. *Problemy razvitiia psikhiki.* Moscow, 1959.

Leont'ev, A. N., and A. R. Luriia. "Psikhologicheskie vozzreniia L. S. Vygotskogo." In *Izbrannye psikhologicheskie issledovaniia,* 4–36. *See* Vygotskii 1956.

Leontovich, M. A., and V. L. Ginzburg. "Nekotorye voprosy zhurnal'no-izdatel' skoi deiatel'nosti v oblasti fiziki." *VAN,* 1961, no. 5: 131–134.

Lepeshinskaia, O. B. *Proiskhozhdenie kletok iz zhivogo veshchestva.* Moscow, 1951.

————. *Proiskhozhdenie kletok iz zhivogo veshchestva i rol' zhivogo veshchestva v organizme.* Moscow, 1950.

————. "Tvorcheskoe znachenie trudov Marksa, Engel'sa, Lenina, Stalina dlia razvitiia nauki." *VF,* 1953, no. 2: 12–38; no. 3: 46–58.

Levin, B. Iu. "Ravitie planetnoi kosmogonii." In *Razvitie,* 300–314. *See* Ambartsumian et al. 1967.

————. *The Origin of the Earth and Planets*, 2d ed. Translated from the Russian by A. Shkorovsky. Moscow, 1958.

Lewis, Robert A. "Government and the Technological Sciences in the Soviet Union: The Rise of the Academy of Sciences." *Minerva* 15, no. 2 (1977): 174–199.

————. *Science and Industrialization in the USSR: Industrial Research and Development, 1917–1940*. New York: Holmes and Meier, 1979.

Lewontin, Richard, and Richard Levins. "The Problem of Lysenkoism." In *Ideology*, 173–205. *See* Rose and Rose 1976.

"Lichnyi sostav Imperatorskoi Akademii nauk 1 aprelia 1868 goda." *ZAN* 13 (1968): 103–115.

Lindener, B. A. "Issledovatel'skaia rabota Akademii nauk SSSR v sovetskoi obstanovke." *NR*, 1925, no. 3: 9–24.

————, ed. *Raboty Rossiiskoi Akademii nauk v oblasti issledovaniia prirodnykh bogatstv Rossii*. Petrograd, 1922.

Liubimenko, V. N., and I. I. Liubimenko. "Ivan Parfen'evich Borodin." In *Iubileinyi sbornik*, 3–38. *See* Iachevskii 1927.

Liubishchev, A. A. "K klassifikatsii evoliutsionnykh teorii." *Problemy evoliutsii* 4 (1975): 206–220.

[Lobachevskii, N. I.] *N. I. Lobachevskii: Sbornik stat'ei*. Moscow, 1943.

[Lomonosov, M. V.] *Mikhail Vasil'evich Lomonosov on the Corpuscular Theory*. Translated, with an Introduction, by Henry M. Leicester. Cambridge, Mass.: Harvard University Press, 1970.

Losskii, N. "Osvobozhdenie nauki ot opeki gosudarstva." *Poliarnaia zvezda*, 1906, no. 6: 408–412.

Lubrano, Linda L. "Scientific Collectives: Behavior of Soviet Scientists in Basic Research." In *The Social Context*, 101–136. *See* Lubrano and Solomon 1980.

————. *Soviet Sociology of Science*. Columbus, Ohio: American Association for the Advancement of Slavic Studies, 1976.

Lubrano, Linda L., and Susan Gross Solomon, eds. *The Social Context of Soviet Science*. Boulder, Colo.: Westview Press, 1980.

Lukirskii, P. I., et al. *Sbornik posviashchennyi semidesiatiletiiu akademika A. F. Ioffe*. Moscow, 1950.

Lunacharskii, A. V. "Kul'turnaia revoliutsiia i nauka." *NR*, 1928, no. 4: 3–7.

————. "Rol' marksizma v nauchnom issledovanii." *NR*, 1925, no. 2: 19–30.

————. *Sobranie sochinenii*. Vol. 6. 1965.

————. *Vospominaniia i vpechatleniia*. Moscow, 1968.

Luppol, I. K. "IV Plenum Tsentral'nogo soveta o rabote Akademii nauk SSSR." *NR*, 1928, no. 7: 14–21.

————. "K vyboram v Akademiiu nauk SSSR." *NR*, 1928, no. 11: 3–9.

Luriia, A. R. "Mozg i psikhika." *Kom*, 1964, no. 6: 107–117.

————. "Psikhologiia v sisteme estestvoznaniia i obshchestvennykh nauk." In

Filosofiia, 216–224. *See* M. A. Markov et al. 1974.

———. "Teoriia razvitiia vysshikh psikhicheskikh funktsii v sovetskoi psikhologii." *VF*, 1966, no. 7: 72–79.

L'vov, V. "Protiv idealizma v fizike." *LG*, November 24, 1948.

———. "Trubadury idealizma v fizike." *LG*, November 20, 1948.

Lysenko, T. D. *Agrobiologiia*, 4th ed. Moscow, 1948.

———. "Novoe v nauke o biologicheskom vide." In *Filosofskie voprosy*, 3–16. *See* Novinskii and Platonov 1951.

———. "Raboty O. B. Lepeshinskoi." In *Vnekletochnye formy zhizni*, 191–193.

McClelland, James C. *Autocrats and Academics: Education, Culture and Society in Tsarist Russia*. Chicago: University of Chicago Press, 1979.

———. "Bolshevik Approaches to Higher Education, 1917–1921." *SR* 30, no. 4 (1971): 818–831.

———. "Utopianism versus Revolutionary Heroism in Bolshevik Policy: The Proletarian Culture Debate." *SR* 39, no. 3 (1980): 403–425.

Maiorov, F. P. "O mirovozzrenii I. P. Pavlova." *VAN*, 1936, no. 3: 17–20.

Makeeva, V. N. "Komitet nauki pri Sovnarkome v 1922–1924 gg." *VAN*, 1966, no. 5: 110–114.

———. "V. I. Lenin i Nauchno-tekhnicheskii otdel VSNKh." *VIE-T*, 1970, no. 1: 46–51.

Maksimov, A. A. "*Bol'shaia Sovetskaia Entsiklopediia:* Estestvenno-nauchnyi otdel." *PZM*, 1930, nos. 10–12: 339–352.

———. "Bor'ba Lenina s 'fizicheskim' idealizmom." In *Velikaia sila*, 187–222. *See* Stepanian 1950.

———. "Lenin i krizis estestvoznaniia epokhi imperializma." *PZM*, 1931, nos. 1–2: 12–44.

———. "Ob otnoshenii Lenina k estestvoispytateliiam." *PZM*, 1931, nos. 9–10: 125–162.

———. "Obsuzhdenie knigi I. V. Kuznetsova 'Printsip sootvetstviia v sovremennoi fizike i ego filosofskoe znachenie.'" *VF*, 1950, no. 2: 378–387.

———. *Ocherki po istorii bor'by za materializm v russkom estestvoznaniiu*. Moscow, 1947.

———. "O znachenii truda I. V. Stalina 'Marksizm i voprosy iazykoznaniia' dlia istorii estestvoznaniia." *Priroda*, 1951, no. 9: 3–11.

Maksimov, A. A., et al., eds. *Filosofskie voprosy sovremennoi fiziki*. Moscow, 1952.

Maksimov, N. A. "Razvitie i dostizheniia fiziologii rastenii v SSSR za 25 let." *Uspekhi sovremennoi biologii* 16, no. 5 (1943): 469–498.

———. Review of *Trudy Instituta genetiki Akademii nauk SSSR*. *VAN*, 1949, no. 3: 105–107.

Malkin, G. V. "Organizatsiia Zapadno-sibirskogo filiala Akademii nauk SSSR." In *Akademiia nauk*, 255–265. *See* Okladnikov et al. 1977.

Mandel'shtam, L. I. *Polnoe sobranie sochinenii*. Vol. 5. 1950.

Mandel'shtam, O. E. "Zametki o naturalistakh." *Puti v neznaemoe* 15 (1980): 442–445.

Man'kovskii, L. "K voprosu o filosofskikh istokakh men'shevistvuiushchego idealizma." *PZM*, 1931, no. 6: 44–72.

Marchuk, G. "Krai bol'shoi nauki." *Kom*, 1977, no. 18: 32–41.

———. "Sibirskomu otdeleniiu Akademii nauk SSSR—dvadtsat' let." In *Oktiabr'*, 635–656. *See* A. P. Aleksandrov et al. 1977.

Markarian, E. S. "Chelovecheskoe obshchestvo kak osobyi tip organizatsii." *VF*, 1971, no. 10: 64–75.

———. *Ocherki teorii kul'tury*. Erevan, 1969.

———. *Voprosy sistemnogo issledovaniia*. Moscow, 1972.

Markov, A. A. "Matematicheskaia logika." In *BSE*, 3d ed. Vol. 15, 479–480.

———. "Matematicheskaia logika." *FE* 3 (1965): 340–342.

Markov, M. A. "O prirode fizicheskogo znaniia." *VF*, 1947, no. 2: 140–176.

Markov, M. A., et al., eds. *Filosofiia i estestvoznanie: K semidesiatiletiiu akademika Bonifatiia Mikhailovicha Kedrova*. Moscow, 1974.

Markova, L. A. "Ob istorii estestvoznaniia kak nauke i ee zadachakh." In *Ocherki*, 126–141. *See* Bibler et al. 1969.

Markovnikov, V. V. "Istoricheskii ocherk khimii v Moskovskom universitete." In *Lomonosovskii sbornik*, 1–281. *See* Markovnikov et al. 1901.

Markovnikov, V. V., et al. *Lomonosovskii sbornik: materialy istorii razvitiia khimii v Rossii*. Moscow, 1901.

Marr, N. Ia. "Kraevedcheskaia rabota." *NR*, 1925, no. 1: 10–18.

"Martovskaia sessiia Akademii nauk SSSR." *UFN* 16, no. 7 (1936): 837–838.

Martynov, I. "O neskromnoi pretenzii pravykh uklonistov." *LP*, October 27, 1929.

"Matematika." In *Istoriia estestvoznaniia v SSSR*, 119–175. *See* Figurovskii 1952.

Materialist. "Komu sluzhit kibernetika." *VF*, 1953, no. 5: 210–219.

Materialy dlia istorii Imperatorskoi Akademii nauk. 10 vols. St. Petersburg, 1885–1900.

"Materialy dlia novogo ustava Imperatorskoi Akademii nauk i sostoiashchikh pri nei muzeev." *ZhMNP* 125, sec. 4 (1865): 141–281.

"Materialy nauchnoi sessii Instituta filosofii Komakademii." *PZM*, 1933, no. 3: 132–161.

Materialy po istorii otechestvennoi khimii: Sbornik dokladov na Pervom Soveshchanii po istorii otechestvennoi khimii, 12–15 maia 1948 g. Moscow-Leningrad, 1950.

Mayr, Ernst, and W. B. Provine, eds. *The Evolutionary Synthesis*. Cambridge, Mass.: Harvard University Press, 1980.

Medvedev, Roy A. *Let History Judge*. Translated by C. Taylor. New York: Vintage, 1973.

Medvedev, Zhores A. *The Medvedev Papers*. Translated by V. Rich. London: Macmillan, 1971.

————. *The Rise and Fall of T. D. Lysenko.* New York: Doubleday, 1971.

————. *Soviet Science.* New York: Norton, 1978.

Meilakh, B. S. *Na rubezhe nauki i iskusstva.* Leningrad, 1971.

————. "Posleslovie." *NM,* 1966, no. 12: 191–198.

Meliukhin, S. T. "Neischerpaemost' svoistv 'elementarnykh' chastits." In *Nekotorye filosofskie voprosy,* 191–222. *See* Kaganov 1957.

Mendel, Gregor. *Opyty nad rastitel'nymi gibridami.* Translated from the German by K. Fliaksberg. St. Petersburg, 1910. (Also published in *Trudy Biuro po prikladnoi botaniki,* 1910, no. 11.)

Mendeleev, D. I. "Kakaia zhe Akademiia nuzhna v Rossii?" *NM,* 1966, no. 12: 176–191.

Menshutkin, B. N. "M. V. Lomonosov kak fiziko-khimik: K istorii khimii v Rossii." *Zhurnal Russkogo fiziko-khimicheskogo obshchestva* 36, sec. 2 (1904): 77–304.

————. *Russia's Lomonosov.* Translated by J. Eyre and E. T. Webster. Princeton, N.J.: Princeton University Press, 1952.

————. *Zhizn' i deiatel'nost' N. A. Menshutkina.* St. Petersburg, 1908.

Merton, Robert K. *The Sociology of Science.* Chicago: Chicago University Press, 1973.

Metal'nikov, S. "Petr Frantsevich Lesgaft: Biograficheskii ocherk." In *Pamiati Petra Frantsevicha Lesgafta,* 35–56.

Metodologicheskie problemy nauki: Materialy zasedaniia Prezidiuma Akademii nauk. Moscow, 1964.

Meyer, Klaus. "Wissenschaftspolitik." In *Kulturpolitik,* 145–189. *See* Anweiler and Ruffman 1973.

————. *Das wissenschaftliche Leben in der UdSSR nach dem Stand von 1961.* Berlin, 1963.

Michurin, I. V. *Sochineniia,* 2d ed. 4 vols. Moscow, 1948.

Mikhailov, A. A., et al., eds. *Astronomiia v SSSR za sorok let: 1917–1957.* Moscow, 1960.

Mikulinskii, S. R. "From the Revolution in the Natural Sciences to the Scientific and Technological Revolution." In *Science,* 73–82. *See* Velikhov, Gvishiani, and Mikulinskii 1980.

————. "Nekotorye problemy organizatsii nauchnoi deiatel'nosti i ee izucheniia." In *Organizatsiia,* 137–149. *See* Beliaev et al. 1968.

————. "O naukovedenii kak obshchei teorii razvitiia nauki." In *Upravlenie.* Vol. 2, 21–36. *See* Gvishiani et al. 1970.

————. "Problemy analiza istorii nauki v trudakh V. I. Vernadskogo." *VF,* 1975, no. 5: 106–118.

————, ed. *Nauka segodnia.* Moscow, 1969.

Mikulinskii, S. R., and N. I. Rodnyi. "Istoriia nauki i naukovedenie." In *Ocherki,* 35–66. *See* Bibler et al. 1969.

Mikulinskii, S. R., and L. A. Markova. "Chem interesna kniga T. Kuna 'Struktura nauchnykh revoliutsii.'" In *Struktura*, 265–282. *See* Kuhn 1975.

Mikulinskii, S. R., and N. I. Rodnyi. "Nauka kak predmet spetsial'nogo issledovaniia." *VF*, 1966, no. 5: 25–38.

Miliukov, P. "Universitety v Rossii." In *Entsiklopedicheskii slovar.* Vol. 68, 788–800. F. A. Brokgauz and I. A. Efron, 1902.

Millar, James R. *The ABC's of Soviet Socialism.* Urbana: University of Illinois Press, 1981.

———, ed. *The Soviet Rural Community.* Urbana: University of Illinois Press, 1971.

Millionshchikov, M. D. "Shirokie gorizonty." *Pravda*, November 13, 1969.

Millionshchikov, M. D., et al., eds. *Nauka i chelovechestvo—1969.* Moscow, 1969.

———. *Nauka i chelovechestvo—1971–1972.* Moscow, 1972.

Mirskii, E. M. "Naukovedenie v SSSR (Istoriia, problemy, perspektivy)." *VIE-T*, 1971, nos. 3–4: 86–98.

Mitin, M. B. *Boevye voprosy materialisticheskoi dialektiki.* Moscow, 1936.

———. "K voprosu o leninskom etape v razvitii dialekticheskogo materializma." *PZM*, 1931, nos. 7–8: 9–32.

———. "Leninskii etap marksistskoi filosofii." *VAN*, 1970, no. 2: 3–12.

———. "Problema material'nogo i ideal'nogo v svete dialekticheskogo materializma." In *Filosofskie voprosy fiziologii*, 13–34. *See* Fedoseev et al. 1963.

———. "Rol' i znachenie raboty tovarishcha Stalina 'O dialekticheskom i istoricheskom materializme' v razvitii marksistsko-leninskoi filosofii mysli." In *Voprosy istorii*, 15–40. *See* Vavilov et al. 1949.

———. "Za peredovuiu sovetskuiu geneticheskuiu nauku." *PZM*, 1939, no. 10: 147–176.

Mitin, M. B., and I. Razumovskii, eds. *Dialekticheskii i istoricheskii materializm.* 2 vols. Moscow, 1932.

Mitin, M. B., et al., eds. *Protiv reaktsionnogo mendelizma-morganizma.* Moscow-Leningrad, 1950.

Mitkevich, V. F. "O pozitsii I. E. Tamma v otnoshenii printsipial'nykh vozzrenii Faradeia i Maksvella." *PZM*, 1933, nos. 6: 278–281.

———. "Oznovnye vozzreniia sovremennoi fiziki." In *Pamiati Karla Marksa*, 221–244. *See* Bukharin and Deborin 1933.

———. "O sovremennoi bor'be materializma s idealizmom v oblasti fiziki." *PZM*, 1938, no. 8: 111–137.

———. "Za Faradee-Maksvellovskuiu ustanovku v voprose o prirode fizicheskikh vzaimodeistviiakh." *SRiN*, 1934, no. 8: 97–103.

Mitriakova, N. M. "Struktura, nauchnye uchrezhdeniia i kadry AN SSSR (1917–1940 gg.)." In *Organizatsiia*, 203–235. *See* Beliaev et al. 1968.

Mochalov, I. I. *Vladimir Ivanovich Vernadskii.* Moscow, 1982.

Modestov, V. "Russkaia nauka v poslednie dvadtsat' piat' let." *Russkaia mysl'*, 1890, no. 5, sec. 2: 73–91.

Modrzhinskaia, E. D., and Ts. A. Stepanian, eds. *The Future Society: A Critique of Modern Bourgeois Philosophical and Socio-Political Conceptions.* Moscow, 1973.

Mokshin, Stepan. "Sovnarkomovskii paek." *Puti v neznaemoe* 8 (1970): 5–20.

Molas, B. N. "Iubileinye dni." *NR*, 1925, no. 3: 38–53.

"Molodezh v nauke." *Pravda*, June 18, 1967.

Molotov, V. M. *31-ia godovshchina Velikoi Oktiabr'skoi Sotsialisticheskoi Revoliutsii.* Moscow, 1948.

Monod, Jacques. *Chance and Necessity.* Translated by A. Weinhouse. New York: Knopf, 1971.

"Moskovskii Nauchnyi institut." *Priroda*, 1917, no. 2: 281–282.

Mostowski, Andrzej. *Thirty Years of Foundational Studies.* New York: Barnes and Noble, 1966.

Moszkowski, Alexander. *Einstein the Searcher.* Translated by H. L. Brose. London: Methuen, 1921.

Muksinov, I. M., and I. L. Davitnidze. "Akademiia nauk SSSR i akademii nauk soiuznykh respublik." In *Organizatsionno-pravovye voprosy*, 208–234. *See* Piskotin et al. 1973.

Müller-Markus, S. "Diamat and Einstein." *Survey*, 1961, nos. 35–39: 68–78.

———. *Einstein und die Sowjetphilosophie.* 2 vols. Dordrecht, Holland, 1969.

———. "Niels Bohr in the Darkness and Light of Soviet Philosophy." *Inquiry* 9 (1966): 73–90.

Naan, G. I. "K voprosu o printsipe otnositel'nosti v fizike." *VF*, 1951, no. 2: 57–77.

Na bor'bu za materialisticheskuiu dialektiku v matematiki: Sbornik statei. Moscow-Leningrad, 1931.

Nagibin, S. "Russkoe Botanicheskoe obshchestvo." *Priroda*, 1917, no. 2: 267–268.

Nalimov, V. V., and Z. M. Mul'chenko. *Naukometriia.* Moscow, 1969.

Narodnoe khoziaistvo SSSR v 1973 g. Moscow, 1974.

Nasledov, D. "Otchetnyi doklad akad. A. F. Ioffe." *FNiT*, 1936, no. 4: 19–23.

"Nauchnaia obshchestvennost' kleimit vragov v sovetskoi maske." *FNiT*, 1936, no. 7: 121–123.

"Nauchnaia rabota v vuzakh." *VAN*, 1968, no. 1: 3–16.

Nauchnaia sessiia posviashchennaia problemam fiziologicheskogo ucheniia I. P. Pavlova, 28 iiunia–4 iiulia 1950 g. Moscow, 1950.

Nauchno-tekhnicheskaia revoliutsiia i sotsial'nyi progress. Moscow, 1972.

"Nauchnye issledovaniia po fiziologii SSSR." *VAN*, 1970, no. 6: 48–54.

"Nauchnye sessii posviashchennye 50-letiiu KPSS." *VAN*, 1953: 21–39.

"Nauchnyi tsenter biologicheskikh issledovanii v Pushchine." *VAN*, 1968, no. 11: 9–18.

Nauka i iskusstvo. Vol. 1, Moscow-Leningrad, 1926.

"Nauka i zhizn'." *Kom*, 1954, no. 5: 3–13.

Neapolitanskaia, V. S., ed. *Perepiska V. I. Vernadskogo s B. L. Lichkovym, 1918–1939*. Moscow, 1979.

———. *Perepiska V. I. Vernadskogo s B. L. Lichkovym, 1940–1944*. Moscow, 1980.

Nedospasov, A. V. "K vysochaishei vershine strany." *VAN*, 1974, no. 7: 114–118.

"Nekotorye voprosy razvitiia sovremennoi fiziki." *VAN*, 1965, no. 2: 3–46.

Nemchinov, V. S. *Izbrannye proizvedeniia*. Vol. 1. Moscow, 1967.

Nesmeianov, A. N. "Ob osnovnykh napravleniiakh v rabote Akademii nauk SSSR." *VAN*, 1957, no. 2: 3–42.

———. "Velikaia sila tvorcheskogo sodruzhestva." *VAN*, 1952, no. 3: 3–21.

———. "Vozmozhnosti nauki bezgranichny." *LG*, 1967, no. 1, p. 11.

Nevskii, V. I. "Restavratsiia idealizma i bor'ba s 'novoi' burzhuaziei." *PZM*, 1922, nos. 7–8: 117–131.

Nikolai Semenovich Kurnakov—osnovopolozhnik fiziko-khimicheskogo analiza. Novosibirsk, 1960.

Nikol'skii, V. K., and N. F. Iakovlev. "Osnovnye polozheniia materialisticheskogo ucheniia N. Ia. Marra o iazyke." *VF*, 1949, no. 1: 265–285.

"Noiabr' 1936 goda." *SRiN*, 1936, no. 9: 3–8.

Novikov, M. M. *Ot Moskvy do N'iu-Iorka*. New York: Chekhov, 1952.

Novinskii, I. I., and G. V. Platonov, eds. *Filosofskie voprosy sovremennoi biologii*. Moscow, 1951.

Novinskii, I. I., et al., eds. *Voprosy dialekticheskogo materializma*. Moscow, 1951.

Novogrudskii, D. "Geokhimiia i vitalizm (O 'nauchnom mirovozzrenii' akad. V. I. Vernadskogo)." *PZM*, 1931, nos. 7–8: 168–203.

"Novye uspekhi sovetskoi nauki." *Bol*, 1948, no. 11: 1–9.

Nuzhin, M. T., et al., eds. *Kazanskii universitet: 1804–1979*. Kazan, 1979.

"Ob akademike Luzine." *VAN*, 1936, nos. 8–9: 7–10.

"Ob itogakh soveshchaniia, posviashchennogo problemam morfologii zhivotnykh i kriticheskoi otsenke 'evoliutsionnoi morfologii' A. N. Severtsova." *VAN*, 1953, no. 6: 62–63.

"Oblispolkom i Leningradsovet—akad. I. P. Pavlovu." *LP*, September 27, 1929.

"Ob odnom filosofskom dispute." *FN*, 1967, no. 4: 133–139.

"Obrashchenie k tovarishchu Stalinu." *VAN*, 1948, no. 9: 19–20.

"Obshchee sobranie Akademii nauk SSSR (1938)." *VAN*, 1938, nos. 9–10: 80–85.

"Obshchee sobranie Akademii nauk SSSR (29–30 iiunia 1962 g.)." *VAN*, 1962, no. 8: 3–13.

"Obshchee sobranie Akademii nauk SSSR (19–20 oktiabria 1962 g.)." *VAN*, 1962, no. 12: 19–50.

"Obsuzhdenie doklada M. V. Keldysha." *VAN*, 1961, no. 7: 40–89.

"Ob filosofskikh voprosakh psikhologii. (K itogam diskussii)." *VF*, 1954, no. 4: 182–193.

Okladnikov, A. P., et al., eds. *Akademiia nauk i Sibir'*. Novosibirsk, 1977.

Okulov, A. F. *Sovetskaia filosofskaia nauka i ee problemy.* Moscow, 1970.

Olby, Robert. *The Path to the Double Helix.* Seattle: University of Washington Press, 1974.

Ol'denburg, S. F., "K dvukhsotletiiu Rossiiskoi Akademii nauk." *NR,* 1925, no. 1: 65–69.

―――. "Nauchno-kul'turnye itogi akademicheskogo iubileia." *NR,* 1925, no. 3: 3–8.

―――. "Voprosy organizatsii nauchnoi raboty." In *Tvorchestvo,* 3–14. *See* Blokh et al. 1923.

Ol'derogge, D. A., and I. I. Potekhin. "Funktsional'naia shkola v etnografii na sluzhbe britanskogo imperializma." In *Anglo-amerikanskaia etnografiia,* 41–66. *See* Potekhin 1951.

Omel'ianovskii, M. E. "Bor'ba materializma protiv idealizma v sovremennoi fizike." In *Voprosy dialekticheskogo materializma,* edited by I. I. Novinskii et al., 143–170.

―――. "Dialekticheskii materializm i problema real'nosti v kvantovoi fizike." In *Filosofskie voprosy,* 5–54. *See* I. V. Kuznetsov and Omel'ianovskii 1958.

―――. "Dialekticheskii materializm i sovremennaia fizika." In *Filosofskie voprosy,* 3–29. *See* I. V. Kuznetsov and Omel'ianovskii 1958.

―――. "Lenin i dialektika v sovremennoi fizike." In *Lenin,* 127–148. *See* Omel'ianovskii 1969.

―――, ed. *Lenin i sovremennoe estestvoznanie.* Moscow, 1969.

Omel'ianskii, V. L. "Ocherednye zadachi sovremennoi mikrobiologii." *Otchet-R,* 1924 (supplement).

Oparin, A. I. "Institut biokhimii." *VAN,* 1937, nos. 10–11: 197–210.

―――. "Problema proiskhozhdeniia zhizni v sovremennom estestvoznanii." In *Filosofskie voprosy sovremennoi biologii,* 267–287. *See* Novinskii and Platonov 1951.

―――. "Problema proiskhozhdeniia zhizni v svete dostizhenii sovremennogo estestvoznaniia." *VAN,* 1953, no. 11: 39–48.

―――. *Zhizn' kak forma dvizheniia materii.* Moscow, 1963.

"O polozhenii na fronte estestvoznaniia. (Rezoliutsiia prezidiuma Komakademii sovmestno s Assotsiatsiei estestvoznaniia Komakademii i estestvennym otdeleniem IKP, F. i E. po dokladu O. Shmidta i sodokladu A. Maksimova)." *VKA,* 1931, no. 1: 23–32.

O polozhenii v biologicheskoi nauke: Stenograficheskii otchet Sessii Vsesoiuznoi Adademii sel'skokhoziaistvennykh nauk imeni V. I. Lenina, 31 iiulia–7 avgusta 1948 g. Moscow, 1948.

"O proekte ustava i shtatov Imperatorskoi Akademii nauk v S. Petersburge." *Chteniia v Imperatorskom Obshchestve istorii i drevnostei rossiiskikh pri Moskovskom universitete,* 1965, no. 2: 142–165.

"O raznoglasiiakh na filosofskom fronte." *VKA,* 1930, no. 42: 20–89.

Orbeli, L. A. "Razvitie biologicheskikh nauk v SSSR za 25 let." In *Iubileinaia sessiia Akademii nauk*, 216–227. *See Iubileinaia sessiia* 1943.

———, ed. *Uspekhi biologicheskikh nauk v SSSR za 25 let: 1917–1942*. Moscow, 1945.

Oreshkin, V. V. *Vol'noe ekonomicheskoe obshchestvo, 1765–1917*. Moscow, 1963.

"Organizatsiia Kirgizskogo filiala." *VAN*, 1973, nos. 7–8: p. 122.

"Organizatsionno-administrativnaia khronika." *VAN*, 1931, no. 3: 49–51.

Orlov, I. "Klaissicheskaia fizika i reliativizm." In *Teoriia otnositel'nosti*, 153–187.

Orlov-Davydov, V. "Biograficheskii ocherk V. G. Orlova." *Russkii arkhiv*, 1908, no. 7: 301–395.

"O sostoianii podgotovki doktorov nauk cherez doktoranturu Akademii nauk SSSR." *VAN*, 1953, no. 6: 64.

Ossowska, Maria, and Stanislaw Ossowski. "The Science of Science," *Minerva* 3 (1964): 78–82.

Ostrovitianov, K. V., et al., eds. *Istoriia Akademii nauk SSSR*. 2 vols. Moscow-Leningrad, 1958–1964.

Otchety o deiatel'nosti Komissii po izucheniiu estestvennykh proizvoditel'nykh sil' Rossii. No. 8. Petrograd, 1917.

"Ot Instituta istorii Akademii nauk SSSR." In *Protiv istoricheskoi kontseptsii*. Vol. 1, 3–4. *See* Grekov et al. 1939.

"Ot redaktsii." In *Materialy po istorii otechestvennoi khimii*, 3–6.

"Ot redaktsii." *VF*, 1950, no. 2: 194–196.

"Ot Rossiiskoi assotsiatsii nauchno-issledovatel'skikh institutov obshchestvennykh nauk (RANION)." *PZM*, 1926, nos. 4–5: 255–261.

"O vragakh v sovetskoi nauke." *Pravda*, July 3, 1936.

"O zhurnale 'Pod znamenem marksizma.'" *PZM*, 1930, nos. 10–12: 1–2.

P., A. "Pervyi s'ezd rossiiskikh biologov imeni I. M. Sechenova." *Priroda*, 1917, nos. 5–6: 706–708.

"Pamiati O. Iu. Shmidta." *VAN*, 1961, no. 12: 117–118.

Pamiati Petra Frantsevicha Lesgafta. St. Petersburg, 1912.

Pankratova, A. "Razvitie istoricheskikh vzgliadov M. N. Pokrovskogo." In *Protiv istoricheskoi kontseptsii*. Vol. 1, 5–69. *See* Grekov et al. 1939.

———. "Sovetskaia istoricheskaia nauka za 25 let i ee zadachi v usloviiakh Velikoi Otechestvennoi voiny." In *Dvadtsat' piat' let*, 3–40. *See* Volgin et al. 1942.

Parrott, Bruce. "The Organizational Environment of Soviet Applied Research." In *The Social Context*, 69–100. *See* Lubrano and Solomon 1980.

Parry, Albert. *The Russian Scientist*. New York: Macmillan, 1973.

———, ed. *Peter Kapitsa on Life and Science*. New York: Macmillan, 1968.

"Patrioticheskii dolg sovetskikh uchenykh." *VAN*, 1962, no. 1: 3–6.

Pauling, Linus. "Fifty Years of Progress in Structural Chemistry and Molecular Biology." In *Twentieth-Century Science*, 281–307. *See* Holton 1970.

Pavinskii, P. P., and A. M. Mostopanenko, eds. *Kvantovaia mekhanika i teoriia otnositel'nosti*. Leningrad, 1980.

Pavlov, A. P., and V. I. Vernadskii. *Proekt Ustava dlia obespecheniia i organizatsii russkikh estestvenno-nauchnykh s'ezdov.* Moscow, 1892.

Pavlov, D. "Iz nauchnoi zhizni provintsii: na severnom Kavkaze." *NiR*, 1921, no. 1: 28–29.

Pavlova, G. E. "Fundamental'nye nauchnye issledovaniia v 1917–1925 gg." *VIE-T*, no. 2: 69–71.

Pavlova, G. E., and A. S. Fedorov, *Mikhail Vasil'evich Lomonosov.* Moscow, 1980.

Pekarskii, P. P. "Ekaterina II i Eiler." *ZAN* 6 (1865): 59–92.

———. *Istoriia Imperatorskoi Akademii nauk v Peterburge.* 2 vols. St. Petersburg, 1870–1873.

———. *Nauka i literatura pri Petre Velikom.* Vol. 2. St. Petersburg, 1862.

"Perevybory v Smolenske." *VAR*, 1929, no. 4: 10–11.

Perel', Iu. G., "O knige B. E. Raikova 'Ocherki po istorii geliotsentricheskogo mirovozzreniia.'" *VF*, 1952, no. 4: 196–198.

Pervaia Moskovskaia obshchegorodskaia konferentsiia proletarskikh kul'turno-prosvetitel'-nykh organizatsii 23–28 fevralia 1918 g. Moscow, 1918.

"Pervaia Vsesoiuznaia konferentsiia marksistsko-leninskikh nauchno-issledovatel'skikh uchrezhdenii." *VKA*, 1928, no. 26: 239–294; no. 27: 289–315.

"Pervoe Vsesoiuznoe soveshchanie obshchestv voinstvuiushchikh materialistov-dialektikov." *PZM*, 1929, no. 5: 129–169.

"Petrogradskoe Fiziko-matematicheskoe obshchestvo." *NiR*, 1921, no. 4: 38–40.

Petrov, F. N. "Ob organizatsii nauchno-issledovatel'skoi raboty v SSSR." In *Nauka i iskusstvo.* Vol. 1, 9–14. Moscow-Leningrad, 1926.

———, ed. *Desiat' let sovetskoi nauki.* Moscow, 1927.

Petrov, M., and A. Potemkin. "Nauka poznaet sebia." *NM*, 1968, no. 6: 238–252.

Petrushevskii, S. "Trudy velikogo fiziologa-materialista." *Bol*, 1949, no. 15: 72–80.

———. "Voinstvuiushchii materializm I. P. Pavlova." *Bol*, 1950, no. 15: 24–35.

Piat' let vlasti sovetov. Moscow, 1922.

"Piatoe zasedanie 14 maia 1921 goda." *IRAN* 15, series 6 (1921): 10–12.

Piskotin, M. I., et al., eds. *Organizatsionno-pravovye voprosy rukovodstva naukoi v SSSR.* Moscow, 1973.

"Pis'ma K. F. Gaussa v S. Peterburgskuiu Akademiiu nauk." *AINT*, 1934, no. 3: 209–238.

Plate, A. F. "V. V. Markovnikov i Akademiia nauk." *TIIE-T* 6 (1955): 292–297.

Platonov, G. V. "Nekotorye filosofskie voprosy diskusii o vide i vidoobrazovanii." *VF*, 1954, no. 6: 116–32.

———. "O nekotorykh voprosakh filosofii estestvoznaniia v svete truda I. V. Stalina 'Ekonomicheskie problemy sotsializma v SSSR.'" *VAN*, 1953, no. 1: 3–15.

Plenum Tsentral'nogo Komiteta KPSS, 22–25 dekabria 1959 goda. Moscow, 1960.

Pliushch, L. N. "Ob osnovakh novoi kletochnoi teorii." *VF*, 1953, no. 4: 185–191.

Plotkin, S. Ia. "Organizatsiia v SSSR issledovanii po istorii estestvoznaniia i tekhniki." *VIE-T*, 1967, no. 23: 3–9.

Podvigina, E. P. "Akademik Nikolai Petrovich Gorbunov. K 90-letiiu so dnia rozhdeniia." *VAN*, 1982, no. 8: 85–97.

Poincaré, Henri. *Mathematics and Science: Last Essays.* Translated by J. W. Bolduc. New York: Dover, 1963.

Pokrovskii, K. "IV-yi Astronomicheskii s'ezd." *Priroda*, 1929, no. 2: 183–185.

Pokrovskii, M. N. "Obshchestvennye nauki v SSSR za 10 let." *VKA*, 1928, no. 26: 3–30.

[————.] "Primechaniia M. Pokrovskogo." *VKA*, 1922, no. 1: 38–39.

Poliakov, I. A. "Otechestvennye biologi v bor'be s mendelizmom." *TIIE* 3 (1949): 3–11.

Polianskii, V. I., and Iu. I. Polianskii, eds. *Istoriia evoliutsionnykh uchenii v biologii.* Moscow-Leningrad, 1966.

Polnoe sobranie uchenykh puteshestvii v Rossii. 7 vols. St. Petersburg, 1818–1825.

Pontriagin, L. S. "Topologiia v Sovetskom Soiuze." *UZMU*, 1946, no. 92: 65–76.

Popov, V. F. "Podgotovka nauchnykh kadrov v SSSR." *SN*, 1939, no. 11: 188–193.

Pospelov, I. "Iz zhizni vysshei shkoly." *Vestnik vospitaniia*, 1914, no. 4–6: 98–114.

"Postanovlenie Prezidiuma Akademii nauk SSSR 26 avgusta 1948 goda po voprosu o sostoianii i zadachakh biologicheskoi nauki v institutakh i uchrezhdeniiakh Akademii nauk SSSR." *VAN*, 1948, no. 9: 21–24.

Potekhin, I. I. "Kosmopolitizm v amerikanskoi etnografii." In *Anglo-amerikanskaia etnografiia*, 17–40. See Potekhin 1951.

————. "Zadachi bor'by s kosmopolitizmom v etnografii." *Sovetskaia etnografiia*, 1949, no. 2: 7–26.

————, ed. *Anglo-amerikanskaia etnografiia na sluzhbe imperializma.* Moscow, 1951.

Pozner, V. "Gegel' i estestvoznanie." *FNiT*, 1932, no. 6: 48–55.

Prazdnovanie dvukhsotletnei godovshchiny rozhdeniia M. V. Lomonosova Imperatorskim Moskovskim universitetom. Moscow, 1912.

"Predislovie." In *Protiv reaktsionnogo mendelizma-morganizma*, 3–5. See Mitin et al. 1950.

Predtechenskii, A. V., and A. V. Kol'tsov. "Iz istorii Akademii nauk v period revoliutsii 1905–1907 godov." *VAN*, 1953, no. 3: 82–89.

Preobrazhenskii, E. A. "Blizhaishie zadachi Sotsialisticheskoi akademii." *VKA*, 1922, no. 1: 5–12.

Presniakov, D. "Ideologiia Sviashchennogo Soiuza." *Annaly*, 1923, no. 3: 72–81.

Prezent, I. I. "O lzhenauchnykh vozzreniiakh prof. N. K. Kol'tsova." *PZM*, 1939, no. 5: 146–153.

"Prilozhenie k protokolu VIII zasedaniia Obshchego sobraniia Rossiiskoi Akademii nauk 18 (5) maia 1918 goda." *IRAN* 12, series 6 (1918): 1387–1399.

Prokof'eva, I. A. "Konferentsiia po ideologicheskim voprosam astronomii, soz-vannaia Leningradskim otdeleniem Vsesoiuznogo Astronomo-geodezi-cheskogo obshchestva (LOVAGO)." *Priroda*, 1949, no. 6: 71–77.

"Protiv Luzina i luzinovshchini." *FNiT*, 1936, no. 7: 123–125.

"Protokol Obshchego sobraniia chlenov Kommunisticheskoi Akademii 29 noia-bria 1924 g." *VKA*, 1925, no. 11: 380–402.

"Protokol sovmestnogo zasedaniia Komissii i Voenno-khimicheskogo komiteta 10 ianvaria 1917 g." In *Otchety o deiatel'nosti Komissii*, 147–155.

Protskaia, D. G. "K voprosu o perezhitkakh proshlogo v otnoshenii k trudu." In *Priroda soznaniia*, 126–130. *See* Borisov et al. 1968.

Pypin, A. N. *Istoriia russkoi etnografii*. Vol. 1. St. Petersburg, 1890.

———. *Obshchestvennoe dvizhenie pri Aleksandre I*, 3d ed. St. Petersburg, 1900.

———. "Russkaia nauka i natsional'nyi vopros v XVIII veke." *Vestnik Evropy* 3, part 1, no. 5: 212–256; part 2, no. 6: 548–600; 4, part 3, no. 7: 72–117.

Rabkin, Yakov M. "'Naukovedenie': The Study of Scientific Research in the Soviet Union." *Minerva* 14, no. 1 (1976): 61–78.

Raboty Rossiiskoi Akademii nauk v oblasti issledovaniia prirodnykh bogatstv Rossii: Obzor deiatel'nosti KEPS za 1915–1921 gg. Petrograd, 1922.

Rachkov, P. A., et al., eds. *O strukture marksistskoi sotsiologicheskoi teorii*. Moscow, 1970.

Radovskii, M. I. *M. V. Lomonosov i Peterburgskaia Akademiia nauk*. Moscow-Len-ingrad, 1961.

———. "N'iuton i Rossiia." *VIMK*, 1957, no. 6: 96–106; 1958, no. 2: 123–124.

Raikov, B. E. *Akademik Vasilii Zuev; ego zhizn' i trudy*. Moscow-Leningrad, 1955.

———. "Iz istorii darvinizma v Rossii." *TIIE-T* 31 (1960): 68–81.

———. *Karl Ber: ego zhizn' i trudy*. Moscow-Leningrad, 1961.

———. *Russkie biologi-evoliutsionisty do Darvina*. 3 vols. Moscow-Leningrad, 1951–1952.

Rainov, T. I. "N'iuton i russkoe estestvoznanie." In *Isaak N'iuton*, 329–344. *See* Vavilov 1943.

———. "Russkoe estestvoznanie vtoroi poloviny XVIII v. i Lomonosov." In *Lomonosov: Sbornik statei i materialov*, 318–388. *See* Andreev and Modza-levskii 1940.

———. "Teoriia i praktika v tvorchestve M. V. Lomonosova." *SRiN*, 1936, no. 9: 9–21.

"Rasshirennoe zasedanie Prezidiuma Akademii nauk SSSR 24–26 avgusta 1948 goda po voprosu o sostoianii i zadachakh biologicheskoi nauki v insti-tutakh i uchrezhdeniiakh Akademii nauk SSSR." *VAN*, 1948, no. 9: 17–208.

Rassokhin, V. P. "Problemy razvitiia systemy nauchnykh organizatsii v SSSR." In *Pravo*, 64–104. *See* Rassudovskii and Rassokhin 1980.

Rassudovskii, V. A. *Gosudarstvennaia organizatsiia nauki v SSSR*. Moscow, 1971.

Rassudovskii, V. A., and M. S. Bastrakova. "Akademiia nauk SSSR—vysshee nauchnoe uchrezhdenie strany." *Sovetskoe gosudarstvo i pravo*, 1974, no. 2: 12–21.

Rassudovskii, V. A., and V. P. Rassokhin, eds. *Pravo i upravlenie nauchnym organizatsiiami*. Moscow, 1980.

"Razvertyvat' kritiku i bor'bu mnenii v nauke." *Pravda*, November 17, 1952.

"Razvitie nauchnykh uchrezhdenii v Rossiiskoi federatsii." *VAN*, 1970, no. 1: 3–6.

Regel', R. E. "Selektsiia s nauchnoi tochki zreniia." *Trudy Biuro po prikladnoi botanike*, 1912, no. 11: 425-540.

"Resheniia Plenuma TsK VKP(b) i zadachi nauchnoi obshchestvennosti." *FNiT*, 1937, no. 5: 3–18.

Reutov, O. A. "K voprosu o formalizme i uproshchenstve v teorii organicheskoi khimii." *VF*, 1950, no. 2: 181–194.

———. "O knige G. V. Chelintseva 'Ocherki po teorii organicheskoi khimii.'" *VF*, 1949, no. 3: 309–317.

"Rezoliutsiia Martovskoi sessii Akademii nauk po otchetnym dokladam akademikov A. F. Ioffe, D. S. Rozhdestvenskogo i S. I. Vavilova." *UFN* 16, no. 7 (1936): 839–846.

"Rezoliutsiia X Plenuma IKKI o tov. Bukharine." *LP*, August 20, 1929.

Riabinin, A. "O deiatel'nosti Russkogo Paleontologicheskogo obshchestva za 1916 god." *Priroda*, 1917, no. 2: 268–270.

Richter, Liselotte. *Leibniz und sein Russenbild*. Berlin, 1946.

Rodkevich, V. Review of F. Bacon, "O printsipakh i nachalakh" (Moscow, 1937). *FNiT* 1937, no. 11: 187–188.

Rodnyi, N. I. *Ocherki po istorii i metodologii estestvoznaniia*. Moscow, 1974.

Rodnyi, N. I., et al., eds. *Uchenye o nauke i ee razvitii*. Moscow, 1971.

Rokhlina, M. "Obshchestvennyi smotr Instituta Eksperimental'noi biologii." *VAR*, 1930, no. 5: 44–48.

———. "V poriadke samokritiki." *VAR*, 1930, no. 2: 57.

Romanov, A. K., L. A. Androsova, and A. F. Felinger. *Nauchnye kadry Sibirskogo Otdeleniia AN SSSR: Metody i rezul'taty statisticheskogo issledovaniia*. Novosibirsk, 1979.

Romashkin, P. S. "O nekotorykh problemakh metodologii v iuridicheskoi nauke." In *Metodologicheskie problemy*, 319–325.

Rose, Hilary, and Steven Rose, eds. *Ideology of/in the Natural Sciences*. Cambridge, Mass.: Schenkman, 1976.

Rostovtsev, M. I. "Russia's Contribution to the World's Science." *Struggling Russia* 1, nos. 38–40 (1919): 602–605.

Rozhdestvenskii, S. V. *Istoricheskii obzor deiatel'nosti Ministerstva narodnogo prosveshcheniia, 1802–1902*. St. Petersburg, 1902.

Rubanovskii, L. M. "Vtoroi zakon termodinamiki." In *Pamiati Karla Marksa*, 269–327. *See* Bukharin and Deborin 1933.

Rubinshtein, S. L. *Bytie i soznanie*. Moscow, 1957.

Ruble, Blair A. "The Expansion of Soviet Science." *Knowledge* 2, no. 4 (1981): 529–553.

Runin, B. *Lichnost' i tvorchestvo*. Moscow, 1966.

——. *Vechnyi poisk*. Moscow, 1964.

Russell, Bertrand. *The Wisdom of the West*. New York: Crescent Books, 1959.

Rutkevich, M. P. "Problemy izmeneniia sotsial'noi struktury sovetskogo obshchestva." *FN*, 1968, no. 3: 44–52.

——. "Sotsial'nye istochniki popolneniia sovetskoi intelligentsii." *VF*, 1967, no. 6: 15–23.

Ryvkina, R. V. "O strukture predmeta sotsiologii kak sistemy nauk." In *Sotsiologicheskie issledovaniia*, 5–45. *See* Ryvkina 1966.

——, ed. *Sotsiologicheskie issledovaniia: Voprosy metodologii i metodiki*. Novosibirsk, 1966.

Sachkov, Iu. V. "Obsuzhdenie filosofskikh voprosov kvantovoi mekhaniki." *VF*, 1959, no. 2: 72–76.

Sakharov, A. D. *Progress, Coexistence and Intellectual Freedom*. New York: W. W. Norton, 1968.

Salov, V. I. "Iz istorii Akademii nauk SSSR v pervye gody Velikoi Otechestvennoi voiny (1941–1943)." *Istoricheskie zapiski* 60 (1957): 3–30.

Samoilov, A. F. "Dialektika prirody i estestvoznanie." *PZM*, 1926, nos. 4–5: 61–81.

Samuelian, Thomas I. *The Search for a Marxist Linguistics in the Soviet Union, 1917–1950*. Ph. D. dissertation, University of Pennsylvania, Philadelphia, 1981.

Satkevich, A. "Leonard Eiler (v dvukhsotuiu godovshchinu dnia ego rozhdeniia)." *Russkaia starina* 132 (1907): 467–506.

Scharff, Ronald. *Wissenschaftsorganisation und Wissenschaftler in der UdSSR. Berichte des Bundesinstituts für osteuropeische und internationale Studien*, vol. 51. Berlin, 1978.

Schlözer, August Ludwig. *Öffentliches und Privat-Leben von ihm selbst beschrieben*. Vol. 1. Göttingen, 1922.

Sechenov, I. M. "Nauchnaia deiatel'nost' russkikh universitetov po estestvoznaniiu za poslednee dvadtsatipiatiletie." *Vestnik Evropy* 6 (1883): 330–342.

VII. Vsesoiuznomu s'ezdu sovetov—Akademiia nauk SSSR. Moscow-Leningrad, 1935.

Segré, Emilio. *From X-Rays to Quarks: Modern Physicists and Their Discoveries*. San Francisco: W. H. Freeman, 1976.

Selunskaia, V. M. "V. I. Lenin i problema preemstvennosti v razvitii nauki." *VMU*, 1970, no. 2: 29–40.

Semenov, N. N. "Budushchee cheloveka v atomnom veke." *VAN*, 1958, no. 11: 94–103.

_____. *Nauka i obshchestvo*. Moscow, 1973.

_____. "Nauka ne terpit sub'ektivizma." *NiZh*, 1965, no. 4: 38–43, 132.

_____. "Nauka segodnia i zavtra." *Izvestiia*, August 9, 1959: 3.

Semkovskii, S. Iu. "'Dialektika prirody' Engel'sa i teoriia otnositel'nosti." *FNiT*, 1935, no. 9: 8–15.

_____. "K sporu v marksizme o teorii otnositel'nosti." *PZM*, 1925, nos. 8–9: 126–169.

_____. "O nekotorykh metodologicheskikh problemakh iadernoi fiziki." *FNiT*, 1935, no. 3: 29–36.

Serebrovskii, A. "Teoriia nasledstvennosti Morgana i Mendelia i marksisty." *PZM*, 1926, no. 3: 98–117.

"Sessii i zasedaniia, posviashchennye godovshchini opublikovaniia genial'nogo proizvedeniia I. V. Stalina 'Marksizm i voprosy iazykoznaniia.'" *VAN*, 1951, no. 8: 36–84.

"Sessiia Akademii nauk SSSR." *SRiN*, 1936, no. 5: 155–162.

"Sessiia Akademii nauk SSSR posviashchennaia istorii otechestvennoi nauki." *VAN*, 1949, no. 2: 3–125.

Setrov, M. I. "Ob obshchikh elementakh tektologii A. Bogdanova, kibernetiki i teorii sistem." *Uchenye zapiski kafedr obshchestvennykh nauk* 8 (1967): 49–60.

Severtsov, A. N. *Sobranie sochinenii*. Vol. 3. Moscow-Leningrad, 1945.

Sh. "V Akademii nauk." *FNiT*," 1936, no. 5: 108–111.

Shabanov, A. "O teorii evoliutsii." *LP*, October 30, 1929.

Shapiro, Joel. *A History of the Communist Academy, 1918–1936*. Ph.D. dissertation, Columbia University, 1976.

Sharov, A. Ia. "Prezidium Akademii nauk SSSR obsuzhdaet rabotu Instituta filosofii." *VF*, 1970, no. 3: 134–139.

Shchapov, A. P. *Sotsial'no-pedagogicheskie usloviia umstvennogo razvitiia russkogo naroda*. St. Petersburg, 1870.

Shcheglov, A. V. "Filosofskie diskussii v SSSR v 20-kh i nachale 30-kh godov." *FN*, 1967, no. 5: 110–118.

Shcherbakov, D. I. "Gor'kii o nauke." *NM*, 1965, no. 1: 272–274.

_____. "Rol' V. I. Vernadskogo v izucheniiu prirodnykh resursov nashei strany." *VIE-T*, 1957, no. 8: 92–95.

Sheinin, Y. *Science Policy: Problems and Trends*. Translated by Yuri Sdobnikov. Moscow, 1978.

Shevshchenko, V. I. "K voprosu ob uchete nauchnykh kadrov." *VAN*, 1931, no. 2: 23–30.

Shinkaruk, V. I. "Integrativnaia funktsiia marksistskoi filosofii v sovremennoi nauke." In *Nauchno-tekhnicheskaia revoliutsiia*, 41–72. *See* Depenchuk 1978.

Shkurinov, P. S. *Pozitivizm v Rossii XIX veka*. Moscow, 1980.

Shmal'gauzen, I. I. *Organizm kak tseloe v individual'nom i istoricheskom razvitii*. Moscow-Leningrad, 1938.

_____. *Puti i zakonomernosti evoliutsionnogo protsessa.* Moscow-Leningrad, 1939.

Shmidt, O. Iu. "Problema proiskhozhdeniia Zemli i planet." *VF,* 1951, no. 4: 120–133.

Shorin, V. G., and A. A. Popova. "Organizatsiia raboty v vuzakh." In *Osnovnye printsipy,* 199–215. See Ananichev et al. 1973.

Shparo, B. "V Akademii nauk SSSR." *VKA,* 1935, no. 5: 39–41.

Shpol'skii, E. V. "Al'bert Einshtein (1879–1955)." *UFN* 57, no. 2 (1955): 177–186.

Shreider, Iu. "Nauka—istochnik znanii i sueverii." *NM,* 1969, no. 10: 207–226.

Shtokalo, I. Z., et al., eds. *Istoriia matematicheskogo obrazovaniia v SSSR.* Kiev, 1975.

Shtorkh, A., and F. Adelung. *Sistematicheskoe obozrenie literatury v Rossii v techenie piatiletiia, s 1801 po 1806 god.* 2 vols. St. Petersburg, 1811.

Shtraikh, S. "I. I. Mechnikov o razvitii estestvoznaniia v Rossii." *Priroda,* 1917, no. 1: 109–116.

Sidorenko, A. V., et al., eds. *50 let sovetskoi geologii.* Moscow, 1968.

Skriabin, G. K., ed. *Akademiia nauk: Personal'nyi sostav.* 2 vols. Moscow, 1974–1975.

Skriabin, G. K., et al., eds. *Ustavy Akademii nauk SSSR.* Moscow, 1974.

Skriabin, K. I. *Moia zhizn' v nauke.* Moscow, 1969.

Skripchinskii, V. V. "O razlichii v vzgliadakh I. V. Michurina i T. D. Lysenko na nekotorye iz osnovnykh problem genetiki." *BMOIP: Otdel biologicheskii,* 1965, no. 4: 105–116.

Smirnov, A. A. *Razvitie i sovremennoe sostoianie psikhologicheskoi nauki v SSSR.* Moscow, 1975.

Smirnov, E. S. "Novye dannye o nasledstvennom vliianii sredy i sovremennyi lamarkizm." *VKA,* 1928, no. 25: 183–197.

Smirnov, I. S. *Lenin i sovetskaia kul'tura.* Moscow, 1960.

Smirnov, N. A. "Zapadnoe vliianie na russkii iazyk v petrovskuiu epokhu." *Sbornik Otdeleniia Russkogo iazyka i slovesnosti Imperatorskoi Akademii nauk* 88, no. 2 (1910).

Smirnov, V. A., and P. V. Tavanets. "O vzaimootnoshenii simvolicheskoi logiki i filosofii." In *Filosofiia i logika,* 5–34. *See* Tavanets and Smirnov 1974.

Smit, N. "Plannovoe vreditel'stvo i statisticheskaia teoriia." In *Na bor'bu* 25–52.

Sobol', S. L. "Iz istorii bor'by za darvinizm v Rossii." *TIIE-T* 14 (1957): 195–226.

Soboleva, E. V. *Bor'ba za reorganizatsiiu Peterburgskoi Akademii nauk v seredine XIX veka.* Leningrad, 1971.

Sochor, Zenovia A. "Was Bogdanov Russia's Answer to Gramsci?" *Studies in Soviet Thought* 22 (1981): 59–81.

Sokolov, N. D. "Soveshchanie po teorii khimicheskogo stroeniia v organicheskoi khimii." *UFN* 45, no. 2 (1951): 277–293.

Sologub, V. "Nauchnaia obshchestvennost' o planakh Akademii nauk." *SN,* 1939, no. 2: 178–180.

Solov'ev, E. *D. I. Pisarev.* Berlin-Petrograd, 1922.

Solov'ev, Iu. I., and N. N. Ushakova. *Otrazhenie estestvennonauchnykh trudov M. V. Lomonosova v russkoi literature XVIII i XIX vv.* Moscow, 1961.

Sominskii, M. S. *Abram Fedorovich Ioffe.* Moscow-Leningrad, 1964.

"Soveshchanie po teorii khimicheskogo stroeniia v organicheskoi khimii." *VAN*, 1951, no. 12: 111–123.

Speranskii, N. *Voznikovenie Moskovskogo Gorodskogo universiteta imeni A. L. Shaniavskogo.* Moscow, 1913.

Spiess, Otto. *Leonhard Euler: Ein Beitrag zum Geistesgeschichte des XVIII Jahrhunderts.* Leipzig, 1929.

The Spiritual Regulation of Peter the Great. Translated by A. V. Muller. Seattle: University of Washington Press, 1972.

Spravochnik dlia uchastnikov prazdnovaniia dvukhsotletnego iubileia Akademii nauk, 1725–1925. Leningrad, 1925.

Spravochnik partiinogo rabotnika. Vol. 4. Moscow, 1963.

Sreznevskii, I. I. "Otchet o pervom prisuzhdenii Lomonosovskoi premii." *ZAN*, 1868, 3, part 2: 195–210.

[Stählin, Karl] *Aus den Papieren Jacob von Stählins: Ein biographischer Beitrag zur deutsch-russischen Kulturgeschichte des 18. Jahrhunderts.* Leipzig, 1926.

Stakhanov, I. P. "Logika 'vozmozhnogo.'" *VF*, 1970, no. 2: 126–127.

Stalin, I. V. "Ekonomicheskii problemy sotsializma v SSSR." *VAN*, 1952, no. 9: 3–61.

———. *Problems of Leninism.* Moscow, 1947.

Staniukovich, T. V. *Etnograficheskaia nauka i muzei.* Leningrad, 1978.

Steklov, V. A. *Galileo Galilei.* Berlin-Petrograd-Moscow, 1923.

———. *Lomonosov.* Berlin-Petrograd, 1921.

———. *Teoriia i praktika v issledovaniiakh Chebysheva.* Petrograd, 1921.

Stepanian, Ts. A., ed. *Velikaia sila idei leninizma.* Moscow, 1950.

Stiazhkin, N. I. "Elementy algebry logiki i teorii semanticheskikh antinomii v pozdnei srednevekovoi logike." In *Logicheskie issledovaniia*, 20–32. *See* Kol'man et al. 1959.

Stoletie neevklidovoi geometrii Lobachevskogo. Kazan, 1927.

Stoletov, V. N. "'Materializm i empiriokrititsizm' Lenina i voprosy biologii." In *Velikaia sila idei*, 223–266.

Stresemann, E. "Leben und Werk von Peter Simon Pallas." In *Lomonosov*, 247–257. *See* Winter et al. 1962.

Subbotin, A. L. "Po sledam 'Novogo Organona.'" *VF*, 1970, no. 9: 97–108.

Sukachev, V. N. "O vnutrividovykh i mezhvidovykh vzaimootnosheniiakh." *BZh* 38, no. 1 (1953): 57–96.

Sukhanov, A. F. "Ustav vysshei shkoly SSSR." *SN*, 1938, no. 2: 82–85.

Sukhomlinov, M. I. *Issledovaniia i stat'i po russkoi literature i prosveshcheniia.* 2 vols. St. Petersburg, 1889.

————. *Istoriia Rossiiskoi Akademii.* 8 vols. St. Petersburg, 1885–1900.

————. *Materialy dlia istorii obrazovaniia v Rossii v tsarstvovanie imperatora Aleksandra I.* 2 vols. St. Petersburg, 1866.

————. "Piatidesiatiletnii i stoletnii iubilei S.-Peterburgskoi Akademii nauk: 1776 i 1826 gg." *Russkaia starina* 8 (1878): 1–20.

————. *Rech' v torzhestvennom sobranii Imperatorskoi Akademii nauk po sluchaiu stoletniego iubileia Aleksandra I.* St. Petersburg, 1877.

Sutton, A. C. *Western Technology and Soviet Economic Development.* 3 vols. Stanford, Calif.: Hoover Institution, 1968–1973.

Suvorov, L. N. "K otsenke leninskogo etapa filosofii marksizma." *FN,* 1968, no. 2: 113–121.

————. "Rol' filosofskikh diskussii 20-30kh godov v bor'be za leninizm protiv mekhanitsizma, formalisticheskikh i idealisticheskikh oshibok v filosofii." In *Leninskii etap,* 38–45. *See* Arkhiptsev et al. 1972.

Suvorov, S. G. "Leninskaia teoriia poznaniia—filosofskaia osnova razvitiia fiziki." *UFN* 39, no. 1 (1949): 3–50.

Suvorov, S. G., and R. Ia. Shteinman. "Za posledovatel'no-materialisticheskuiu traktovku osnov mekhaniki." *UFN* 40, no. 3 (1950): 407–439.

Sverdlov, V. "Piat' let raboty VARNITSO." *FNiT,* 1933, no. 1: 12–19.

Taitslin, I. S. "Nauchnye kadry RSFSR." *Nauchnoe slovo,* 1930, no. 10: 10–31.

Takhtadzhian, A. L. "Tektologiia: Istoriia i problemy." *Sistemnye issledovaniia* 3 (1971): 200–277.

Tamm, I. E. "O rabote filosofov-marksistov v oblasti fiziki." *PZM,* 1933, no. 2: 220–231.

————. "Rol' vedushchei nauki estestvoznaniia pereidet v otnositel'no nedalekom budushchem ot fiziki k biologii." *Tekhnika-molodezhi,* 1957, no. 9: 13.

Tarasevich, L. A. "Estestvoznanie i meditsina." In *Istoriia Rossii XIX veka.* Vol. 6, 285–308.

Tataevskii, V. M., and M. I. Shakhparonov. "Ob odnoi makhistskoi teorii v khimii i ee propagandistakh." *VF,* 1949, no. 3: 176–192.

Tatishchev, V. M. *Izbrannye proizvedeniia.* Edited by S. N. Valk. Leningrad, 1979.

————. "Razgovor o pol'ze nauk i uchilishch." *Chteniia v Imperatorskom Obshchestve istorii i drevnostei rossiiskikh pri Moskovskom universitete* 140, no. 1 (1887): 1–171.

Tavanets, P. V., ed. *Problems of the Logic of Scientific Knowledge.* Translated by T. J. Blakely. Dordrecht, Holland: Reidel, 1970.

Tavanets, P. V., and V. A. Smirnov, eds. *Filosofiia i logika: Sbornik statei.* Moscow, 1974.

Teoriia otnositel'nosti i materializm. Leningrad-Moscow, 1925.

Terent'ev, P. "Zhizn' pod luchom matematiki." *Pravda,* December 11, 1966.

Terletsky, Ya. P. "On the Problem of the Spatial Structure of Elementary Particles." In *Philosophical Problems,* 140–149. *See* I. V. Kuznetsov 1968.

Terpigorev, A. M. "O vysshei shkole." *SRiN*, 1936, no. 7: 13–17.

Thomas, J. R., and U. M. Kruse-Vaucienne, eds. *Soviet Science and Technology.* Washington, D.C.: George Washington University Press, 1977.

Thomas, L. L. "Some Notes on the Marr School." *SR*, 1957, October: 322–348.

Timiriazev, A. K. "Eshche raz o volne idealizma." *PZM*, 1938, no. 4: 124–152.

———. *Nauka v Sovetskoi Rossii.* Moscow, 1922.

———. "Novaia neudachnaia popytka primirit' teoriiu otnositel'nosti s dialekticheskim materializmom." *Voinstvuiushchii materialist* 1925, no. 4: 243–253.

———. Review of A. Einshtein, "O spetsial'noi i vseobshchei teorii otnositel'-nosti." *PZM*, 1922, nos. 1–2: 70–73.

———. "Teoriia 'kvanta' i sovremennaia fizika." *PZM*, 1923, nos. 2–3: 98–120.

———. "Teoriia otnositel'nosti Einshteina i dialekticheskii materializm." *PZM*, 1924, nos. 8–9: 142–157; nos. 10–11: 92–114.

———. "Volna idealizma v sovremennoi fizike na Zapade i u nas." *PZM*, 1933: 94–123.

Timiriazev, K. A. "Probuzhdenie estestvoznaniia v tret'ei chetverti veka." In *Istoriia Rossii XIX veka*, 1–30.

———. *Nauka i demokratiia: Sbornik statei, 1904–1919 gg.* Moscow, 1963.

———. *Sochineniia.* Vol. 8. Moscow, 1939.

Timofeev-Resovskii, N. V. "Elementarnye iavleniia evoliutsionnogo protsessa." In *Filosofiia i teoriia evoliutsii*, 114–120. *See* Il'in et al. 1974.

Timoshenko, S. *Vospominaniia.* Paris, 1963.

Tiurin, I. V. *Achievements of Russian Science in the Province of Chemistry of Soils.* Leningrad, 1927.

Tokin, B. "Doklad." In *Protiv mekhanisticheskogo materializma*, 8–34. *See* Bondarenko 1931.

———. "Teoreticheskaia biologiia i ee dostizheniia." *FNiT*, 1934, no. 9: 47–56.

Tolstoi, D. A. *Akademicheskii universitet v XVIII stoletii.* St. Petersburg, 1885.

Tolstov, S. P. "Krizis burzhuaznoi etnografii." In *Anglo-amerikanskaia etnografiia*, 3–16. *See* Potekhin 1951.

Tolstykh, V. I. "Galileo protiv Galilea." In *Nauka*, 324–357. *See* Tolstykh 1971.

———. "Nauka i nravstvennaia otvetstvennost' uchenogo." *VF*, 1967, no. 4: 77–86.

———, ed. *Nauka i nravstvennost'.* Moscow, 1971.

Topchiev, A. V. "Sovetskaia nauka v pervom godu semiletki." *VAN*, 1959, no. 1: 3–10.

Trapeznikov, S. "Leninizm i sovremennaia nauchno-tekhnicheskaia revoliutsiia." In *Nauchno-tekhnicheskaia revoliutsiia*, 5–29. *See* Trapeznikov et al. 1972.

Trapeznikov, S., et al. *Nauchno-tekhnicheskaia revoliutsiia i sotsial'nyi progress.* Moscow, 1972.

Travin, I. S. "Sovremennye tsentry intensivnogo vidoobrazovaniia rastenii." *BZh*, 1945, no. 6: 245–250.

Triska, J. F., and P. M. Cocks, eds. *Political Development in Eastern Europe.* New York: Praeger, 1977.

Trofimuk, A. "Strategiia nauchnogo nastupleniia." *Kom,* 1977, no. 18: 41–44.

Troshin, D. M. "Marksizm-Leninizm ob ob'ektivnom kharaktere zakonov nauki." In *Voprosy,* 216–241. See Andreev et al. 1952.

———. "V plenu sub'ektivizma: idealisticheskii tupik estestvoznaniia v burzhuaznykh stranakh." *Priroda,* 1953, no. 9: 39–46.

———. "Znachenie truda I. V. Stalina 'Marksizm i voprosy iazykoznaniia dlia estestvennykh nauk.'" In *Voprosy,* 208–229. See G. F. Aleksandrov et al. 1951.

Trotsky, L. "Kul'tura i sotsializm." *NM,* 1927, no. 1: 166–184.

———. *Marxism and Science.* Miradana, Ceylon, 1949.

———. *Sochineniia.* Vol. 21. Moscow-Leningrad, 1927.

Trubetskoi, S. *Sobranie sochinenii.* Vol. 1. Moscow, 1907.

Tseitlin, Z. "O 'misticheskoi' prirode svetovykh kvant." *PZM,* 1925, no. 4: 74–101.

Tsitsin, N. V. "I. V. Michurin i sovremennaia biologiia." *VF,* 1955, no. 5: 94–111.

Tugarinov, V. P. *O tsennostiakh zhizni i kul'tury.* Leningrad, 1960.

Tur. "Pod kupolom Akademii." *LP,* February 8 and 12, 1929.

Turbin, N. V. "Darvinizm i novoe uchenie o vide." *BZh* 37 (1952): 798–818.

Tymanskii, G. S. "Lenin i nauka." *Priroda,* 1934, no. 1: 7–18.

Ugrinovich, D. M. "O strukture marksistskoi sotsiologicheskoi teorii." In *O strukture,* 5–19. See Rachkov et al. 1970.

Ul'ianovskaia, V. A. *Formirovanie nauchnoi intelligentsii v SSSR: 1917–1937 gg.* Moscow, 1966.

Ul'ianskaia, B. "Vtoraia Vsesoiuznaia konferentsiia VARNITSO." *SRiN,* 1933, no. 3: 180–184.

"Uluchshit' rabotu s nauchnymi kadrami." *VAN,* 1953, no. 2: 3–15.

"Usilit' revoliutsionnuiu bditel'nost'!" *VAN,* 1938, nos. 8–9: 1–6.

Utkina, N. F. *Estestvennonauchnyi materializm v Rossii XVIII veka.* Moscow, 1971.

Vagnov, G. "Nauka i filosofskii nigilizm." *LG,* February 15, 1967: 12.

"V Akademii nauk." *FNiT,* 1936, no. 5: 108–111.

Val'den, P. A. *Nauka i zhizn'.* 2 vols. Petrograd, 1922.

Valeskal, P., and V. Sverdlov. "Itogi Ob'edinennogo plenuma SNR i VARNITSO." *FNiT,* 1933, nos. 4–5: 1–7.

Valk, S. N. et al., eds. *Ocherki po istorii Leningradskogo universiteta.* Vol. 1. Leningrad, 1962.

VARNITSO: tseli i zadachi. Moscow, 1931.

Vasetskii, G. "O knige A. A. Maksimova 'Ocherki po istorii bor'by za materializm v russkom estestvoznanii.'" *Bol,* 1948, no. 14: 91–96.

"Vasilii Robertovich Vil'iams (1863–1939)." In *Liudi russkoi nauki*. Vol. 2, 785–794. *See* I. V. Kuznetsov 1948.

Vavilov, S. I. "Akademiia nauk v razvitii otechestvennoi nauki." In *Voprosy*, 41–59. *See* Vavilov et al. 1949.

————. "Mikhail Vasil'evich Lomonosov." In *LRN*:mat, 9–25.

————. "Neskol'ko predlozhenii ob organizatsii nauchno-issledovatel'skoi raboty." *FNiT*, 1935, no. 7: 39–40.

————. "Ocherk razvitiia fiziki v Akademii nauk SSSR za 220 let." In *Fiziko-matematicheskie nauki*, 3–29. *See* Ioffe 1945.

————. *Sobranie sochinenii*. Vol. 3. Moscow, 1959.

————. "Staraia i novaia fizika." In *Pamiati Karla Marksa*, 207–220. *See* Bukharin and Deborin 1933.

————. *Tridtsat' let sovetskoi nauki*. Moscow-Leningrad, 1947.

————. "Vstupitel'noe slovo." In *Voprosy*, 9–14. *See* Vavilov et al. 1949.

[————.] "Vstupitel'noe slovo prezidenta Akademii nauk SSSR akademika S. I. Vavilova." In *Nauchnaia sessiia*, 5–8.

[————.] "Vstupitel'noe slovo prezidenta Akademii nauk SSSR akademika S. I. Vavilova." *VAN*, 1947, no. 1: 23–25.

————, ed. *Isaak N'iuton*. Moscow, 1943.

————, ed. *Materialy k istorii Akademii nauk SSSR za sovetskie gody (1917–1947)*. Moscow-Leningrad, 1950.

Vavilov, S. I. et al., eds. *Voprosy istorii otechestvennoi nauki*. Moscow-Leningrad, 1949.

"Vazhneishii istoricheskii dokument." *PZM*, 1930, nos. 10–12: 3–14.

V., E. "Podgotovka nauchnykh kadrov i Akademiia nauk." *VKA*, 1929, nos. 35–36: 381–85.

Vekua, I. N. "Vysshaia shkola v nauchnom tsentre Sibiri." *VAN*, 1964, no. 6: 12–20.

Velikaia sila idei leninizma: Sbornik statei. Moscow, 1950.

Velikhov, E. P., J. M. Gvishiani, and S. R. Mikulinskii, eds. *Science, Technology and the Future*." Oxford: Pergamon Press, 1980.

Vernadskii, V. I. "The First Years of the Ukrainian Academy of Sciences (1918–1919)." *Annals of the Ukrainian Academy of Arts and Sciences in the United States*. 11, nos. 1–2 (1964–1968): 3–31.

————. *Mysli o sovremennom znachenii istorii znanii*. Leningrad, 1927.

————. *Nachalo i vechnost' zhizni*. Petrograd, 1922.

————. "Ob ispol'zovanii khimicheskikh elementov v Rossii." *Russkaia mysl'*, 1916, no. 1, sec. 2: 73–88.

————. *Ocherki i rechi*. 2 vols. Moscow, 1922.

————. "O gosudarstvennoi seti issledovatel'skikh institutov." In *Otchety o deiatel'nosti Komissii*, 156–161.

————. "O nauchnoi rabote v Krymu v 1917–1921 gg." *NiR*, 1921, no. 4: 5–12.

————. *Pamiati akademika K. M. fon-Bera.* Leningrad, 1927.

————. *Pis'ma o vysshem obrazovanii v Rossii.* Moscow, 1913.

————. "Raboty po istorii znanii." In *Akademiia nauk,* 155–163. *See* Fersman 1927.

————. "Radievye instituty." *Russkaia mysl',* 1911, no. 2, sec. 2: 251–256.

————. "Sovremennoe sostoianie nauchnogo tvorchestva." *Priroda,* 1927, nos. 7–8: 644–646.

————. *Stranitsy avtobiografii.* Moscow, 1981.

————. "Tri resheniia." *Poliarnaia zvezda,* 1906, no. 14: 163–173.

Veselov, M. G. "Nauchnaia deiatel'nost' V. A. Foka." In *Kvantovaia mekhanika,* 7–25. *See* Pavinskii and Mostopanenko 1980.

Veselovskii, K. S. *Istoricheskoe obozrenie trudov Akademii nauk na pol'zu Rossii.* St. Petersburg, 1865.

————. "Otchet": 1861. *ZAN* 1 (1862): 1–45.

————. "Otchet": 1865. *ZAN* 7 (1866): 1–48.

————. "Otchet": 1869. *ZAN* 17 (1870): 1–24.

————. "Otnosheniia imp. Pavla k Akademii nauk." *Russkaia starina* 94 (1898): 5–18, 225–246.

————. "Petr Velikii, kak uchreditel' Akademii nauk." *ZAN* 21 (1872): 20–30.

V., F. "Problemy teoreticheskoi fiziki na sessii." *FNiT,* 1936, no. 4: 30–31.

"Vil'iams, Vasilii Robertovich." In *BSE,* 2d ed. Vol. 4, 81–84. 1951.

Vil'nitskii, M. B. "Za posledovatel'no-materialisticheskuiu traktovku printsipa otnositel'nosti." *VF,* 1951, no. 1: 183–186.

Vklad akademika A. F. Ioffe v stanovlenie iadernoi fiziki v SSSR. Leningrad, 1980.

Vladimirskii, V. "III. Vserossiiskii s'ezd zoologov, anatomov i gistologov." *Priroda,* 1928, no. 2: 174–178.

Vladislavskii, L. A., and V. I. Kuraev. "Obsuzhdenie knigi P. V. Kopnina 'Filosofskie idei V. I. Lenina i logika.'" *VF,* 1970, no 7: 116–129.

Vnekletochnye formy zhizni: Sbornik materialov. Moscow, 1952.

Volgin, V. P. "Akademiia nauk na novom etape." *VAN,* 1935, no. 4: 3–26.

————. "Akademiia nauk k 15 godovshchine oktiabria." *FNiT,* 1932, nos. 11–12: 25–30.

————. "Akademiia v 1933 g." *VAN,* 1934, no. 3: 27–36.

————. *Akademiia nauk SSSR za chetyre goda, 1930–1933.* Leningrad, 1934.

————. "Nashi zadachi i perspektivy." *VAN,* 1934, nos. 11–12: 14–20.

————. "Reorganizatsiia Akademii nauk." *VAN,* 1931, no. 1: 3–11.

————. "Sovetskaia vlast' i nauchnye rabotniki za 10 let." *NR,* 1927, no. 11: 17–25.

Volgin, V. P., et al., eds. *Dvadtsat' piat' let istoricheskoi nauki v SSSR.* Moscow-Leningrad, 1942.

Volkov, V. A. "Novye dokumenty o literaturnom nasledstve D. I. Mendeleeva i ego sem'e." *VIE-T,* 1969, no. 4: 138–141.

Vol'kovich, S. I., et al., eds. *Dmitrii Nikolaevich Prianishnikov: zhizn' i deiatel'nost'*. Moscow, 1972.

"80-letie akad. I. P. Pavlova: Postanovlenie Soveta narodnykh komissarov Soiuza SSSR." *LP*, September 27, 1929.

Vremennik Obshchestva sodeistviia uspekham opytnykh nauk i ikh prakticheskikh primenenii imeni Kh. S. Ledentsova. No. 2. Moscow, 1912.

"Vsesoiuznoe soveshchanie po filosofskim voprosam estestvoznaniia." *VF*, 1959, no. 2: 59–88.

"V tsentral'nom biuro VARNITSO." *FNiT*, 1932, nos. 4–5: 125–126.

Vucinich, Alexander. "The Peasants as a Social Class." In *The Soviet Rural Community*, 307–324. See Millar 1971.

———"Russia: Biological Sciences." In *The Comparative Reception*, 227–255. See Glick 1974.

———"Soviet Physicists and Philosophers in the 1930's: Dynamics of a Conflict." *ISIS* 71, no. 257 (1980): 236–250.

Vucinich, W. S. "The Structure of Soviet Orientology: Fifty Years of Change and Accomplishment." In *Russia and Asia*, pp. 52–134. See W. S. Vucinich 1972.

Vucinich, W. S., ed. *Russia and Asia: Essays on the Influence of Russia on the Asian Peoples*. Stanford, Calif.: Hoover Institution, 1972.

Vygodskii, M. "Problemy istorii matematiki s tochki zreniia metodologii marksizma." In *Na bor'bu za materialisticheskuiu dialektiku*, 161–182.

Vygotskii, L. S. *Izbrannye psikhologicheskie issledovaniia*. Moscow, 1956.

"Vystupleniia uchenykh i resheniia po organizatsionnym voprosam." *VAN*, 1962, no. 3: 19–22.

Werskey, P. G. Introduction to *Science*, xi–xxix. See Bukharin et al. 1971.

———. *The Visible College: The Collective Biography of British Scientific Socialists of the 1930's*. New York: Holt, 1979.

Wiener, Norbert. *The Human Use of Human Beings: Cybernetics and Society*. New York: Avon, 1973.

Wienert, Helgard. "The Organization and Planning of Research in the Academy System." In *Science Policy*, 187–287. See Zaleski et al. 1969.

Winter, E. et al., eds. *Lomonosov, Schlözer, Pallas: Deutsch-russische Wissenschaftsbeziehungen im 18. Jahrhundert*. Berlin, 1962.

Zagoskin, N. N. *Istoriia Imperatorskogo Kazanskogo universiteta za pervuiu sto let ego sushchestvovaniia*. Vol. 3. Kazan, 1904.

Zaleski, E., et al. *Science Policy in the USSR*. Paris, 1969.

"Zamechaniia russkikh universitetov na proekt novogo ustava Akademii nauk." *Otechestvennye zapiski*, 1866, no. 4: 185–197.

"Za patrioticheskuiu sovetskuiu nauku." *VAN*, 1949, no. 4: 3–14.

Za povorot na fronte estestvoznaniia. Moscow-Leningrad, 1931.

"Za protsvetanie peredovoi biologicheskoi nauki." *VAN*, 1948, no. 9: 3–16.

"Za rastsvet peredovoi biologicheskoi nauki." *Bol*, 1948, no. 15: 1–9.

"Za razvitie peredovoi sovetskoi fiziki, za razvertyvanie bor'by mneniia v nei." *UFN* 48, no. 4 (1952): i–viii.

"Za reshitel'nuiu perestroiku form i metodov organizatsii teoreticheskoi raboty." *PZM*, 1931, no. 6: 1–14.

"Zasedaniia otdelenii Akademii nauk SSSR, posviashchennye 70-letiiu so dnia rozhdeniia I. V. Stalina." *VAN*, 1950, no. 1: 75.

"Za svobodnuiu, tvorcheskuiu nauchnuiu kritiku." *VAN*, 1950, no. 8: 10–20.

Zavadovskii, B. M. "Nauka v Sovetskoi Rossii. (Vpechatleniia o rabotakh peter-burgskikh fiziologicheskikh laboratorii)." *Krasnaia nov'*, 1921, no. 4: 128–147.

———. "The 'Physical' and 'Biological' in the Process of Biological Evolution." In *Science*, 69–80. *See* Bukharin et al. 1971.

Zavadovskii, M. M. "Darvin i ego nasledstvo." *FNiT*, 1938, no. 3: 38–53.

———. "Osnovnye etapy v istorii eksperimental'noi biologii (zoologii) v Rossii." *UZMU*, 1946, no. 103: 47–60.

Zbarskii, B. "Na povorote: IV. S'ezd fiziologov." *VAR*, 1930, no. 6: 67–69.

Zhdanov, A. A. *Vystuplenie na diskussii po knige G. F. Aleksandrova 'Istoriia zapad-noevropeiskoi filosofii', 24 iunia 1947 g.* Moscow, 1951.

Zhdanov, Iu. "Ob'edinenie, koordinatsiia, dvizhenie vpered." *NiZh*, 1978, no. 5: 33–38.

Zhigalova, L. V. "K istorii sozdaniia institutov Nauchno-tekhnicheskogo otdela VSNKh." In *Organizatsiia*, 187–202. *See* Beliaev et al. 1968.

Zhukov-Verezhnikov, N. N., I. N. Maiskii, and L. A. Kalinichenko. "O nekle-tochnykh formakh zhizni i razvitie kletok." In *Filosofskie voprosy*, 318–334. *See* Novinskii and Platonov 1951.

Zinov'ev, A. A. "Logical and Philosophical Implications." In *Problems*, 91–159. *See* Tavanets 1970.

———. "O logike mikrofizike." *VF*, 1970, no 2: 126–135.

———. "O printsipakh determinizma i indeterminizma." *VF*, 1970, no. 9: 60–63.

Zubov, V. P. *Istoriografiia estestvennykh nauk v Rossii.* Moscow, 1956.

Zvorykin, A. A. "Nauka i nauchno-tekhnicheskii progress." In *Osnovnye print-sipy*, 51–67. *See* Ananichev et al. 1973.

———. "O sovetskom prioritete." *Bol*, 1948, no. 22: 23–42.

Zvorykin, A . A. and E. I. Rabinovich. "Organizatsionnaia struktura nauki na sovremennom etape." In *Osnovnye printsipy*, 87–118. *See* Ananichev et al. 1973.

INDEX

Khvol'son, O. D., 66
Kiev University, 191
Kirchhoff, Gustav Robert, 340
Kirghiz Branch of the Academy of
Sciences, 203
Kirghiz philosophers, 342
Kirghiz republic, 292
Kirpotin, V. Ia., 229
Kol'man, E. Ia., 150, 164–165, 173,
194, 196, 202, 237, 251, 319,
385–386
Kolmogorov, A. N.,* 74, 136, 173,
180, 193, 249, 273
Kol'tsov, N. K., 69, 76, 161, 174, 175,
191, 193, 209, 234, 251, 356
Komarov, V. L., 127, 166, 171, 177,
186, 204, 207, 208, 233
Kommunist, 258, 327
Kon, I. S., 328
Konstantin Konstantinovich, Grand
Duke, 59
Konstantinov, F. V.,* 167, 257, 341
Kopnin, P. N., 341
Korol, Alexander, 301, 402, 403
Korolenko, V. G., 59
Kosygin, A. N., 283
Kotel'nikov, S. K.,* 27, 30, 38
Kovalevskii, A. O., 26, 48, 53, 55
Krasheninnikov, S. P.,* 19, 38
Krasnaia Pakhra Center, 291
Krylov, A. N.,* 204, 205
Krylov, I. A., 229
Krzhizhanovskii, G. L.,* 130, 149, 180
Kudriavtsev, P. S., 236
Kugel, S. A., 400, 401
Kuhn, Thomas S., 2, 3, 235, 368
Kunstkamera, 7
Kupffer, A. T.,* 45, 51
Kurchatov, I. V.,* 201
Kurnakov, N. S.,* 92, 105
Kursk magnetic anomaly, 105, 106
Kuznets metallurgical complex, 200
Kuznetsov, B. G., 217, 349

Kuznetsov, I. V., 223, 224, 229, 236,
248, 251, 255, 330, 395

Labor camps, 172
Le Harpe, Frédéric Cezar, 32
Lamarck, Jean Baptiste, 215
Lamarckism, 153, 164, 175, 195, 258
Landau, L. D.,* 138, 141, 142, 148,
159, 165, 174, 201, 223, 251, 359
Landsberg, G. S.,* 193
Lapirov-Skoblo, M. Ia., 121
Laplace, Pierre Simon, 2, 164, 221
Lappo-Danilevskii, A. S., 60, 62,
97–98
Latvian Academy of Sciences, 206
Laue, Max von, 3, 368, 374, 389
Lavoisier, Antoine, 37, 93, 236
Lav'rentev, M. A.,* 286
Lavrov, P. L., 52
Lazarev, P. P.,* 117
Learned societies, 27, 34, 57, 58, 72,
77–81, 86, 111, 118
Lebedev, P. N., 63, 64, 94
Leclerc, N. G., 23
Lecourt, D., 391
Ledentsov, Kh. S., 63
Leibniz, Gottfried Wilhelm, 7, 8, 13,
229, 332, 360
Lenin, V. I., 94, 95, 107, 166, 170,
230, 242, 361; and Academy of Sci-
ences, 103, 114–115; on bourgeois
science, 75, 119–120; Deborin on,
131, 151, 158–159; and dialectical
materialism, 154; Frenkel' on, 156;
in history of Marxist thought, 153;
against Mach, 159; on mathematics, 154, 250; on "objectivism," 155;
reflection theory of, 88
Leningrad Institute of Nuclear Physics in Gachina, 291
Leningrad Museum of Anthropology
and Ethnography, 80–81
Leningradskaia pravda, 124

Designer: U.C. Press Staff
Compositor: Publisher's Typography
Printer: Vail-Ballou
Binder: Vail-Ballou
Text: Palatino
Display: Palatino

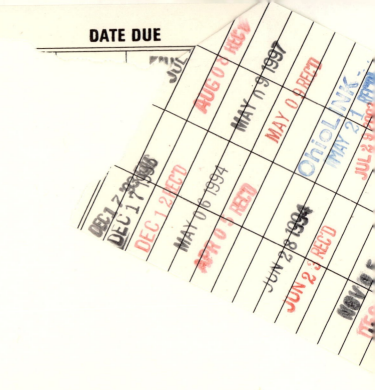